INTERNATIONAL SERIES IN

NATURAL PHILOSOPHY

General Editor: D. TER HAAR

VOLUME 90

SOLAR NOISE STORMS

Some other titles of interest

In the International Series in Natural Philosophy:

BATTEN:
Binary and Multiple Systems of Stars

BOWLER:
Gravitation and Relativity

GLASBY:
The Nebular Variables

KAPLAN & TSYTOVICH:
Plasma Astrophysics

PACHOLCZYK:
Radio Galaxies

REDDISH:
Stellar Formation

STROHMEIER:
Variable Stars

ZHELEZNYAKOV:
Radioemission of the Sun and Planets

Other Titles

BEER:
Vistas in Astronomy (Review Journal)

CLARK & STEPHENSON:
Historical Supernovae

HEY:
The Radio Universe, 2nd edition

HILLAS:
Early Solar Physics

SOLAR NOISE STORMS

by

E. O. ELGAROY

Institute of Theoretical Astrophysics
University of Oslo

PERGAMON PRESS

OXFORD · NEW YORK · TORONTO · SYDNEY · PARIS · FRANKFURT

U.K.	Pergamon Press Ltd., Headington Hill Hall, Oxford OX3 0BW, England
U.S.A.	Pergamon Press Inc., Maxwell House, Fairview Park, Elmsford, New York 10523, U.S.A.
CANADA	Pergamon of Canada Ltd., 75 The East Mall, Toronto, Ontario, Canada
AUSTRALIA	Pergamon Press (Aust.) Pty. Ltd., 19a Boundary Street, Rushcutters Bay, N.S.W. 2011, Australia
FRANCE	Pergamon Press SARL, 24 rue des Ecoles, 75240 Paris, Cedex 05, France
WEST GERMANY	Pergamon Press GmbH, 6242 Kronberg-Taunus, Pferdstrasse 1, Frankfurt-am-Main, West Germany

First edition 1977

Library of Congress Cataloging in Publication Data

Elgarøy, Øystein, 1929-
Solar noise storms.
(International series in natural philosophy, v. 90)
Bibliography: p.
Includes indexes.
1. Solar noise storms. I. Title.
QB539.N6E43 1976 523.2 76-17598
ISBN 0-08-021039-2

In order to make this volume available as economically and rapidly as possible the author's typescript has been reproduced in its original form. This method unfortunately has its typographical limitations but it is hoped that they in no way distract the reader.

Printed in Great Britain by A. Wheaton & Co., Exeter

CONTENTS

PREFACE

Less than 20 years ago a comprehensive exposition of the results from the whole field of radio astronomical investigations could be covered in a single book. This is no longer possible because of the rapid growth this branch of science has seen. As a result of the development monographs, proceedings from symposia and meetings and review articles have become more specialized. The present treatise illustrates this trend in so far as it is concerned with one section only of solar radio astronomy.

Noise storms is the type of emission with which solar radio astronomers are most familiar. In spite of this the phenomenon is still not well understood. One can say this is a highly unsatisfactory state of affairs, but one can also take a positive attitude and regard noise storms as a challenging subject for further research.

An impressive number of papers deal directly or indirectly with noise storms and are published in various journals and books. Time may have come to collect the different fragments of knowledge and experience in a single volume so as to be able to "survey the landscape". Considering what has been done, one is in a better position to see what should be done. It is the hope of the author that the observations, interpretations and theories presented on the pages which follow would be of value for the planning of new experiments and that the reader be inspired to new and fresh attacks on the problems.

The noise storm is a natural phenomenon which has been of very great value for small observatories. Useful work could in many instances be undertaken with equipment of moderate costs. This period is running out, but sensible experiments can still be undertaken at places with limited financial resources. This is probably one of the reasons why investigations of noise storms have been conducted by a large number of small groups and even scientists bound to work almost in isolation. In this respect one is tempted to regard noise storms as a "kind" gift to the radio astronomers.

One day in the summer of 1955 the author arrived at the office of Professor Rosseland to take up a position as research assistant at the Institute of Theoretical Astrophysics and the recently opened Oslo Solar Observatory. The only instructions he got by the experienced astrophysicist was to supervise some routine work at the observatory and "find a problem which interests you and go ahead". This opportunity to choose a subject which challenged the interest was very much appreciated, and the noise storm phenomenon soon became almost a passion. Since then, although other problems also have claimed time and interest, the author has been monitoring what happened in the field of solar noise storms and also been engaged in work on storms. The present book is a result of this interest.

It is always difficult to include and give credit to all contributions which deserve attention. An attempt has been made to give an exposition of the subject which is not too biased. In some cases pertinent papers have been difficult to acquire, in others a language barrier has been difficult to surmount. Some selection has always to be made, and, of course, the present collection reflects to some extent the personal experience and taste of the author. Others would probably have written a somewhat different book on the same subject.

The author wishes to express his sincere thanks to Dr. J. L. Steinberg, Paris Observatory, Meudon; to Dr. A.D. Fokker, Dr. T. de Groot, R. van Hees, A. Kattenberg and W. Weber, Solar Radio Astronomy Group, University of Utrecht; and to Professor V.V. Zheleznyakov and Dr. V.V. Zaitsev of the Radiophysical Research Institute, Gorki, for reading the manuscript, or parts of it, and for critical remarks and valuable suggestions.

It is a pleasure to extend my gratitude to Berit Schumann-Olsen and Kristin Sæther for careful typing of the manuscript and the final typing of the book.

Øystein Elgarøy.

INTRODUCTION

Intense bursts of radio emission have been whirling out from the sun since times immemorial – escaping detection until modern times. In the 1930s several scientists and radio amateurs observed hissing sounds in their short-wave receivers preceding fade-outs which accompany strong solar flares, but they were not aware of the true nature of the phenomenon.

On February 27 and 28, 1942, severe daytime noise jamming was experienced by British anti-aircraft radar at wavelengths between 4 and 8 m, and Hey recognized that the directions of maximum interference recorded by the operators appeared to follow the sun. Since an exceptionally active sunspot crossed the central meridian on February 28, Hey concluded that the must be radiating electromagnetic waves directly, and he assumed that the active sunspot region was the likely source. Because of the war this discovery was not published in scientific journals until 1946 (Hey, 1946).

In 1946 solar radio astronomy started to grow. Hey was able to discern two forms of solar radio emission: the long-lasting noise storm associated with the sunspot, and the intense bursts accompanying the solar flares (Hey, 1973). Enhanced solar radio emission was also observed by Reber (1946), who gave a somewhat more poetical description of his findings than usually appear in present-day papers. According to Reber "the audible effect in headphones was much like wind whistling through the trees when no leaves are on the limbs. Occasionally great swishes occurred above the rapidly varying background. No snaps or crackling sounds could be heard which might be interpreted as lightning or sparking discharges of any kind."

Other pioneering observations of solar storm radiation were carried out by Martyn (1946), Ryle and Vonberg (1946) and McCready, Pawsey and Payne-Scott (1947). A new and exciting field of research was opened.

The intensity of the visible light from the sun shows very small variations. But this is not true for solar radio waves. The intensity of the emitted radio waves may increase by several orders of magnitude in the course of minutes and stay at an enhanced level for days. At metric wavelengths large numbers of brief flashes of radiation clearly demonstrate that a "radio noise storm" rages in the solar atmosphere. The term "noise storm" was coined by Allen (1947) because of a strong similarity between some records of solar radio noise emission and records of geomagnetic storms. Later on, Wild and McCready (1950) classified different types of solar radio emission according to their spectral behaviour, and noise storms became known as type I emission.

Basic information on solar radio waves was published by Kundu (1965) and Zheleznyakov (1970).

From single-frequency records alone it may be difficult to discriminate between noise storms and some other types of enhanced solar radio emission. Certainly, in papers appearing before 1960 some confusion may be encountered. In particular little or no distinction was made between noise storms and large bursts of radio emission correlated with proton flares and presently designated "type IV bursts". But it is rather easy to determine the type of emission when radio spectra are available, and confusion should not now occur.

Noise storms constitute the most prominent part of the solar radio emission in the metre-wavelength band and the radio bursts associated with the storms represent the most transient phenomenon encountered in solar physics as yet. Both observationally and theoretically, noise storms pose challenging problems to the astrophysicist, and progress in these questions are of cardinal importance for the understanding of the physics of the solar atmosphere.

ACKNOWLEDGEMENTS

For permission to use, free of charge, previously published illustrations I am indebted to: Astronomical Society of Australia (Figures 3.6, 3.12, 3.17, 3.18, 8.5, 8.6, 11.4, 11.5); Astronomical Society of Japan (Figures 3.3, 3.13, 3.15, 11.1, 11.2, 11.3); Centre National de la Recherche Scientifique (Figures 1.2, 1.3, 1.8, 1.9, 1.10, 1.11, 3.1, 4.4, 4.11, 4.12, 4.13, 4.14); Commonwealth Scientific and Industrial Research Organization (Figures 1.4, 1.5, 1.6, 2.7, 2.17, 3.9, 3.11, 6.5, 6.8, 6.11, 6.12, 6.13, 6.15, 6.16, 6.18, 6.19, 6.21, 12.1, 12.2); D. Reidel Publishing Company (Figures 1.12, 1.13, 1.14, 2.8, 2.10, 2.11, 2.15, 2.20, 3.4, 3.5, 3.16, 3.19, 3.20, 3.21, 3.22, 5.7, 5.10, 6.20, 7.5, 7.6, 8.2, 8.7, 8.8, 8.9, 9.1, 10.1, 10.2, 13.2, 13.3, 13.4); Gordon and Break Science Publishers (Figure 13.1); John Wiley and Sons Inc. (Figure 1.6); Macmillan Journals Ltd. (Figure 1.15); North-Holland Scientific Publishers (Figures 5.1, 5.2, 5.3, 5.4); Plenum Publishing Corporation (Figure 3.14); Springer-Verlag (Figures 1.16, 1.18, 4.5, 4.6, 4.7, 4.8, 4.9, 4.10, 5.5, 5.6, 5.8, 5.9, 5.11, 5.12, 5.13, 6.4, 6.6, 6.7, 6.9, 6.10, 8.1, 8.3, 8.4, 9.2); The University of Chicago Press (Figure 11.6).

1. RELATIONS BETWEEN NOISE STORMS AND OPTICALLY OBSERVABLE FEATURES ON THE SUN

1.1. THE ONSET OF NOISE STORMS

A noise storm may start abruptly, but a more gradual development may also take place. The commencement of several noise storms was studied by Fokker (1960). Using radiometer records obtained at 200 MHz, Fokker found that the first phase of the noise storms in the sample differed quite clearly. There may be a sudden increase of burst activity and a gradual rise of continuum intensity, a sudden increase of base level and a gradual increase of burst activity, or a slow increase of both base level and burst activity, just to mention some examples. Sometimes a new storm is triggered while another is in progress, causing a noise storm enhancement.

Some storms occur in restricted frequency ranges, but quite commonly noise storms cover a frequency band of more than 100 MHz. The starting frequencies of noise storms were studied by Malville (1961). As is seen from Fig. 1.1, the noise storm is

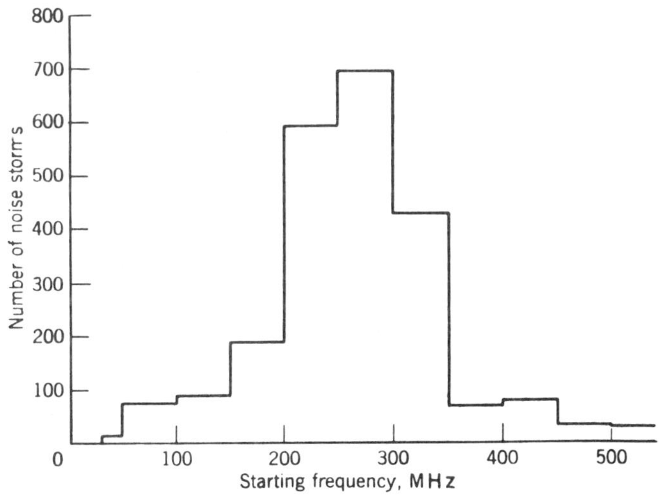

Fig. 1.1. Distribution of upper frequencies of noise storms (Malville, 1961).

a typical metre-wave phenomenon. Some storms start in one frequency region and drift slowly to another band in the course of a day.

1.2. RELATION BETWEEN NOISE STORMS AND FLARES

It is an important question whether particular solar phenomena are associated with the onset of noise storms. In this connection a possible relationship between noise storms and solar flares has been sought for. Several investigations have been undertaken leading to different conclusions. The main reason for this is that it is not possible to obtain reliable results when the observational material is insufficient. For good results to be obtained it is necessary to dispose of: (1) high quality flare data, (2) positional observations of the noise storm sources (two-dimensional), and (3) dynamic spectra or radiometer recordings of good sensitivity. In several investigations the requirements (1), (2) or (3) were only partly satisfied. Thus contradictory results emerged.

The first to study the relation between noise storms and flares were Dodson, Hedeman and Owren (1953) and Dodson (1958) who found that 97% of the storms observed at 200 MHz were associated with flares. Most flares just preceded or coincided with the commencement of noise storms at 200 MHz. Fokker (1960) investigated individual cases and concluded that some storms undoubtedly were initiated by flares whereas others occurred without any preceding flare activity.

Smith and McIntosh (1962) arrived at somewhat different results. They studied the correlation between noise storms and flares during the International Geophysical Year, for which good observational data were available. A noise storm was considered to be associated with a flare if it occurred within 15 min after the start of the flare. Correction for chance associations were introduced. If t is the total observation time in minutes, the number of 15 min periods is $T = t/15$. Let F be the number of flares and S the number of noise storms observed in this time.

The number of chance associations N_c becomes $N_c = \frac{FS}{T}$ provided flares and noise storms are uniformly distributed in time. The strength of the association between storms and flares may be expressed by a coefficient of association Q defined in the following way:

$$Q = \frac{N_o(T-F-S+N_c) - (F-N_o)(S-N_o)}{N_o(T-F-S+N_o) + (F-N_o)(S-N_o)},$$

where N_o is the number of observed associations. When Q=1 there is complete association, Q=0 means independence of the two parameters, and Q=-1 demonstrates complete disassociation. The results of the analysis undertaken by Smith and McIntosh are given in Table 1.1.

TABLE 1.1

Association between flares and noise storms (Smith and McIntosh, 1962)

		F	N_o	N_c	Q	Flares: Apparent (%)	Flares: True (%)	Noise storms: Apparent (%)	Noise storms: True (%)
Sacramento Peak	All flares	3748	330	319	0.03	9	0.3	38	3.1
T = 10 747	$R \leq 0.7$	(1919)	(189)	(163)	(0.10)	(10)	(1.4)		
S = 914	Flares $\geq$ 1	849	120	72	0.31	14	5.6	14	6.6
	$R \leq 0.7$	(294)	(48)	(25)	(0.37)	(16)	(7.8)		
	Flares $\geq$ 2	181	46	15	0.58	25	17	6	4.4
	$R \leq 0.7$	(51)	(21)	(4.3)	(0.77)	(41)	(33)		
McMath - Hulbert List	Flares $\geq$ 1	1701	214	143	0.24	13	4.2	12	4.9
T = 22 595	Flares $\geq$ 2	159	32	13	0.47	20	12	2.3	2.3
S = 1905									

In Table 1.1 the apparent percentage of flares that were associated with noise storms was obtained by taking the ratio of the number of observed coincidences to the total number of flares observed during the time interval considered, i.e. $100N_o/F$. The true percentage is given by $100(N_o-N_c)/F$. In the percentages for noise storms, an S replaces F in the denominator.

When subflares are included, it is seen from Table 1.1 that there is no statistically significant association between noise storms and flares. But the association coefficient increases when only flares of higher importance are considered. But, nevertheless, Table 1.1 shows that only a small fraction of the total number of noise storms observed appear to be truly associated with flares.

Smith and McIntosh used two lists of flare data. The one was a compilation of IGY flares by Dodson and Hedeman (1960), the other consisted of all flares observed at Sacramento Peak during the IGY. In the Sacramento Peak flare data an intensity scale was used ranging from 10 (plage) to 40 (very brightest flare). Since a statistical study of the distribution of flares of different intensity classes showed a sharp decrease in the number of flares brighter than those of class 18, the material was divided in two parts. It was then found that faint flares showed a poor association with storms. The situation was somewhat better when bright flares were considered, then even subflares showed some correlation. The correlation also improved with flare importance, and for bright flares about every third flare of a class greater than or equal to 2 was followed by a noise storm.

Smith and McIntosh also attempted to answer the question whether certain centres of activity were more likely to produce noise storms than others. It is seen from Table 1.2 that seven selected plage regions showed high values for the coefficient of association, and the table supports the suggestion that some centres produce noise storms more efficiently than others.

Considering the whole material, Smith and McIntosh concluded that the association between flares, exclusive of subflares, and noise storms was real, albeit large fractions of both phenomena occurred independently of each other.

By means of the interferometer at Nancay, France, Le Squeren (1963) determined the position of noise storm sources. Between March 1959 and September 1961 the east - west and the north - south branches of the interferometer were in operation

Table 1.2

Association between flares and noise storms in selected plage regions (Smith and McIntosh, 1962)

McMath plage No.	F	S	N_O	NS	N_C	Q	Flares Apparent (%)	Flares True (%)	Noise storms Apparent (%)	Noise storms True (%)
4083	12	50	5	8	0.86	0.83	42	34	16	14
4124	31	64	6	7	2.8	0.45	19	10	11	7
4125	17	61	6	6	1.6	0.71	35	26	10	7
4314	8	75	4	4	1.25	0.70	50	34	5	4
4476	30	77	8	8	3.7	0.47	27	14	10	6
4548	12	31	3	3	0.54	0.77	25	20	10	8
4659	31	61	5	5	2.6	0.37	16	8	8	4

NS = number of noise storms associated with flares.

at 169 MHz, and two-dimensional position determinations were performed. It was then possible to determine the optical centre that was associated with the noise storm and look for associations between flares in this centre and noise storms or storm enhancements.

In the material Le Squeren included only noise storms with duration ≥ 1 h and the increase in flux was not to be less than 3×10^{-21} W m^{-2} Hz^{-1}. The same criteria were applied in the selection of storm enhancements. A total of 109 events were listed. In 100 cases it was possible to associate a flux increase on 169 MHz with a flare when the flare was allowed to occur within 2 h before to 1 h after the flux increase occurred. In 94 of the cases the flare was situated in the associated optical centre; in 4 cases the flare appeared in a nearby optical centre, and the association was less certain. In the two last cases the flares occurred outside plage regions.

Some associations may of course occur just by chance. To test this a superposed epoch diagram was constructed. Taking the time of onset of a noise storm as 0, the number of flares per hour in the associated optical centre in the 5 h preceding and following the zero point was determined. The result is seen in Fig. 1.2

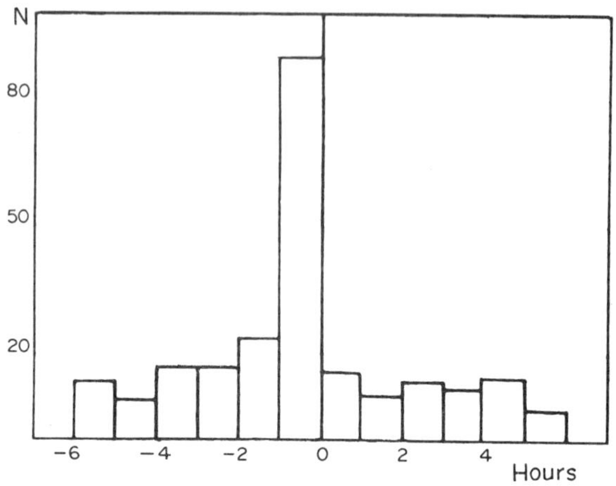

Fig. 1.2. Number of flares in the associated optical centre in the hours preceding and following starts or enhancements of noise storms (Le Squeren, 1963).

According to Fig. 1.2 the number of flares per hour is roughly constant, but a significant increase occurs in the last hour which precedes the storm. Thus a physical connection between flares and storm emission appears plausible.

In Fig. 1.3 the distribution of time intervals which separate the flare and the onset of noise storm emission is plotted. When the flare precedes the storm, the time interval is positive. The figure confirms the results seen in Fig. 1.2. In five cases the flare follows after the beginning of the noise storm, then the association may be by chance or the timing of the noise storm or the flare start may be in error. When strong flares are considered, the time intervals T_o are less dispersed than occurs when all flares are included. Limiting the material to strong flares, the average value of T_o becomes about 30 min. If it is assumed that a perturbation travels from the flare region and initiates a noise storm higher up in the corona, a time delay of 30 min corresponds to a velocity of propagation of about 150 km s^{-1}. (The storm emission is then assumed to be generated at a height of approximately 0.4 $R_\odot$ above the photo-

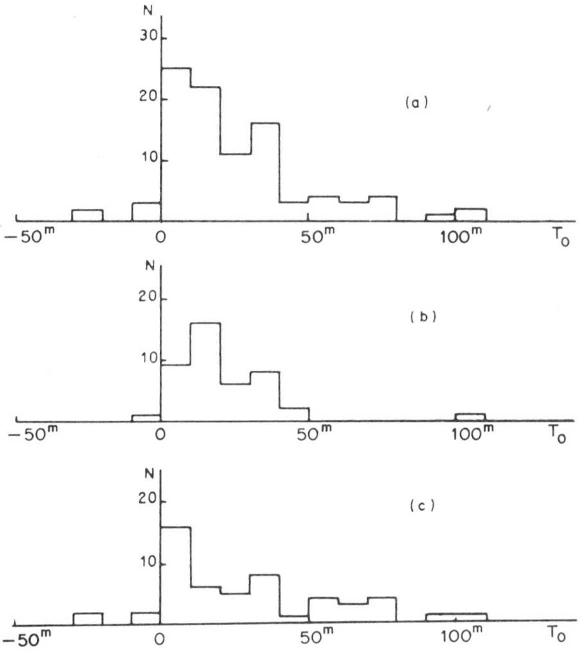

Fig. 1.3. Distribution of time intervals T_o between start of flares and start of noise storm.
(a) all cases; (b) cases in which the flare was of importance 2 or 3; (c) cases in which the flare was of importance 1^-, 1 or 1^+
(Le Squeren, 1963).

sphere.) This is less than the velocity of type II burst exciters which is about 1000 km s^{-1} and it is also less then the velocity of moving type IV bursts which is between 200 and 1000 km s^{-1}. But it is comparable with the velocity of the coronal irregularities reflecting radar signals (James, 1966) and the radial velocity of disturbances generating chains of type I bursts (Hanasz, 1966; Elgarøy and Ugland, 1970).

There seems to be a statistical relation between the intensity of the noise storm and the importance of the associated flare. Le Squeren found that flux enhancements larger than 10^{-20} W m^{-2} Hz^{-1} were twice as frequently associated with flares of importance 2 or 3 as storms with flux enhancements below this limit.

The above evidence favours the view that noise storms are connected with the occurrance of flares. But all flares are not associated with noise storms, and other characteristics than flare importance determine whether a noise storm is triggered or not. It seems very reasonable to assume that the structure of the corona and the magnetic field strength above the active region must be taken into account in this connection.

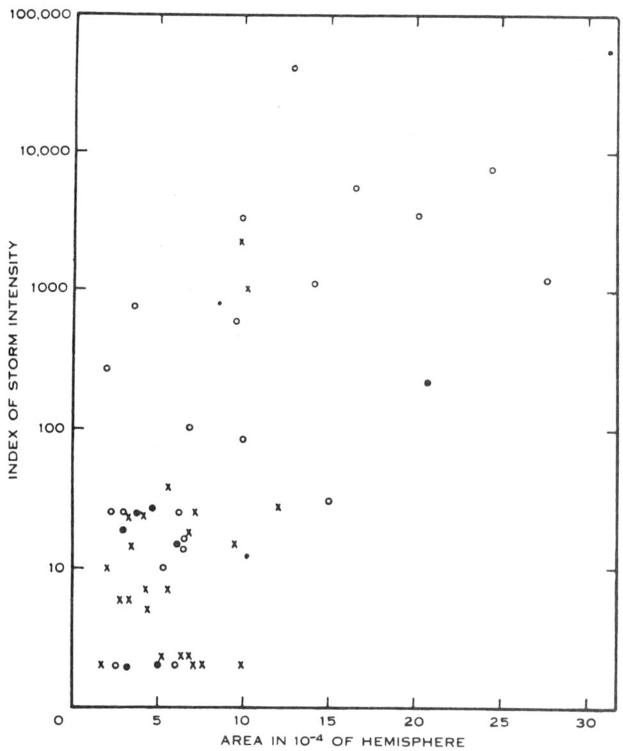

Fig. 1.4. Radio storm intensity, sunspot group area and limb characteristics: o, loops and/or streamers; •, east and west limbs contradictory; x, quiescent, or no prominences (Wild and Zirin, 1956).

The importance of the latter point was demonstrated by Wild and Zirin (1956). Prominence surveys from Climax and Sacramento Peak were collected and the visible activity on the solar limb was compared with radio data. An arbitrary index of storm intensity was arrived at in the following way: the daily values in excess of 10^{-21} W m^{-2} Hz^{-1} flux at 62, 80, 98, 175 and 200 MHz were added for each day of the storm, being limited in all cases to within 5 days of CMP. Then the values on all five frequencies were added to give an index of the intensity of the storm. Storm intensity is plotted against spot group area in Fig. 1.4. Each entry indicates whether the limb appearance was positive, negative or contradictory. The predominance of loops in the large storms is clearly marked. Since the loops were observed 7 days before CMP when the storms usually culminate, the correlation is surprisingly good. The same data are presented in a somewhat different way in Fig. 1.5.

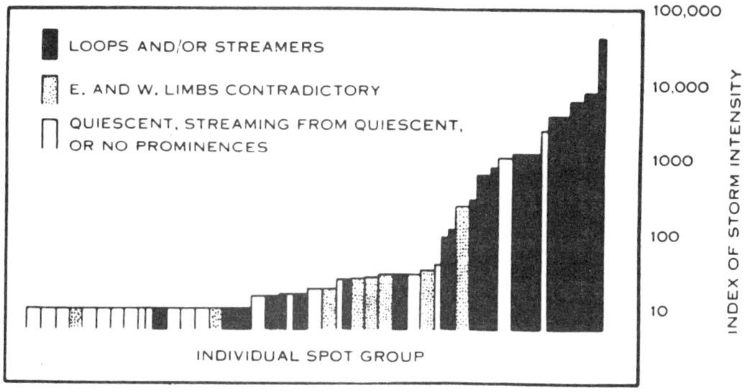

Fig. 1.5. Spot groups arranged in order of increasing radio intensity and shaded according to limb characteristics (Wild and Zirin, 1956).

Wild and Zirin concluded that there was a close connection between the ordering of the coronal material by the magnetic field, producing loops and streamers, and the production of radio noise storms.

1.3. NOISE STORMS AND SUNSPOTS

Right from the beginning it was found that noise storms were associated with sunspots. Many statistical studies have been made to correlate radio emissivity with optical properties of spots. One of the earliest was performed by Payne-Scott and Little (1951) who used observations of storms at a frequency of 97 MHz. It was found that the necessary condition for a sunspot group to give rise to a noise storm at some stage was that it should contain a sufficiently large spot. As is seen from Fig. 1.6, there is a rough line of demarcation at groups containing no spots of area larger than 400 millionths of the solar disk. When a group contains a large spot, a storm is observed in two out of three cases.

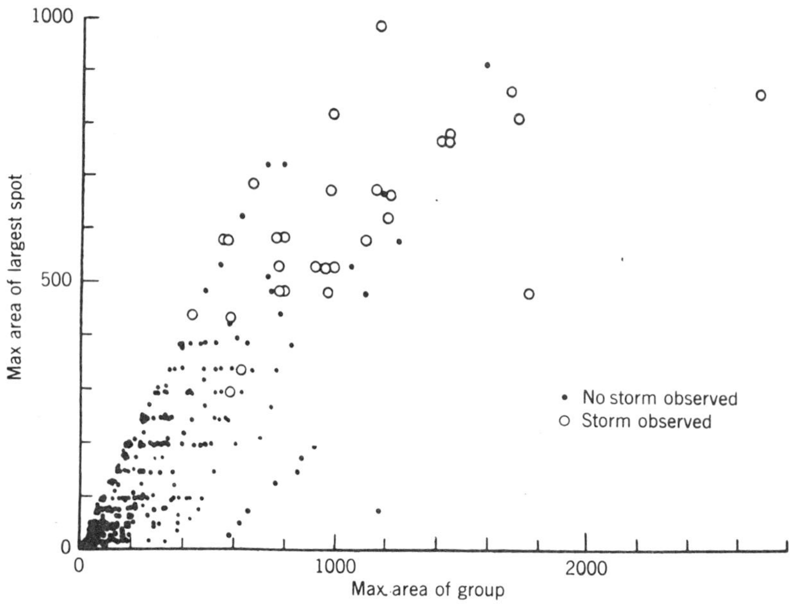

Fig. 1.6. The occurrence of noise storms related to maximum area of associated sunspot group and maximum area of largest spot in the group (Payne-Scott and Little, 1951, after Kundu, 1965).

The same trend was found by Le Squeren (1963). The probability that a photospheric centre was associated with a noise source in the corona increased rapidly when the area of the largest spot in the group exceeded 450 millionths of the visible hemisphere. When the area exceeded 750 millionths the probability of being associated with a storm centre.was about

80%. But there was poor correlation between sunspot area and maximum radio flux in a storm.

Maltby and Steen (1960) investigated the relation between noise activity and the largest magnetic field strength observed in the associated group and found only a slight association. The magnetic flux of the largest spot in the group appeared to be a better indicator than the magnetic field strength.

Sunspots may be classified according to the Zürich scheme. The essential features in each class appear from Fig. 1.7 and are the following:

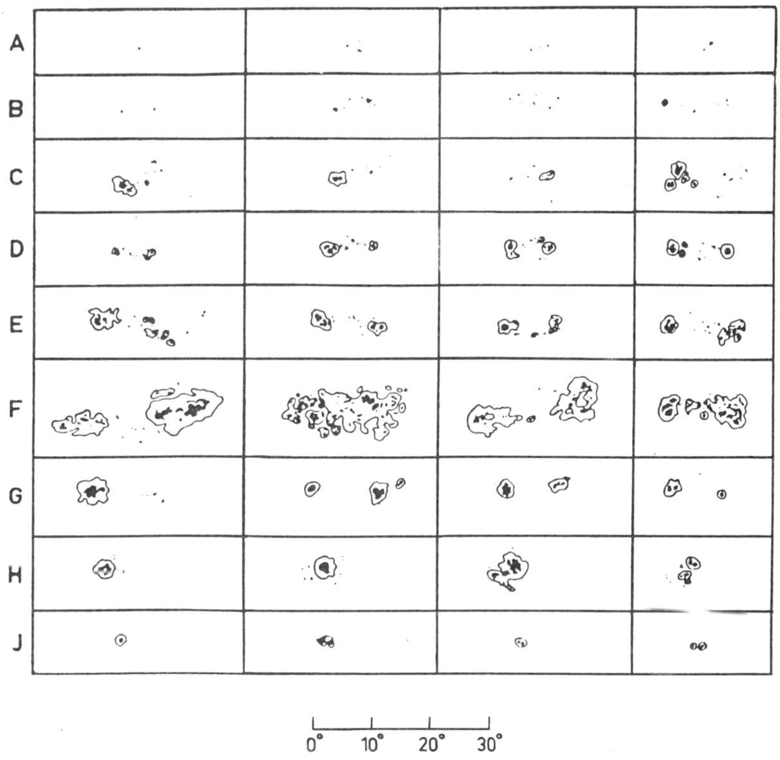

Fig. 1.7. The Zürich classification of sunspot groups. Four examples of each class are shown. The scale at the bottom indicates degrees of heliographic longitude.

A One or several small spots without penumbra. Short lifetime.

B Bipolar group without penumbra.

C Bipolar group as B, but at least one spot has penumbra.

D Bipolar group; more than one spot shows penumbra.

E Large bipolar group with complicated structure.

F Very large bipolar or complex group. At least 15° in longitude.

G Large bipolar group but without small spots between the large ones.

H Unipolar spot with penumbra. Diameter >2.5 degrees.

J Unipolar spot with penumbra. Regular with diameter <2.5 degrees.

Because two-dimensional interferometer data were available, Le Squeren (1963) could point out the optical centre associated with different noise storms with good accuracy. It was then attempted to relate noise storms to the different types of sunspots according to the Zürich classification. For 42 noise storms the distribution shown in Fig. 1.8 was found. The first thing to recognize from Fig. 1.8 is that no spots of types A, B or J were associated with storms. Secondly, it is seen that the

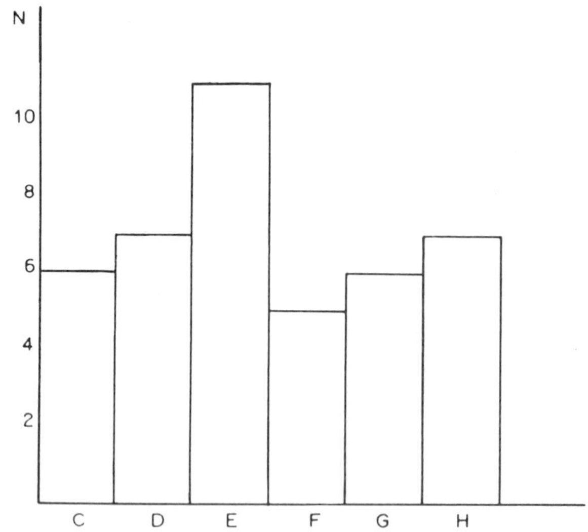

Fig. 1.8. Distribution of different sunspot groups associated with noise storm centres (Le Squeren, 1963).

distribution is almost uniform for spots from type C to type H. But it should be borne in mind that the statistics is not very good as the number of storms entering the sample is small. Thus Fig. 1.8 only indicates that roughly the same number of storms were associated with spots in the pre-maximum and in the post-maximum phase. A comparison with the number distribution of sunspots in the different classes was not undertaken. When this is done, there is no significant discrepancy between the results of Le Squeren and those obtained by Fokker (1960). Fokker found that E, F and G spots showed the highest probability of being associated with noise storms. Both investigations also agree on the point that noise storms are normally associated with large, magnetically bipolar sunspot groups.

1.4. RELATIVE POSITION OF NOISE STORM CENTRES AND OPTICAL CENTRES OF ACTIVITY

The position of the sources producing the enhanced noise storm continuum emission has been determined at different frequencies by means of interferometers. Improved techniques and larger instruments have led to results which show an increasing amount of details and reveal quite complex phenomena. In the present chapter the more simple aspects of the problem are considered.

Two dimensional position determinations of the noise sources were carried out at 169 MHz by Le Squeren (1963). When the positions of the sources were plotted on optical maps of the photosphere, the sources always occurred in the vicinity of important optical centres of activity. But there was no definite geometrical relation between the position of the optical centre and the position of the noise source. This is clearly seen from Fig. 1.9.

Totally more than 200 noise storm centres were studied by Le Squeren and in the majority of the cases it was found that the centres were not radially above the optical centre. In addition it occurred that storm centres which were visible for

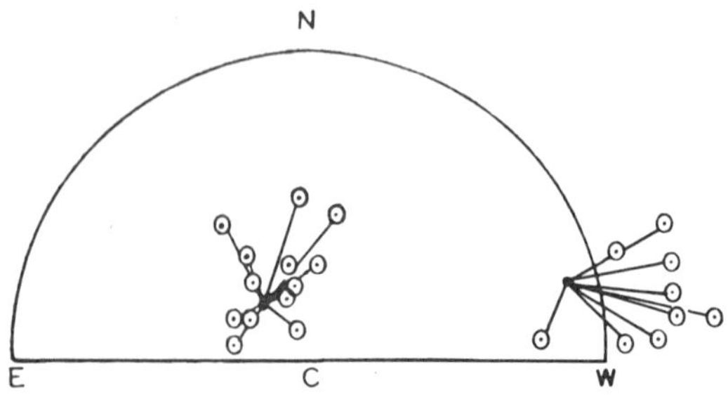

Fig. 1.9. Relative position of noise storm centres (o) and their associated optical centres. Optical centres were chosen which had positions near to the points marked ● (Le Squeren, 1963).

several days changed in position relative to the optical centre. Quite considerable changes might take place, as appears from Fig. 1.10. Considerable variations in the behaviour of individual storm centres thus seem to be allowed.

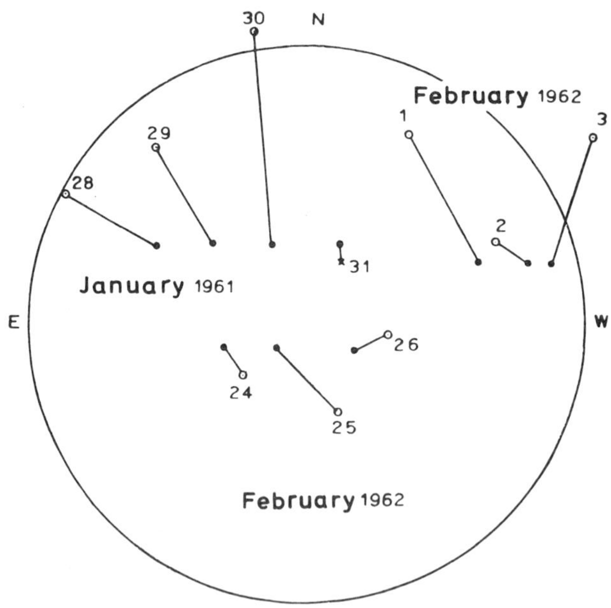

Fig. 1.10. Three examples of shifts in relative position between storm centres (o) and sunspots (●) (Le Squeren, 1963).

Different storm centres show a rather individual behaviour regarding position relative to optical centres of activity. But when the average properties of a large number of storm centres are determined, a more systematic picture emerges. Figure 1.11 was constructed by Le Squeren and shows that the "average" radio centre and optical centre are not radially aligned. The radio centre is displaced towards the pole and towards the north - south diameter of the sun. The deviation towards the pole is the larger of the two effects.

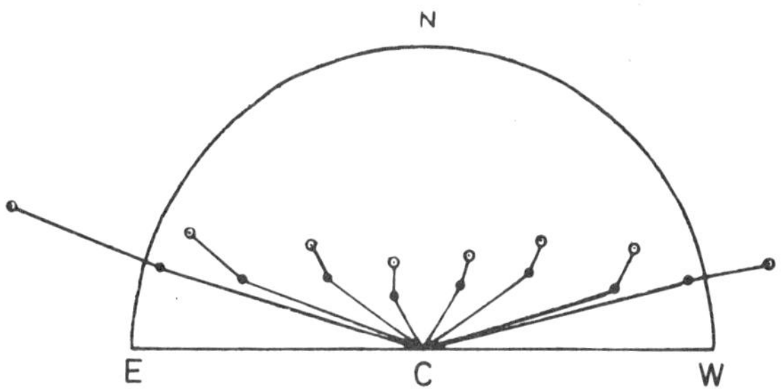

Fig. 1.11. Average position of noise storm centres (o) and associated optical centres (•) at different solar longitudes (Le Squeren, 1963).

The complicated relationship which exists between noise storm centres and optical centres of activity introduces difficulties when it comes to determine the height in the solar corona at which the storm centres occur.

1.5. THE HEIGHT OF THE STORM CENTRES

From interferometer observations the altitude of the storm centres above the photosphere may be determined in three different ways (Kundu, 1965):

(a) Direct method. When it is assumed that the radio noise centre is situated radially above the optical centre, it is a matter of simple geometry to determine the height of the source above the solar photosphere. But this method may give results which have large errors because the assumption of radial alignment of optical and radio centre does normally not hold.

(b) Measurement of the speed of rotation. The apparent speed of rotation of a storm source increases with the height of the source above the photosphere. When the daily displacement of the source is known, the height may be found. Best results are obtained when the source is near the meridian. It is then not necessary to make assumptions about the relative position of the optical centre and the storm centre.

(c) Altitude at the limb. When the associated optical centre has been identified, the height of the storm source may be determined when the optical centre approaches the solar limb. Deviations from radial alignment of the optical centre and the radio centre may here be less important. But refraction effects should not be neglected, in particular at lower frequencies.

The altitude of the noise storm sources has been determined at different observing frequencies.

At 97 MHz Payne-Scott and Little (1951) found that the height of noise storms was between 0.3 and 1 $R_\odot$. But the observational sample was small and the results must be regarded as uncertain.

At 169 MHz Le Squeren (1963) found that the average source height near the solar limb was 0.5 $R_\odot$. But there was a considerable scatter in the values which showed heights varying from 0.2 $R_\odot$ to about 1.7 $R_\odot$. A systematic variation in the source heights was present. The average height for central sources amounted to about 0.15 $R_\odot$ whereas it had increased to 0.5 $R_\odot$ for sources near the limb.

At 201 MHz Suzuki (1961) determined the height of the noise sources by assuming radial alignment with the optical centre, except for sources within a distance of 0.3 $R_\odot$ from the meridian for which the height was determined from their speed of rotation. The upper limit of noise storm heights found by Suzuki was 0.65 $R_\odot$. The average source height varied from 0.3 $R_\odot$ for central sources to about 0.5 $R_\odot$ for limb sources.

At 255 MHz Fokker (1960) carried out different height determinations. The results were the following:

Average distance from sub-terrestrial point	Height ($R_\odot$)	Number of cases
18°	0.32 ± 0.06	59
33°	0.26 ± 0.06	50
39°	0.44 ± 0.03	19
53°	0.45 ± 0.04	21
73°	0.36 ± 0.04	18

The source heights at distances of 18° and 33° were determined from the apparent shift in daily position whereas the remaining heights were obtained by assuming radial alignment of the optical and the radio sources. It appears from the results of Fokker that there is no important difference in height between sources near the centre of the solar disk and sources near the limb.

Fokker also compared the source position at 255 MHz with the 169 MHz position determined at Nancay. In general the same regions which produced excess noise at 169 MHz also produced noise at 255 MHz. When one single noise centre only was active, the root mean square deviation between the positions at the two frequencies was 0.13 $R_\odot$. This is very little. Occasionally differences between the two sources became apparent at the limb. Then the 255 MHz position was found to be systematically nearer to the limb than the 169 MHz position. The ratio between the derived heights in some selected cases ranged between 1.1 and 2.3. The average ratio was 1.5.

At 408 MHz Clavelier (1967) found an average height of 0.1 $R_\odot$ for noise storm sources. There was a tendency for limb sources to be situated somewhat higher than sources near the centre of the solar disk, but the effect was far less pronounced than at 169 MHz.

The investigations described above do not give unambiguous results, but they indicate that the average height of the noise storm source in the corona increases with decreasing frequency and that the source height is larger at the limb than near the centre of the solar disk.

More definite evidence of increasing source height with decreasing frequency was obtained by Gergely and Kundu (1975), who investigated six decametric storms and found average source heights of 0.7 and 1.1 $R_{\odot}$ at frequencies of 50 and 40 MHz respectively.

If it is supposed that the storm emission occurs at the electron plasma frequency, the observed source heights are considerably larger than those to be expected if the electron density follows the Baumbach - Allen model (Allen, 1947) for the quiet corona. Better agreement is found when it is assumed that the electron density amounts to 10 times the standard value for the quiet corona as obtained from the Baumbach - Allen equation. Enhanced electron densities are found in the corona above active regions. Here the corona is very inhomogeneous, and large deviations from the average coronal density may occur.

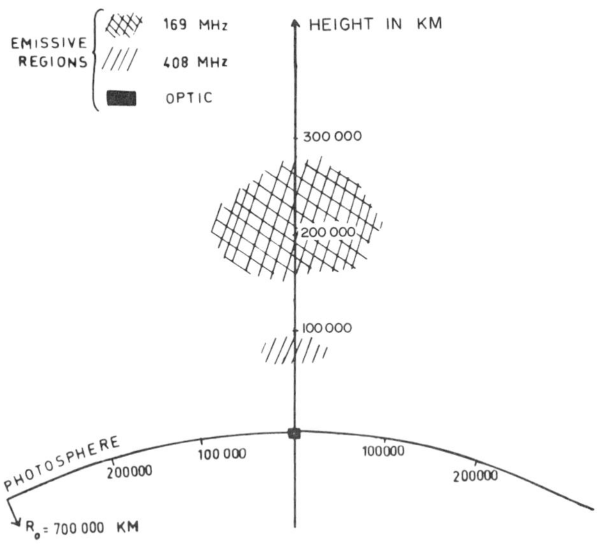

Fig. 1.12. Structure of simple noise storm sources at two frequencies (Clavelier, 1968).

The variation of the source height from the centre of the sun to the limb is most easily explained as a propagation effect of the emitted radio waves. Ray trajectories in a spherically symmetric corona were calculated by Jaeger and Westfold (1949). At a given frequency rays from near the centre of the solar disk may escape from deeper layers than those near the limb.

A model of a simple noise storm source at two frequencies is depicted in Fig. 1.12.

1.6. DOUBLE AND MULTIPLE SOURCES

When observations were undertaken with radio interferometers of poor spatial resolution, noise storm centres appeared single. But instruments with resolving power of a few minutes of arc revealed that the structure may be more complicated.

At 169 MHz Le Squeren (1963) detected a few cases in which a single region of optical activity seemed to be associated with two noise centres. This finding was firmly established by Clavelier (1967) who observed a number of double centres at 408 MHz with an interferometer having a resolution of 1.7′ arc. It seems likely that observations with instruments of still higher resolving power might reveal more cases of double or multiple source structure. A schematic picture of the structure of a double source as proposed by Clavelier is shown in Fig. 1.13.

It is tempting to associate double centres with a bipolar magnetic structure. Bipolar structure of noise storm sources was detected by Kai (1970) from observations undertaken at 80 MHz with the Culgoora radioheliograph. The two components were spatially separated by 3′ arc and strongly circulary polarized in opposite senses (Fig. 1.14).

New observations undertaken simultaneously at two frequencies (80 and 160 MHz) nicely confirm the results of Kai (Dulk and Nelson, 1973).

Bipolar structure is also a feature of other categories of solar radio emission: the slowly varying component, microwave bursts and stationary type IV bursts. Frequently, stationary type IV bursts are almost indistinguishable from noise storm sources.

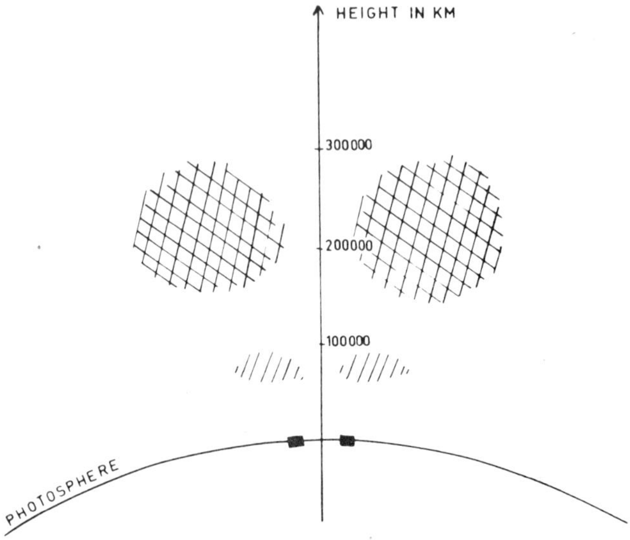

Fig. 1.13. Schematic structure of a double noise storm source at two frequencies (Clavelier, 1968).

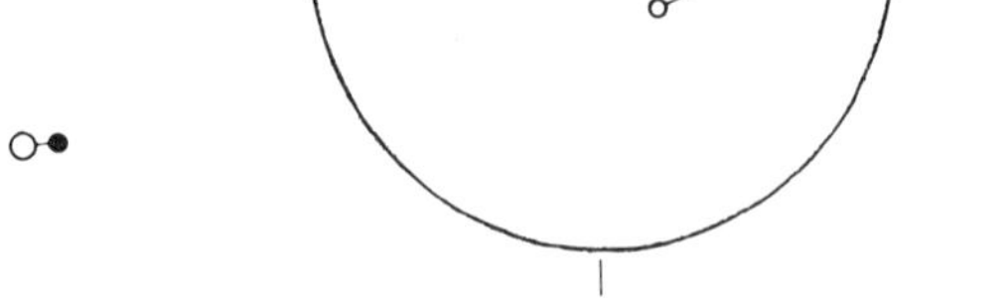

Fig. 1.14. Distribution of observed bipolar sources on the sun's disk. Dots and open circles represent RH and LH polarization sources respectively. The stronger sources are marked by the larger dots or open circles. (Kai, 1970.)

1.7. TYPE I BURST SOURCES

In the regions where the noise storm continuum emission is generated, quite localized sources of type I bursts are seen (Daigne, 1968; Kai *et al.*, 1968; Gnezdilov, 1970).

Records from the Nancay east - west multi-lobe interferometer were analysed by Daigne who found that the angular distribution of type I bursts may be very narrow, as illustrated in Fig. 1.15a. Here the mean positions of the type I bursts coincide with the position of the continuum. This seems to be a general property. In other cases there was evidence for the existence of two separate burst sources (Fig. 1.15b). Then the diameter of the continuum source was large.

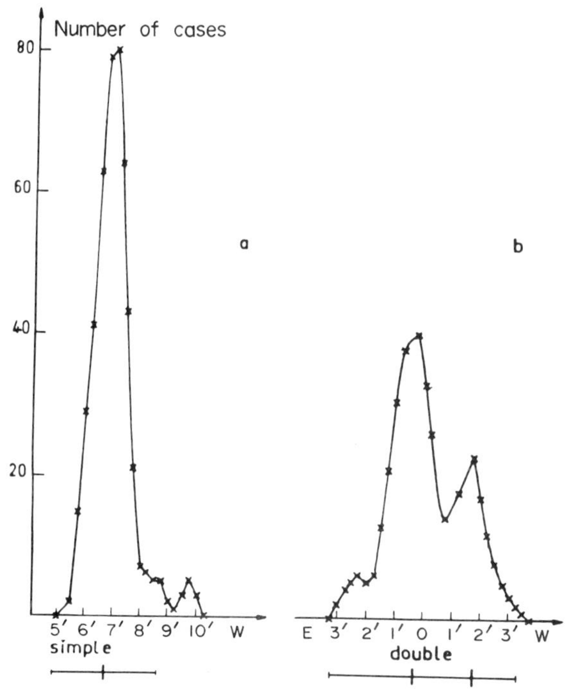

Fig. 1.15. (a) The east - west distribution of all type I bursts during a given noise storm. The mean positions of type I bursts are indicated by the crosses and the solid line represents approximately the continuum source. This noise storm shows one "simple" burst source; (b) this noise storm shows a "double" burst source (Daigne, 1968).

Dividing the total distribution into successive periods of 4 min, Daigne found diameters near to 1.2′ arc for all cases of simple centres.

The space - time properties of individual type I bursts were determined by Bougeret (1973b). From observations made at 169 MHz with the Nancay radio heliograph, detailed studies of the type I burst sources were possible. It was shown by Bougeret that the brightness distribution may assume a variety of shapes. These may be classified in three main groups according to the evolution in time of the peak brightness distribution:

stationary (i)	the peak brightness shows no change in position with time
drifting (d)	the peak moves as a function of time
split (s)	split in position and time

The different types are illustrated in Fig. 1.16. It appears to be a continuity between the three groups. Stationary bursts comprise the largest group; 65% of the bursts may be stationary, 15% drifting and 0.4% splitted. About 20% of the bursts are complex and difficult to classify.

In a particular case when all three burst types occurred, Bougeret found that each burst type was produced within a narrow region, a "quasi point region". Three different sources were resolved and each of them radiated bursts of one definite space - time pattern. The observational experience shows that one may speak of (i), (d) and (s) sources in the noise storm region.

During a three-month period of observation Bougeret found the following frequency of occurrence of sources emitting the different species of bursts: (i) sources: 70%; (d) sources: 24%; and (s) sources: 6%. The east - west distribution of the different sources was determined. The number emitting (i) bursts decreased rapidly from the centre to the limb, but (d) sources appeared preferentially on the limbs. The (s) sources occurred at longitudes larger than 10′ arc, and the second component was always closer to the centre of the solar disk than the first one.

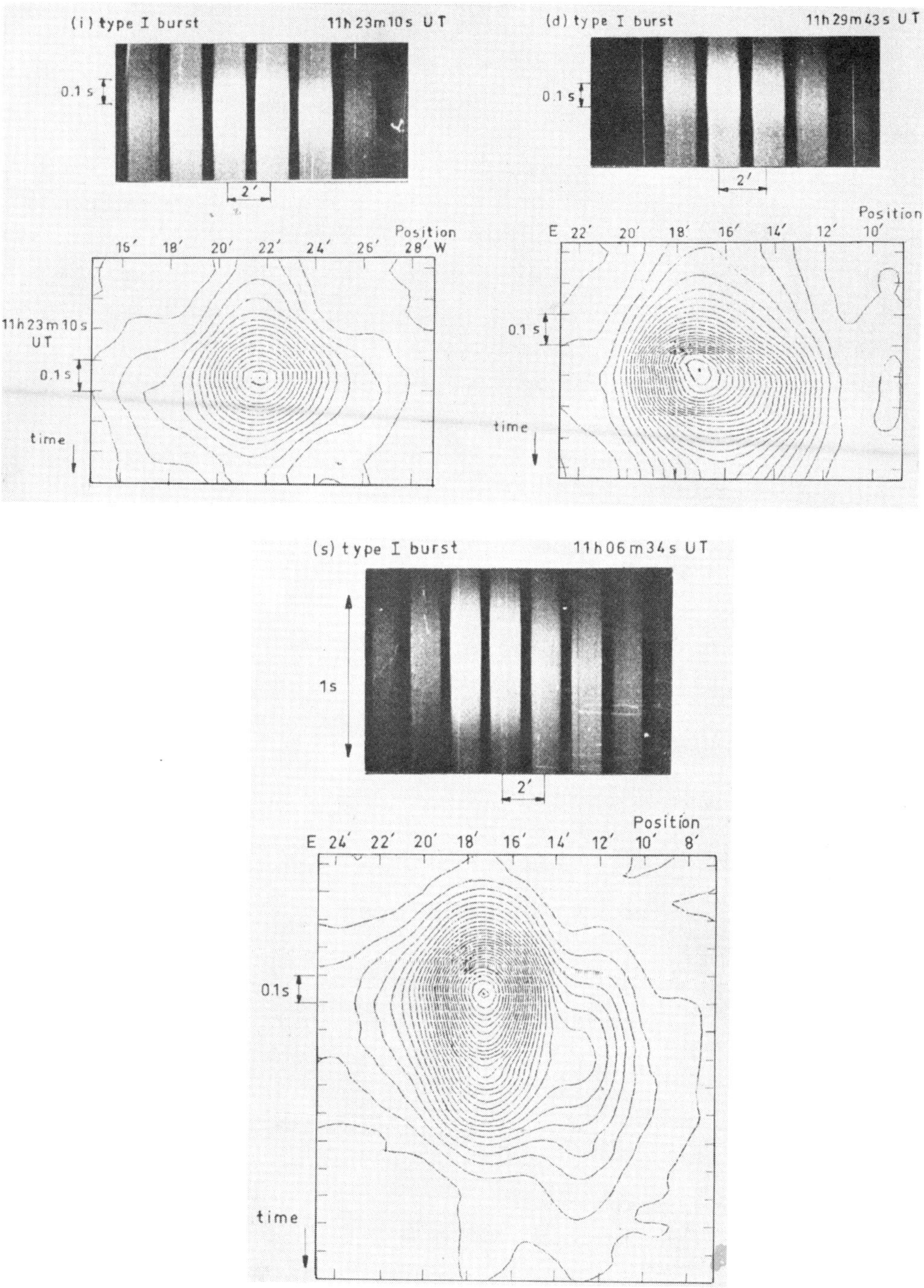

Fig. 1.16. The three different burst shapes observed on November 18, 1968: (i) stationary type I burst; (d) drifting type I burst; and (s) splitted type I burst. A photographic enlargement of the recording film and an isophote plot is given in each case. (Bougeret, 1973b.)

1.8. VARIATIONS IN THE SOURCE POSITION

Variations in the apparent position of noise storm continuum sources were found by Ryle (1952), Owren (1954) and Fokker (1957). The apparent position of the source seems to fluctuate around a mean value. At 200 MHz the fluctuations showed an amplitude of about 1′ arc. Further studies of variations in source positions were made at 200 MHz by Maltby (1958). The effect is shown in Fig. 1.17.

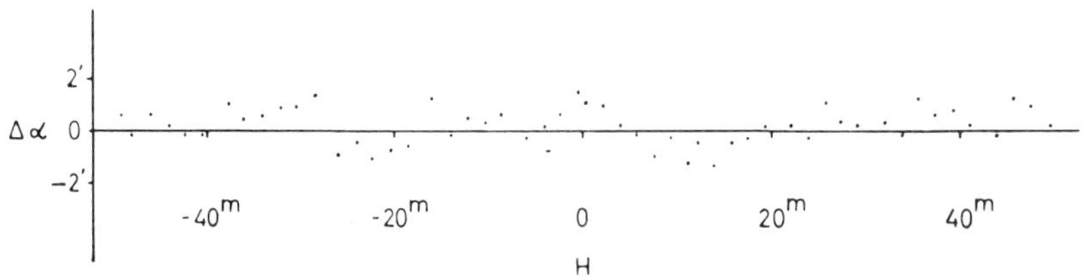

Fig. 1.17. The difference $\Delta\alpha$ between the observed and computed right ascension plotted against the hour angle H, July 13, 1958 (Maltby, 1958).

It is seen from the figure that the source position fluctuates around its mean value. Changes in position of 2′ arc or more may occur. The largest changes took place over time intervals of several minutes. The r.m.s. deviation in position during one hour was determined for 137 hours of observation. Large differences in the magnitude of the fluctuations in position were found between different days and at times on the same day. The upper and the lower limits of the r.m.s. deviation in position during intervals of one hour were 1.2 and 0.2′ arc. The average value for all one-hour intervals was $0.64 \pm 0.20'$ arc. The magnitude of the variation in source position showed poor correlation with the intensity of the radiation; only a slight tendency existed for larger positional fluctuations to occur when the intensity of the radiation was small. When the polarization of the radiation was taken into account, Maltby found the following results:

$$\sigma = 0.76 \pm 0.17' \text{ arc for } P \leq 0.9$$
$$\sigma = 0.50 \pm 0.11' \text{ arc for } P > 0.9,$$

where σ = r.m.s. deviation in position and P the degree of polarization. No real increase in the positional fluctuations was found as the sources moved towards the solar limb. Such an effect has been found by Little and Payne-Scott (1951), but they observed at a considerably lower frequency. If the fluctuations are caused by scattering effects, differences depending on the observing frequency are expected.

Oscillations have also been found in the positions of the type I burst sources. Daigne (1968) determined the mean position of type I bursts during time intervals of 4 min and detected periodic motions with periods of approximately 25 min and amplitudes as large as several minutes of arc at a frequency of 169 MHz. Sometimes these oscillations were superposed on a linear motion of several tens of kilometres per second directed towards the east.

An interesting case of oscillatory motion was described by Bougeret (1973b). The mean values and r.m.s. deviations of the east - west distribution of the positions of bursts produced in two noise centres 40′ arc apart over successive two-minute intervals were calculated. As is seen from Fig. 1.18, the two centres showed a correlated behaviour.

It appears likely that the positional fluctuations are caused by irregularities in the atmosphere of the earth. The possibility that variations in position at distant parts of the solar disk are highly correlated seems remote. Cross correlation of observations from different cites would help to answer the question of the origin of the fluctuations.

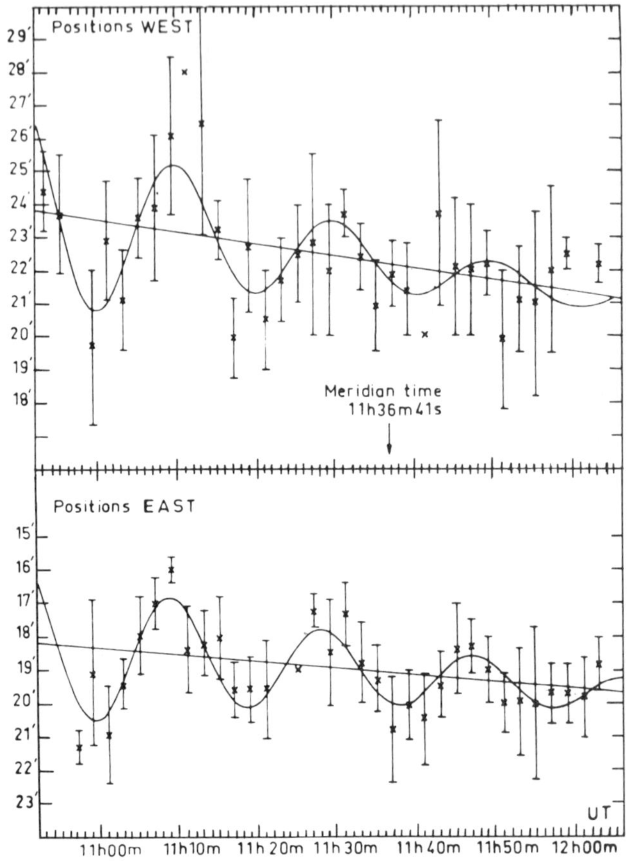

Fig. 1.18. Positional variations of type I bursts produced in two different noise centres on October 18, 1968. Damped sinusoids have been fitted independently for each centre. (Bougeret, 1973b.)

2. THE SPECTRUM OF NOISE STORMS

2.1. BACKGROUND CONTINUUM

Noise storms may occur between 50 and 500 MHz. The high frequency limit is normally between 200 and 300 MHz. In a relatively wide frequency band there is enhanced radiation from the sun consisting of a background continuum with short-lived, narrow band storm bursts (type I bursts) superposed. The spectrum of the background continuum may change with time, but it rarely shows steep variations with frequency. At times of very great burst activity, irregular "ripples" over a wide frequency band might be seen (Wild, 1951).

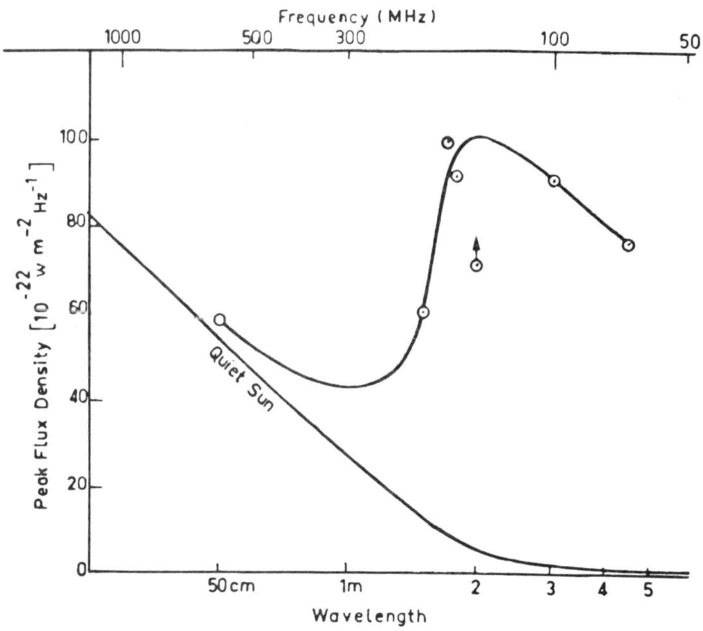

Fig. 2.1. The average spectrum for the recorded peak intensities during 28 storms in 1957/8. The quiet sun spectrum is shown for comparison. (Smerd, 1964.)

Using single-frequency observations undertaken at different observatories, Smerd (1964) determined the average spectrum of 28 isolated noise storms observed during 1957/8 (the International Geophysical Year). The average spectrum of the peak intensities is shown in Fig. 2.1.

At higher frequencies the storm emission is of roughly the same intensity as the radiation from the quiet sun. But at lower frequencies the contribution from the quiet sun is almost negligible as compared with the noise storm continuum. In Fig. 2.2 the average spectrum of the peak intensities above the quiet sun level is given. Two curves show the actual averaged intensity and that normalized for each storm with respect to its value at 200 MHz. At lower frequencies the curves differ significantly. This suggests that storms of different magnitude reached their peak intensities at different frequencies. Maximum intensity in

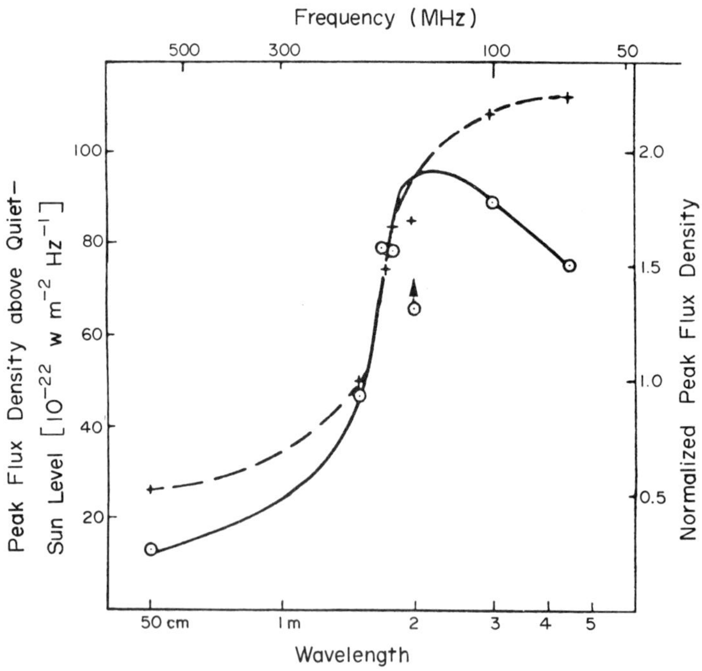

Fig. 2.2. Average spectrum of 28 noise storms. The solid curve corresponds to the average peak flux density and the dashed curve to the average peak flux densities normalized with respect to their values at 200 MHz. (Smerd, 1964.)

the 28 storms studied by Smerd showed a distribution as given in Fig. 2.3. The storm maximum was most likely to occur near 100 MHz. Dividing the material in three groups, 4 storms with the highest intensity at 67 MHz, 9 storms with maximum intensity at 100 MHz and 15 storms whose maxima were in the range 150 - 200 MHz, Smerd found that the actual average intensities and those

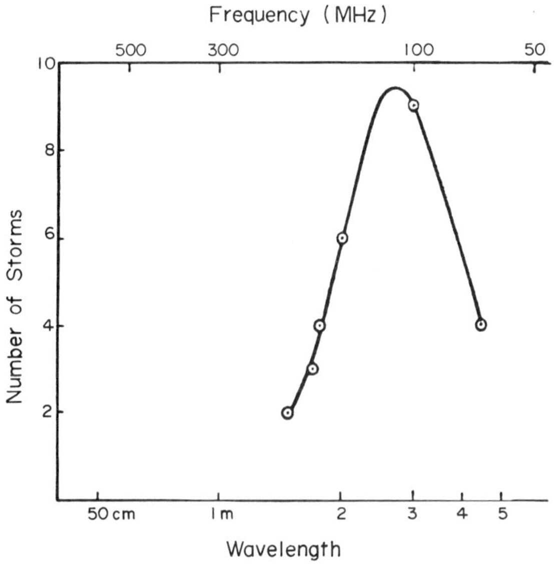

Fig. 2.3. The distribution of the 28 storms according to the frequency of maximum intensity (Smerd, 1964).

normalized with respect to the intensity at 200 MHz resulted in very similar spectra. He therefore concluded that by subtracting the quiet sun contribution and grouping the storms according to the frequency at which the storm reaches its maximum, representative storm spectra are obtained (Fig. 2.4). It then appears that the stronger the storm the lower the frequency of the storm maximum and the wider the storm spectrum.

The spectrum decreases on the high-frequency side. This makes it incompatible with thermal emission unless gyro resonance in strong magnetic fields accounts for the peak. Attempting to fit the relation $S \alpha f^n$ (where S is flux density, f is the frequency and n is the spectral index) to the spectrum at frequencies

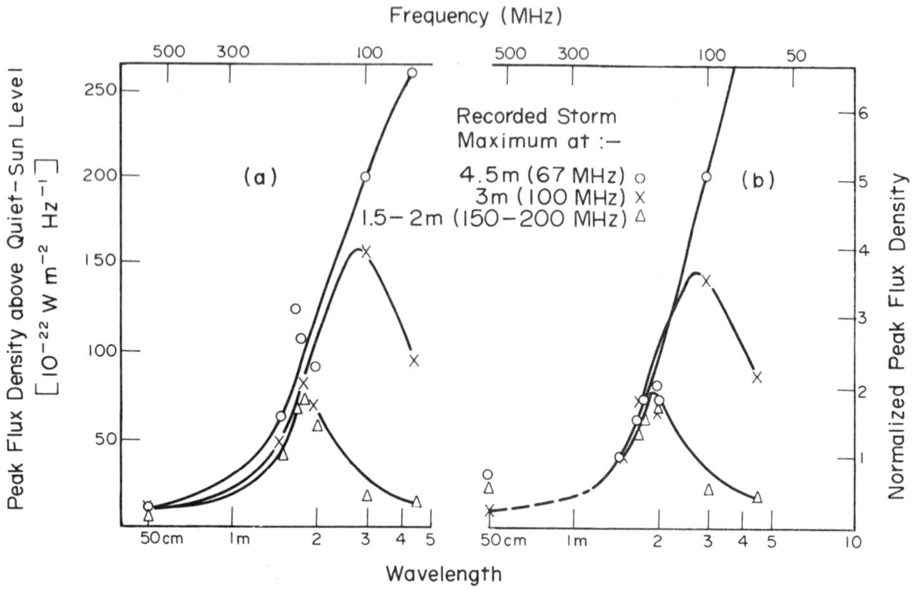

Fig. 2.4. Storm spectra grouped according to frequency at which the greatest intensity was recorded: (a) average peak flux densities; (b) average peak flux densities normalized with respect to their value at 200 MHz (Smerd, 1964).

below the peak intensity, it seems that $n > 2$. This supports the conclusion that the storm radiation cannot be explained as thermal bremsstrahlung. It seems likely that the storm emission is of non-thermal nature.

Many questions regarding the noise storm continuum spectrum are still not answered. Few investigations of the properties of the continuum spectrum have been published in recent years. A matter of particular interest is the evolution in time of the continuum spectrum and the relation of the continuum spectrum to other observable properties of the sun. The low-frequency end of the noise storm spectrum is still unknown. It is also a matter of fact that during storms the continuum emission shows fluctuations, but the spectral properties of these have not been investigated.

2.2. STORM BURSTS

Storm bursts are emitted in very large numbers during noise storms. In very variable storms it may be difficult to separate single bursts. It happens that the bursts are so closely connected in time and frequency that they overlap. Another source of confusion is that more than one active centre may be present at the sun at the same time. These factors have the effect to make it more difficult to determine the properties of individual bursts and to relate variations in the physical properties of bursts to special features of the active region where the source is situated.

Spectral observations using a receiver of high resolution in time and frequency were undertaken by Elgarøy (1961) in order to investigate the fine structure of solar radio emissions. The receiver was well suited for the observation of storm bursts, and an extended programme aiming at such investigations evolved. Observations are now undertaken at different frequencies in the region from 130 to 550 MHz with three different spectrographs. Other investigations have been undertaken with multi-channel receivers (de Groot, 1966; Dröge, 1967). Spectrographic observations of bursts at lower frequencies have been undertaken amongst others by Warwick and Dulk (1969), Ellis (1969) and Boischot *et al.* (1971). A more complete list of instruments is given in Table 2.1.

Studying the fine structure of type I radiation, Elgarøy (1961) found that storm bursts, classified as type I from observations with wide-band radio spectrographs, might have quite different spectral properties. Therefore a subclassification was introduced using the following denotations:

Type I(s)	stable burst, no frequency drift.
Type I(d)	burst with frequency drift from higher to lower frequencies.
Type I(r)	burst with reverse frequency drift, i.e. from lower to higher frequencies.
Type I(c)	curved burst, irregular frequency drift.
Type I(o)	burst with oscillating mid-frequency.

TABLE 2.1

High-resolution instruments for solar observations (de la Noë, 1972)

Authors	Instrument	Frequency range (MHz)	Frequency resolution (KHz)	Time resolution	Polarization
Elgarøy (1961)	Spectrographs	170 - 140 340 - 310	300	22 ms	
Wild (1967)	Interferometer (2′ arc)	80	1000	0.5 ms	L, R
Vinokur (1968)	Interferometer (3,4' arc)	169	500	100 ms	
de Groot and van Nieuwkoop (1970)	Spectrograph multi-channel	320 - 160	900 KHz/channel channel distance 2.7 MHz	10 ms	L, R
Markeev and Chernov (1970)	Spectrograph	220 - 190	250	20 ms	L, R
Tarnstrom and Philip (1972)	Spectrograph	255 - 235 190 - 180 155 - 135	1000	10 ms in 1967 5 ms in 1968	
Ellis and McCulloch (1966)	Spectrographs	230 - 8 28 - 24	200 50	5 ms 5 ms	
Ellis and McCulloch (1967)	Spectrographs	60 - 46 46 - 36 36 - 28 28 - 24	50 30 30 30	20 ms	

Table 2.1 Continued

Ellis (1969)	Spectrographs:				L, R
	before 1967	34 - 24	20		
		34 - 28	20	20 ms	
		34 - 28	25		
		60 - 28			
	after 1967	28 - 27	2		
		35 - 26	25	40 ms	
		50 - 35	25	then	
		200 - 25	50	20 ms	
Yoh and James (1967)	Fixed-frequency receivers	38,15)(38,10 38,20)(38,30 38,25) or 38,40)(38,40)	16	Not indicated	
Warwick and Dulk (1969)	Spectrograph	37 - 24	60	100 ms	Linear, L, R
Boischot *et al.* (1971)	Spectrograph	80 - 40 40 - 20	100 15	20 ms	L, R
Sastry (1969)	Fixed frequency receiver	25	13	10 ms	
Sastry (1971, 1972)	Multi-channel	Around 25 MHz separated by 100 or 200 KHz	6,5 or 13	15 ms	L, R 1972 only
Mosier and Fainberg (1972)	Multi-channel	Around 40	100	20 ms	
Barrow and Saunders (1972)	Fixed-frequency receivers	22 and 18	2, 4	5 ms	L, R

Type I(f) narrow-band burst with fast frequency drift (Type I(fd) and Type I(fr).

Type I(x) dubious case, not possible to classify.

Examples of the most important types are given in Fig. 2.5. The most rarely occurring is type I(o) and later experience has shown that this type perhaps not deserves a special designation.

A few remarks concerning some of the subclasses may be necessary.

Type I(s) bursts. This is a "real" storm burst with narrow bandwidth, short duration and a stable centre frequency. Whether there is fine structure within such bursts is a question that is not yet settled. Observations with short time constant ($\tau < 0.001$ s) at 159 and 318 MHz have not yet revealed sub-structure in such bursts (Peersen, 1967). Perhaps observations of very high sensitivity might reveal interesting features. On many dynamic records there are definitely indications of fine structure (Elgarøy, 1961).

Type I(d). Storm bursts with frequency drift occur. In many cases no signs of fine structure are seen, indicating that one is dealing with a single event. Also bursts with frequency drift in the reverse direction (from lower to higher frequencies) occur. The frequency drift velocity may be about 5 MHz s^{-1}, which is intermediate between the drift velocity of type II and type III bursts.

Type I(fd). Some bursts have a rapid frequency drift of about the same order of magnitude as is characteristic of type III bursts, i.e. 80-100 MHz s^{-1}. But they have a smaller band-width and extend over a smaller frequency range and they occur during type I storms. Such bursts may constitute a variant of the type III class of solar bursts or an "intermediate" case.

De Groot (1960, 1966) observed on frequencies between about 268 and 396 MHz with a seven-channel receiver. Each channel was 1 MHz wide, the spacing between the channels was 2 MHz and the time constant amounted to 0.01 s.

From the recordings, de Groot constructed contour diagrams and studied the dynamic spectra of the bursts. He concluded that weak, single bursts showed no frequency drift and that the time

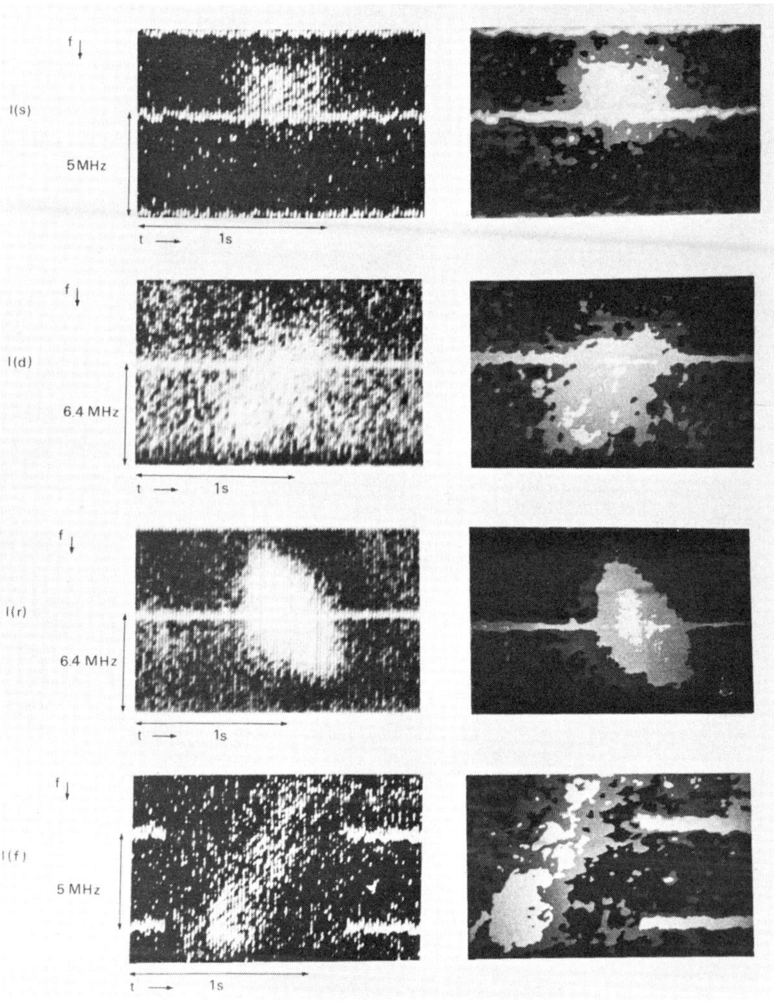

Fig. 2.5. Dynamic spectra of type I bursts of different classes. Photographs to the left, contour diagrams to the right.

and the frequency profiles of a single burst were independent. Thus the two-dimensional intensity distribution may be written

$$I(f,t) = I_1(f)I_2(t).$$

According to de Groot the frequency drift should be a property of complex bursts and not of single bursts. But this view is no longer maintained, and it is definitely not supported by further swept frequency observations.

Observations on low frequencies by Ellis (1969) and by Warwick and Dulk (1969) support the results of Elgarøy, namely that frequency drift may be observed in single storm bursts.

Ellis found bursts of narrow bandwidth that had a frequency - time slope of 1 - 2 MHz s^{-1}. They occurred during noise storms in groups of 10 - 20 bursts, but sometimes they also occurred isolated. In many cases they appeared to be paired. Warwick and Dulk (1969) observed the band from 24 to 37 MHz with high time and frequency resolution. Here they found narrow-band short-lasting bursts drifting in either direction ($df/dt \simeq 0.5$-2 MHz s^{-1}) and bursts without frequency drift.

The frequency of occurrence of the different burst types during noise storms was determined by Elgarøy. Later experience has shown that there may be large differences between different noise storms. But the results in Table 2.2 may be taken as characteristic for a normal type I storm.

TABLE 2.2

Distribution of type I bursts of different subclasses

Date (1959)	Percentage							Number of bursts
	I(s)	I(d)	I(r)	I(fd)	I(fr)	I(c)	I(o)	
August 13	45	13	15	6.5	12	6.5	2	174
September 10	65	10	12	2	3	6	2	189

If the material is divided in bursts without frequency drift, bursts with negative frequency drift and bursts with positive frequency drift, one finds the results shown in Table 2.3.

TABLE 2.3

Distribution of type I bursts with and without frequency drift

Date	Number of bursts			Percentage		
(1959)	I(s)	I(d-fd)	I(r-fr)	I(s)	I(d+fd)	I(r+fr)
August 13	78	34	47	49	21	30
September 10	123	22	29	71	13	16
September 11	238	37	35	77	12	11

As appears from the tables, the class I(s) comprises most bursts. But the other types of bursts occur frequently enough to justify a further division. One should get a wrong and too simplified picture of the storm phenomenon if the other classes were neglected.

It is also seen from the tables that there is an excess of bursts with reverse frequency drift as compared with bursts of positive drift. In this connection it may be mentioned that an investigation of burst-chains undertaken by Elgarøy and Ugland (1970) gives a similar result for chains.

The distribution of bursts in the different subclasses may change from one storm to another and changes may also be noted during the same storm.

High-resolution observations in the frequency range 190 - 220 MHz were performed by Chernov *et al.* (1972). They were also able to measure the degree of polarization of type I bursts as a function of time. Bursts from two noise storms were analysed; the first occurring between May 17-23, 1969, and the other between June 7-13, 1969.

Dividing the observed type I bursts into the classes I(s), I(d) and I(r) as introduced by Elgarøy (1961), Chernov *et al.*

found the following distribution for the noise storm in May 1969: the percentage of type I(s) bursts was 80% on May 17 and decreased steadily and amounted to 51% on May 21. There was a corresponding increase in the relative number of type I(r) and I(d) bursts. During the total period the number of type I(r) bursts was twice as great as that of type I(d) bursts.

During the storm the average duration of the bursts decreased monotonously from 0.46 s on May 17 to 0.23 s on May 21. For the whole storm the average duration of the bursts amounted to 0.37 s and the average bandwidth was 4.6 MHz.

During the June storm the average duration of the type I bursts showed an increase from day to day. For the storm as a whole the average duration amounted to 0.45 s which was longer than for the storm in May, whereas the bandwidth was approximately the same, i.e. 4.7 MHz.

Chernov *et al.* arrived at the following conclusions from their study: Noise storm bursts possess definite spectral polarizational and other characteristics which vary during the development of the storm. This regards in particular the relative number of drifting bursts, but there is also a systematic variation of both duration and polarization of isolated bursts.

Noise storms differ from one another as to the average value of a number of burst parameters and show different characteristic changes with time. This is illustrated, for example, by the observation that more short-lived type I bursts occurred in the May storm than in the June storm, and that in these storms the duration of the type I bursts changed with time in opposite manners.

Considerable variations may be found in the behaviour of the polarization of the bursts. During May 17 and 18, when the bursts were comparatively weakly polarized, most of the bursts showed a changing degree of polarization. As for type I(s) bursts, the most typical was maximum polarization at the moment of maximum intensity, on the one hand, and the steady decrease of polarization on the other hand. During the June storm the polarization did not change with time for practically all the bursts, even when it was relatively small. Further polarization observations are desirable.

As the source of the May storm was near the eastern limb and the noise storm in June was generated over the central part of the disk, one may suppose that the differences of storms are conditioned by the effects of heliographic longitude. However, it is possible that certain regularities in the variation of the parameters and the noise storm structure are connected directly with the process of storm development, and the pecularities of those changes, distinguishing one storm from another, are connected with the development of the corresponding active regions. Properties such as the intensity and the configuration of magnetic field would seem to govern the characteristics of a particular noise storm. The final answer may only be obtained after analysis of a great number of noise storms in different stages of development and generated in different active regions.

As regards polarization, one property was common to both storms and may be generally valid: a gradual increase of the degree of polarization of individual bursts and chains at the initial stage of storm development. After the degree of polarization of the burst component attained 100%, it remained at the same level to the end of the storm. In this case the degree of polarization was 100% in the course of chains and individual type I bursts independently of peculiarities in their spectrum or time profile.

Sometimes great changes in the burst properties occur during a noise storm. One such case was described by Eckhoff (1966), who determined the frequency drift of bursts in a storm occurring on August 15, 1961. It was found that most bursts had no or very small frequency drift velocities, but it also appeared that some had a clear-cut frequency drift. Eckhoff found that in this storm neither positive nor negative frequency drifts dominated any parts of the periods which were examined. Plotting the relative number of bursts of different frequency drift velocities against drift velocity, a dominating maximum at zero drift velocity was found, but also two smaller maxima around ± 17.5 MHz s^{-1} occurred. During a time interval of 4 min the average burst duration decreased considerably, and simultaneously the distribution of frequency drift velocities showed an unusual behaviour. Bursts with high frequency drift velocity [(df/dt > 90 MHz s^{-1})] then were most numerous. The properties of bursts of this kind, denoted flash bursts (Elgarøy, 1962) or more commonly spike bursts (de Groot, 1962), have been treated extensively by Tarnstrom and Philip (1972). The difference between flash bursts and normal storm bursts is illustrated in Fig. 2.6.

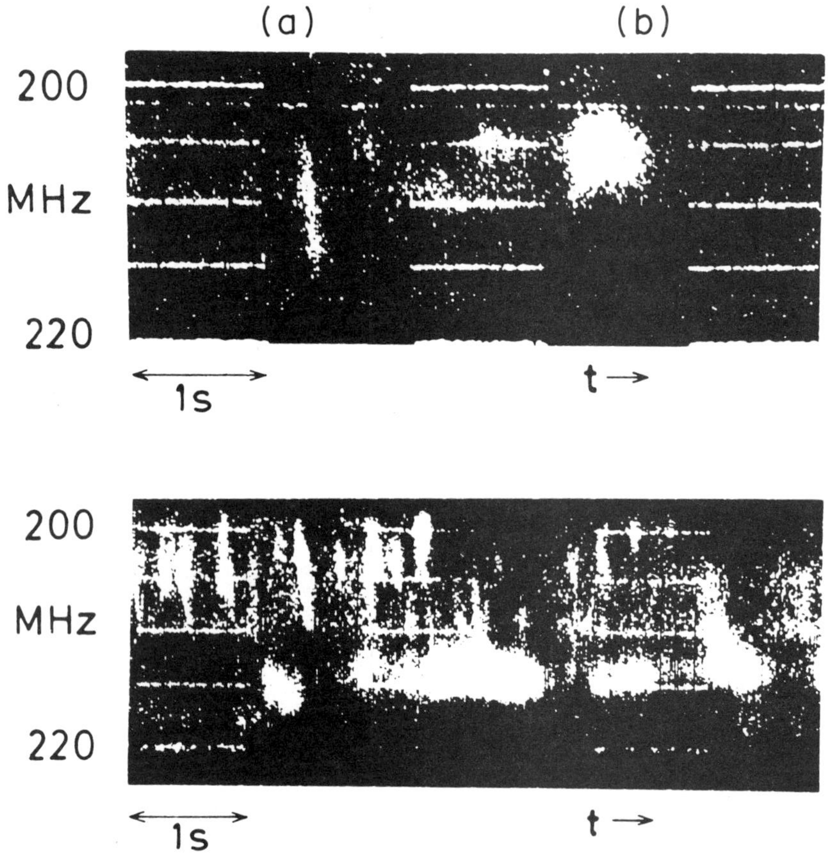

Fig. 2.6. Above: (a) Flash burst; (b) ordinary type I burst. Below: dynamic spectrum showing flash bursts and ordinary type I bursts.

2.3. DURATION OF STORM BURSTS

For a long time it was known that the duration of storm bursts was short. Reber (1946) reported that on 480 MHz he had observed a radio storm on the sun and that there were variations from second to second. In 1951 Wild published the first results in which one could be sure that storm bursts were observed. He measured the lifetime between points of quarter – maximum intensity and found the curve shown in Fig. 2.7. There was a large dispersion of lifetimes. Because of technical difficulties the lower end of the distribution curve could not be determined.

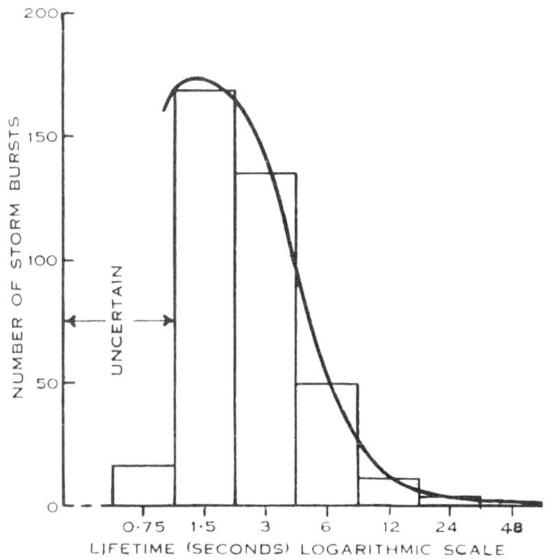

Fig. 2.7. Histogram showing the distribution of storm bursts with lifetime in the frequency region 70 - 130 MHz (Wild, 1951).

It seems that many of the long duration bursts, on using equipment of higher resolving power, may be resolved into several bursts. The bursts of longest lifetime in Wild's histogram are probably composite bursts.

The distribution of burst durations on lower frequencies may probably be more in accordance with the results obtained by Sastry (1969). Figure 2.8 shows that most bursts observed at

25 MHz have a duration t less than 2s. The average duration of the bursts with $t < 2$ s was $\bar{t} = 1.01$ s. Sometimes bursts of extremely short durations, i.e. of the order of 0.2 s, were recorded.

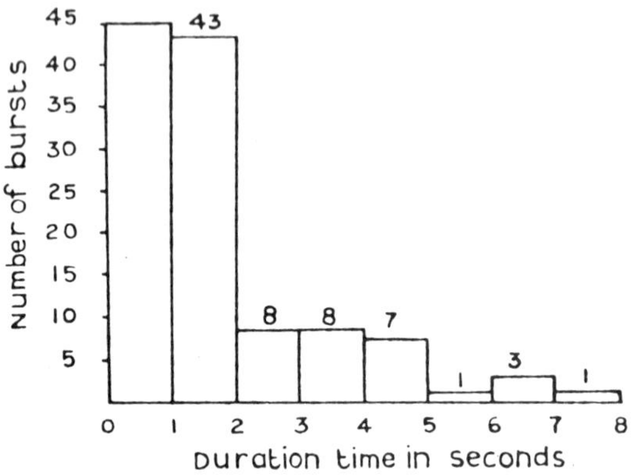

Fig. 2.8. Duration of bursts at 25 MHz (Sastry, 1969).

Similar results were found by Ellis and McCulloch (1967) and by Yoh and James (1967). The lower limit of the burst duration in their cases was about 0.2 s, the average duration was 1.2 s and the longest durations were about 3 - 3.5 s. A paper by Warwick and Dulk (1969) describes similar observations and results which agree with the others.

At 100 MHz Vitkevich and Gorelova (1961) found an upper limit of 2 s to the burst duration, an average value of ≈ 0.5 s and a lower limit which was somewhere between 0.1 and 0.2 s.

At 150 MHz Ugland (1966) found values of t from 0.1 s to about 1 s, the average value being ≈ 0.5 s. Using a high-resolution radiometer on 159 MHz, Peersen (1970) found a lower limit somewhat less than 0.1 s, and the longest bursts had durations of about 1.2 s. The average was about 0.3 s. On 200 MHz Elgarøy and Hauge (1958) found no bursts with $t > 0.8$ s (two-channel observations). This is in agreement with the results of Eckhoff (1966). During about an hour's observations on August 15, 1961, almost 3000 bursts were recorded with the sweep receiver in the 200 - 220 MHz range. The distribution of durations was as shown in Fig. 2.9. Using an isodensitometer bursts with t > 0.8 s were shown to be compound.

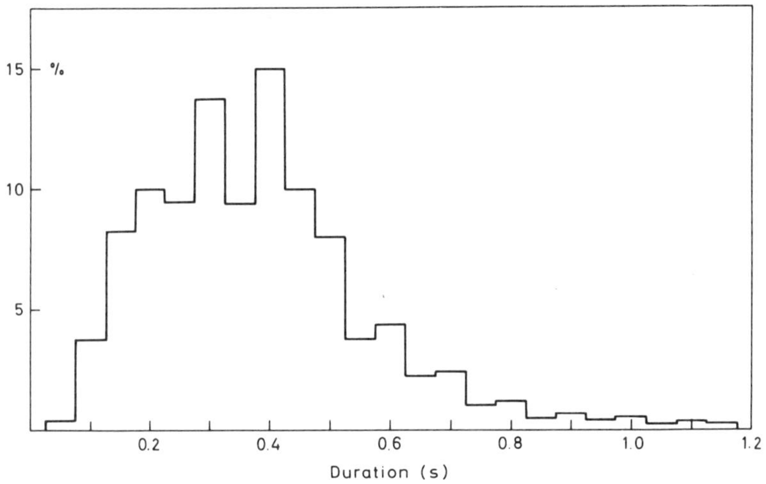

Fig. 2.9. Histogram showing the relative number of bursts of different durations recorded during the period 06.40 UT to 07.50 UT, August 15, 1961 (Eckhoff, 1966).

On 318 MHz Peersen (1970) found an average duration of 0.2 s, a lower limit of about 0.05 s and an upper limit of about 0.5 s in a sample comprising 584 bursts.

On 400 MHz good data were obtained by de Groot (1966). His results appear in Fig. 2.10.

There appears to be a maximum at 0.18 s, a property which de Groot usually found in 400 MHz storms. The rising part of the distribution curve is relatively steep while there is a long tail towards longer lifetimes. In general, de Groot found a mean value of about 0.18 s, which may be rounded off to 0.2 s. The shape of the tail of the distribution may be influenced by clustering of single bursts or by mixed populations of bursts.

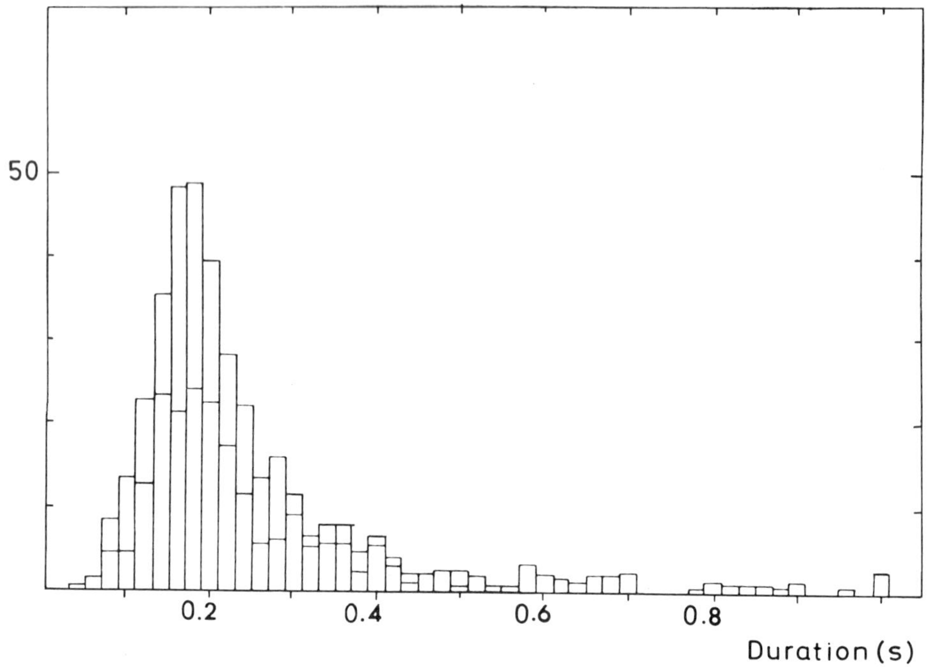

Fig. 2.10. Histogram of the distribution of the duration of single storm bursts at 400 MHz (de Groot, 1966).

de Groot constructed histograms for four different periods during a storm occurring on April 9, 1957. The results are shown in Fig. 2.11.

There is an indication that the average duration decreased gradually. But according to de Groot this was not a general property in his observational material. However, it happens that the most common lifetime of storm bursts decreases down to 0.1 s in particular noise storms fragments.

A variation of this type was very clearly demonstrated by Eckhoff (1966).

It is seen from Fig. 2.12 that between 0.730 UT and 07.40 UT on August 15, 1961, a clear change in the distribution of burst durations occur. During this period the maximum burst duration decreases considerably and a large number of bursts with very short durations occur. The same phenomenon has also been noted

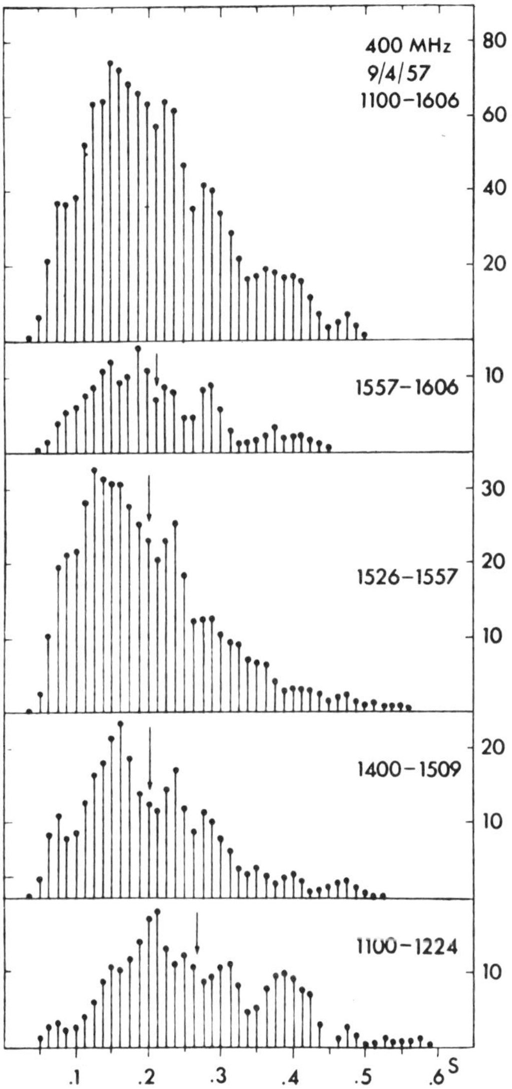

Fig. 2.11. Histograms of burst durations at different times during a noise storm. The distribution for the storm as a whole is at the top of the figure. (de Groot, 1966.)

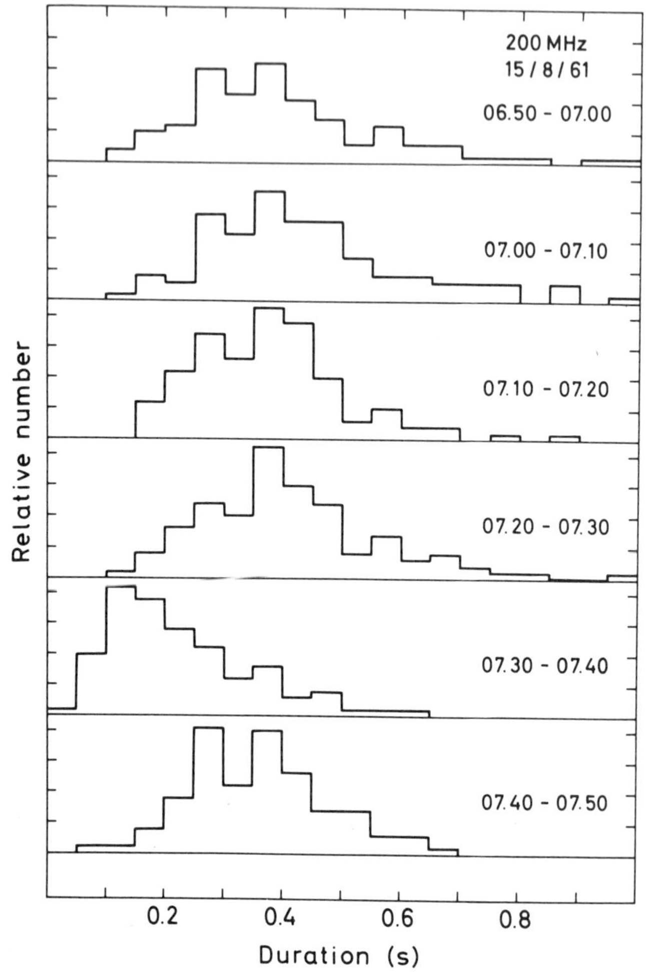

Fig. 2.12. The distribution of burst durations for successive periods of time. (After Eckhoff, 1966.)

by de Groot (1966) who found that the most common lifetime of storm bursts can decrease down to 0.1 s in particular noise storm fragments. It is possible that the very short-lasting bursts ("flash bursts") are caused by a different mechanism as compared with normal storm bursts. Such bursts occur on all wavelengths were storm bursts are found, but their duration seem to depend very little on the wavelength. They may be more frequent in some storms than in others and they may at times occur in large clusters. The difference in duration between flash bursts and storm bursts clearly appears from Fig. 2.13.

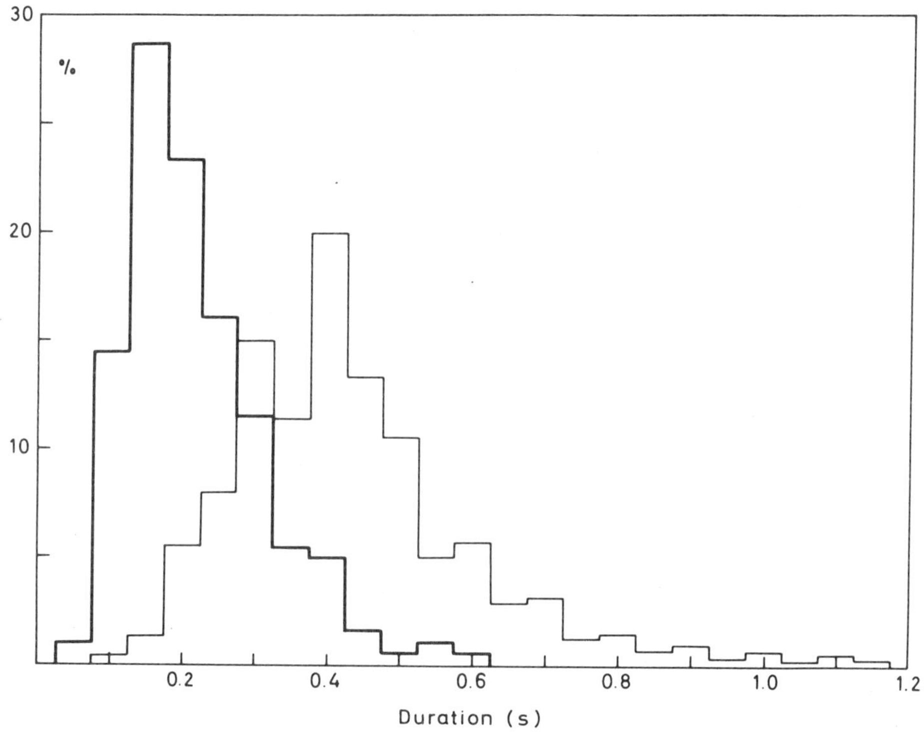

Fig. 2.13. Histograms showing the relative number of bursts of different durations for flash bursts and storm bursts (Eckhoff, 1966).

Elgarøy and Hauge (1958) determined the average value of the burst duration on 9 days in the period from April to August 1957. The observations were made at 200 MHz, and showed that the average duration varied between 0.21 and 0.33 s. Thus small changes occur in the burst duration from one storm to another.

It would be valuable to investigate whether there is any variation of the burst duration with the position of the source on the solar disk. Kai (1962) claimed that the average duration of 200 MHz storm bursts was longer at the limb than near the centre. But these results need confirmation. As yet no such investigations have been reported.

The results described in the present chapter may be summarized in a diagram showing the average burst duration and the dispersion in burst duration as a function of frequency (Fig. 2.14).

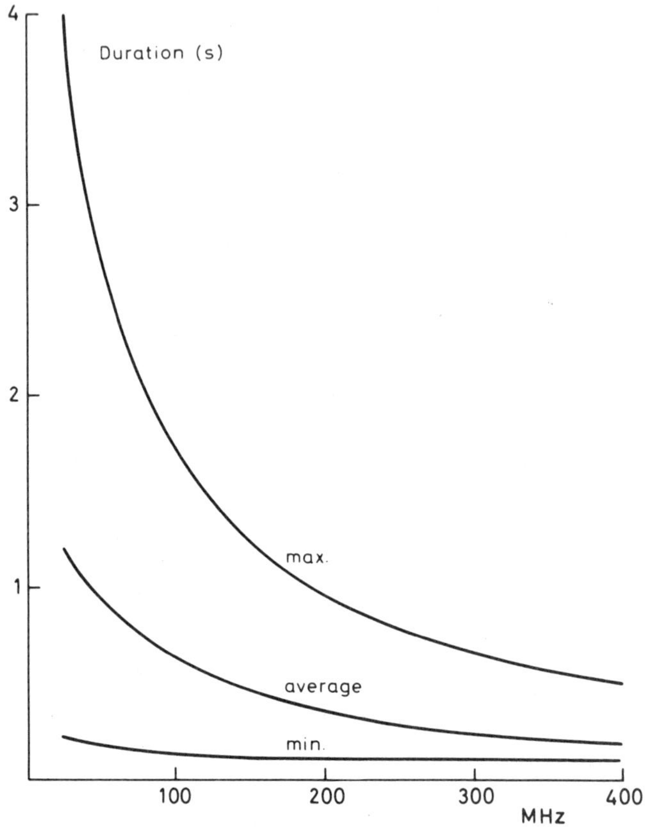

Fig. 2.14. Average burst duration at different frequencies. The upper and the lower limits of the duration are indicated.

It appears that the lower limit of burst duration is independent of observing frequency from 400 MHz down to about 150 MHz, and then increases somewhat. The average duration of the bursts increases more steadily with decreasing frequency, and the same seems to be the case with the upper limit of burst durations. It is a general property at all frequencies that the distribution curve of the number of bursts versus duration shows a steep increase from small values and up to the maximum, and then a longer tail towards longer durations.

The relative dispersion in the distribution of burst durations is almost constant. Choosing typical values for t_{max}, $t_{average}$ and t_{min} from Fig. 2.14, one gets

$$\frac{t_{max} - t_{min}}{t_{average}} \simeq 2.$$

The behaviour of the lower limit of the burst duration as a function of frequency may inspire some speculations. It is possible that this part is determined by flash bursts, and that these show little variation with frequency.

2.4. THE BANDWIDTH

The bandwidth of type I bursts is small. Simultaneous records separated by only a few megahertz are often widely different.

The first determinations of the bandwidth were made from two-channel observations from which a statistical value of the bandwidth could be derived (Blum, 1953; Vitkevich, 1955; de Jager and van't Veer, 1958; Elgarøy and Hauge, 1958; Fokker, 1960). But more reliable values are found from sweep-frequency observations or multichannel observations.

On 400 MHz de Groot (1959) found that about 50% of the bursts had bandwidths less than 10 MHz, the average value being 6 MHz. The other bursts had larger bandwidths (>12 MHz) and may perhaps not be storm bursts. It seems that the bandwidth of 400 MHz storm bursts may vary from about 3 - 4 MHz to about 9 - 10 MHz.

Several histograms of burst bandwidths in the frequency region 274 - 330 MHz have been published by de Groot (1960, 1966). Some are shown in Fig. 2.15. As is seen from the figure, the histograms of burst bandwidths on the different frequencies

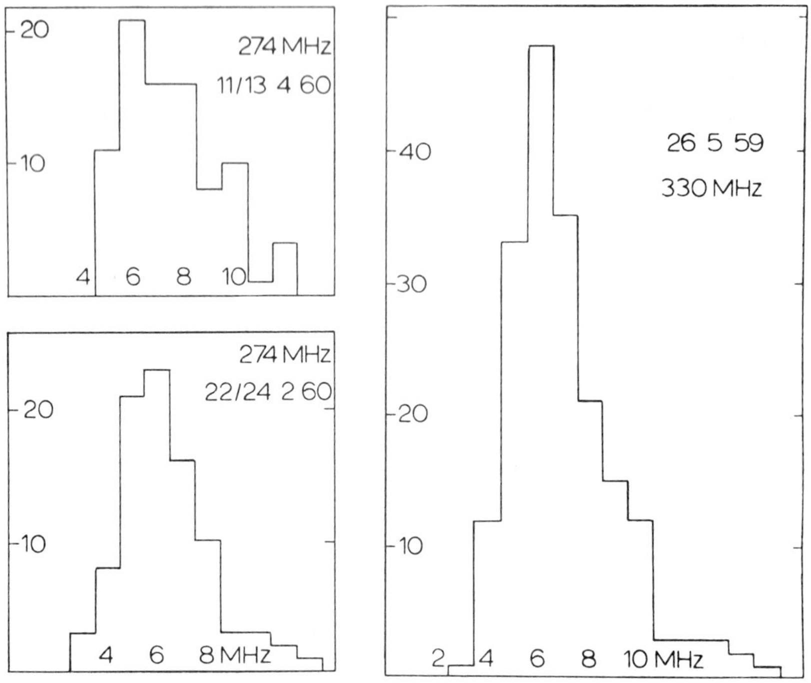

Fig. 2.15. Histograms of the bandwidth of type I bursts at 274 and 330 MHz (de Groot, 1966).

around 300 MHz are very similar. The average bandwidth usually is between 6 and 8 MHz. The maxima in the histograms are around 6 MHz, and almost all bursts fall within the interval 4 - 10 MHz. de Groot attempted to investigate if there was any correlation between the half-widths and the wavelength within the frequency range 274 - 400 MHz, but the results were inconclusive because of the considerable variations in the characteristic parameters of different noise storms at different wavelengths. A considerable drawback was that he had not simultaneous observations at different frequencies.

Using swept-frequency records in the range 315 - 335 MHz, Elgarøy and Ugland (1966) determined bandwidths of type I bursts and found that they occurred within the range from 4 to 15 MHz with maxima in the distributions between 6 and 8 MHz. Typical average values were 8.4, 9.1, 7.6 and 7.2 MHz. This is in good agreement with the results obtained by de Groot.

In the 200 MHz range the bandwidth has been determined on swept-frequency records by Elgarøy (1961). Some of the results are shown in Fig. 2.16.

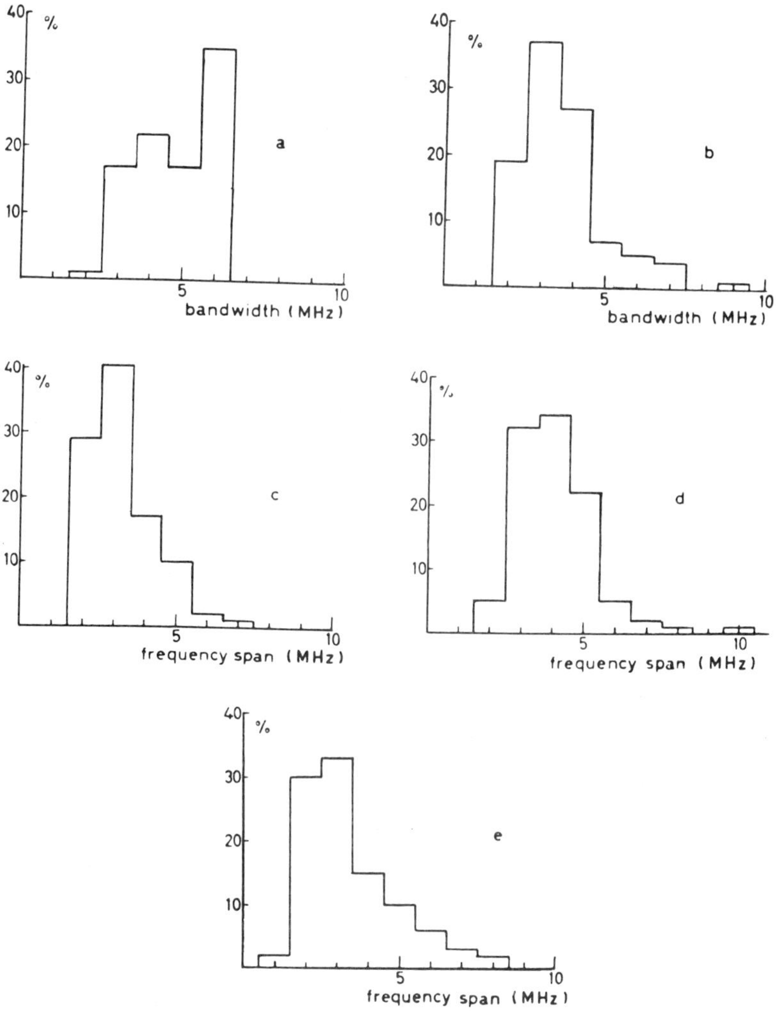

Fig. 2.16. Histograms showing: (a) the distribution of bandwidths as measured from A-scans; (b) the distribution of bandwidths measured from photometer curves; (c), (d) and (e) the distribution of visually determined frequency spans on September 10, 11 and 12, 1959 (Elgarøy, 1961).

It is seen that the bandwidths are confined to the region from 2 to 6 MHz. The average bandwidth may vary from one storm to another and from day to day as is illustrated in Table 2.4.

TABLE 2.4

Average bandwidths of storm bursts in the 200 MHz range

Date (1959)	ΔB (MHz)	Date (1959)	$\Delta B'$ (MHz)
August 28	4.2	September 10	3.2
October 15	3.2	September 11	4.0
October 16	4.1	September 12	3.4
November 13	2.7		

Around 150 MHz bandwidths were determined by Elgarøy and Ugland (1966). The bandwidths were typically 4, 3.8 and 3.5 MHz.

In the region 70 - 130 MHz, Wild (1951) found quarter-power bandwidths as shown in Fig. 2.17. In order to convert to half-power bandwidths the values in the figure should be divided by 1.4 (assuming a gaussian frequency profile). It is then seen

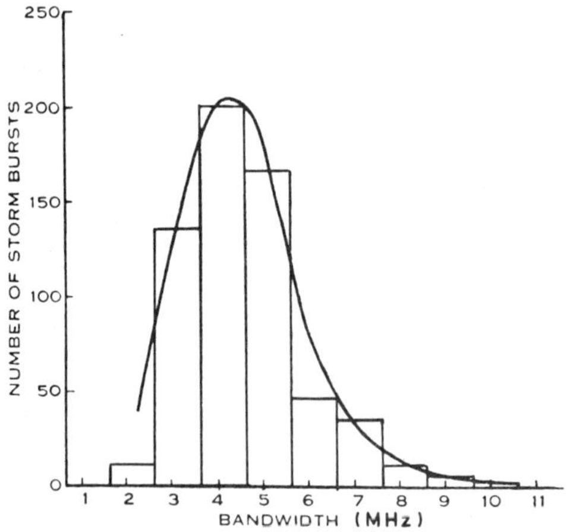

Fig. 2.17. Histogram showing the distribution of recorded storm bursts with bandwidth (Wild, 1951).

that the bandwidths occur between ≈1.5 and ≈5 MHz. The most common bandwidth becomes 3 MHz. Measurements in the frequency range 40 - 150 MHz by Vitkevich (1957) yielded bandwidhts of 4 - 7 MHz.

On 38 MHz, Yoh and James (1967) reports narrowband, highly polarized, short duration bursts with bandwidths of 200 kHz or less. According to Ellis (1969), fast drift type I bursts have bandwidths <100 kHz on frequencies between 30 and 50 MHz. Drift pair bursts have larger bandwidhts, between 0.2 and 0.8 MHz.

On the other hand, Warwick and Dulk (1969) reported that a "typical type I burst in the frequency range 24 - 37 MHz had a bandwidth of 5 MHz". This is at variance with other findings, and more information at frequencies ≈25 MHz is desirable.

Using presently available material, Fig. 2.18 has been constructed.

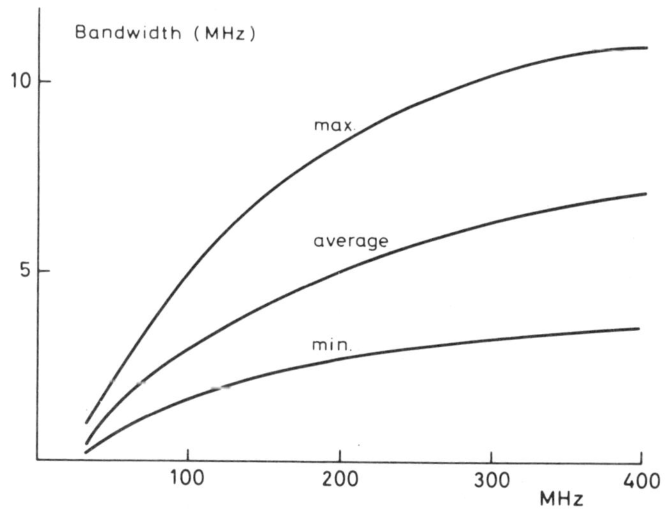

Fig. 2.18. Bandwidth of type I bursts at different frequencies. Upper and lower limits are indicated together with average values.

As is seen from Fig. 2.18, there may be a variation in the burst bandwidth with frequency. One should, however, consider the fact that the figure is constructed by use of rather inhomogeneous data. Not all values, for example, are half-power bandwidths. More definite information was obtained when simultaneous sweep-frequency observations were made between 145 - 165 MHz and between 315 - 335 MHz by Elgarøy (1967). The results of simultaneous bandwidth determinations are shown in Fig. 2.19.

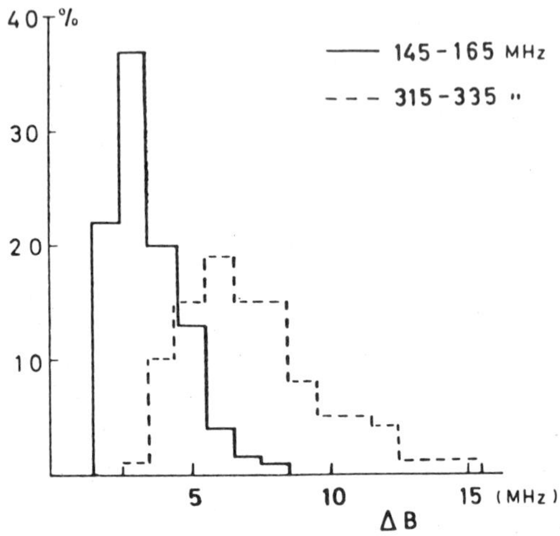

Fig. 2.19. Histograms of burst bandwidths at two different frequencies (Elgarøy, 1967).

Here it is quite clearly seen that there is a variation of burst bandwidth with frequency in the sense that the bandwidth decreases with decreasing frequency.

From Fig. 2.18 it seems that there is a small or no change in bandwidths between 250 and 400 MHz, and then a somewhat more pronounced variation when lower frequencies are approached.

Vitkevich (1955) used two-channel records to make a statistical determination of burst bandwidths on 200 MHz. He divided the material in two groups – one comprising intense bursts (amplitudes above 50% of the average burst level), the other weaker bursts. For strong bursts Vitkevich found a bandwidth of 12.3 MHz, for weak bursts he found 6 MHz. This is in disagreement with the results of Wild (1951) who found no definite correlation between amplitude and bandwidth. Vitkevich has not commented on his first findings in a later contribution on burst properties (1957). In accordance with the results of Wild, de Groot (1966) found no correlation between bandwidth and maximum intensity of bursts in his material. Moreover, he found no correlation between bandwidth and duration (Fig. 2.20).

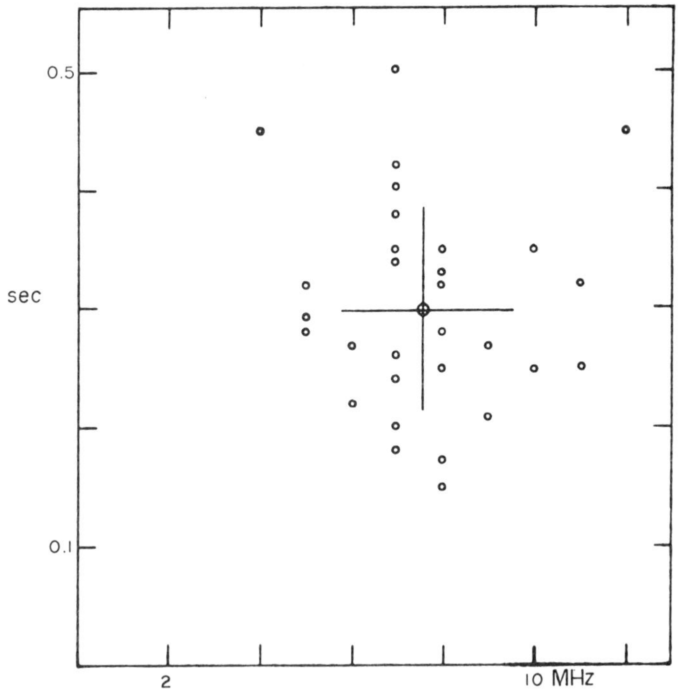

Fig. 2.20. Bandwidth-duration diagram of short storm with frequency drifted narrow band bursts on September 2, 1958 (de Groot, 1966).

Eckhoff (1966) arrived at the same conclusion as de Groot: no correlation between bandwidth and duration except for a weak tendency for short-duration bursts to have smaller bandwidths than the long-lasting ones.

3. POLARIZATION OF NOISE STORM EMISSION

3.1. POLARIZATION OF CONTINUUM

The polarization properties of noise storms have been studied by several authors. In general it is found that both background continuum and bursts are strongly circularly polarized (Wild, 1951). But the sense of polarization, right- or left-handed, may change from one storm to another. By combining polarization measurements at 200 MHz with interferometer determinations of source positions at 169 MHz, Le Squeren (1963) was able to determine the relation between the properties of the associated centre of optical activity and the polarization of the storm emission.

Firstly, there was a connection between the latitude of the optical centre and the sense of polarization of the noise storm continuum as appears from Table 3.1. About 90% of the sources at northern latitudes showed left-handed polarization, whereas at southern latitudes a corresponding percentage of the sources were polarized in the right-handed sense.

TABLE 3.1

Latitudes of optical centre and sense of circular polarization of storm continuum

Latitude	No. of centres	Left-handed	Right-handed	Mixed
North	59	52	6	1
South	21	2	16	2

For a given source of storm radiation, the sense of polarization does not change during the passage over the solar disk (Fokker, 1965a). Some of the centres (10%) in the sample did not follow the trend. A closer study of these centres showed that they had a complex structure, and in a couple of cases the following spot had a stronger magnetic field than the leading one. Also, some of the centres were associated with more than one storm during the passage across the disk, and it was observed that the sense of polarization of the storms were both normal and abnormal. The sense of polarization is such that if the polarity of the leading spot is negative (south polarity), the associated storm was right-handed. This corresponds to emission in the ordinary magneto-ionic mode if the field in the corona at the source has the same polarity as the sunspot with the strongest magnetic field (leading spot).

Payne-Scott and Little (1951) found the opposite behaviour of that given in Table 3.1. This is in accordance with the suggestion that the sense of polarization is determined by the magnetic field of the leading sunspot. The polarity of the leading sunspot in a group is opposite in the two hemispheres and reverses after each 11-year cycle. A reversal occurred at sunspot minimum between the times when the observations of Payne-Scott and Little and those of Le Squeren were carried out.

The results of Payne-Scott and Little and Le Squeren gained support from Smerd (1964) when he investigated storms occurring during the International Geophysical Year.

Some storms show mixed polarization. Then usually more than one centre of activity is present and two source regions may be observed. But in a few cases one noise active region only was seen with the 169 MHz Nancay interferometer.

About 8% of the storms observed by Le Squeren were unpolarized. Unpolarized storms have properties similar to those of polarized storms, but they seem to be less directive than polarized storms. The distribution in longitude of polarized and unpolarized storms as determined by Le Squeren is shown in Fig. 3.1.

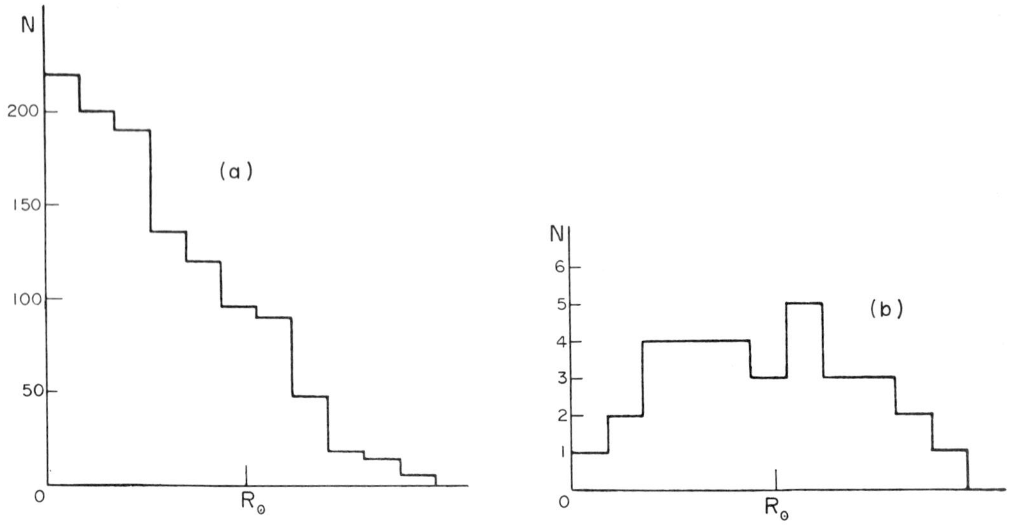

Fig. 3.1. Distribution of the distances of the sources to the north - south diameter of the sun for (a) polarized storms, and (b) unpolarized storms (Le Squeren, 1963).

Suzuki found quite similar results at 200 MHz, as is evident from Fig. 3.2, in which the material has been divided in three classes, i.e. strongly polarized, partially polarized and unpolarized emission. Since the unpolarized storms are less directive than the polarized ones, it may be suggested that unpolarized storms occur at larger heights in the corona. This was confirmed when Le Squeren determined the height of 11 unpolarized limb sources. Their average height was 1.65 $R_\odot$ while for polarized storms it was 1.5 $R_\odot$.

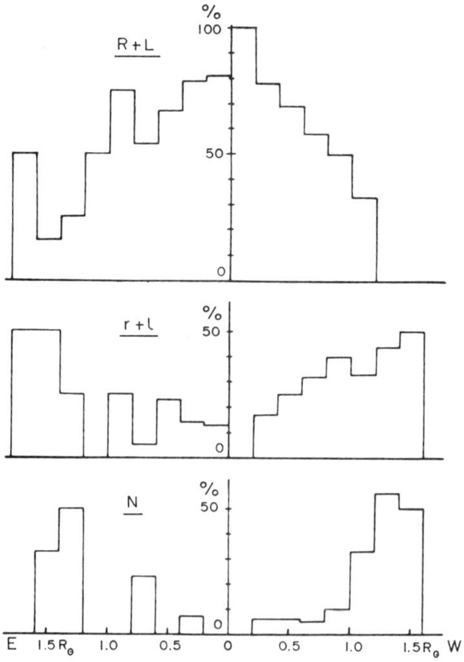

Fig. 3.2. Distribution of different classes of noise storms on the solar surface: *(top)* strongly polarized; *(middle)* partially polarized; *(below)* unpolarized (Suzuki, 1961).

3.2. POLARIZATION OF STORM BURSTS

The polarization of type I bursts is in most cases the same as the polarization of the background continuum. There is a good correlation between the degree of polarization of the continuum and the bursts (Tsuchiya, 1963). It was found by Tsuchiya that the majority of the bursts originating within $\pm 0.5\ R_{\odot}$ from the central meridian had a degree of polarization between 80 and 100%. Kai (1962) showed that the degree of polarization decreased for sources near the limb (Fig. 3.3). The occurrence of completely unpolarized type I bursts with the same properties as those of polarized ones was noted by de Groot (1960). As was found for continuum sources, also the height of the burst sources increases towards the limb (Morimoto and Kai, 1961).

Using spectrograms from Utrecht and polarimetric registrations at 237 MHz, Zlobec (1972) investigated the polarization properties of type I bursts with frequency drift. He found the same polarization for bursts with direct frequency drift, reverse drift and for bursts in chains. There was no relation between duration and bandwidth of the bursts and polarization except that bursts with large bandwidth lasting less than 0.1 s generally were totally polarized. At the other extreme bursts lasting more than 0.6 s were generally non-polarized. However, in another paper Zlobec (1973) found a correlation between frequency drift and polarization for partially polarized type I bursts. Successive bursts with negative drifts showed a decreasing degree of polarization, whereas the opposite behaviour was seen for bursts with positive frequency drift.

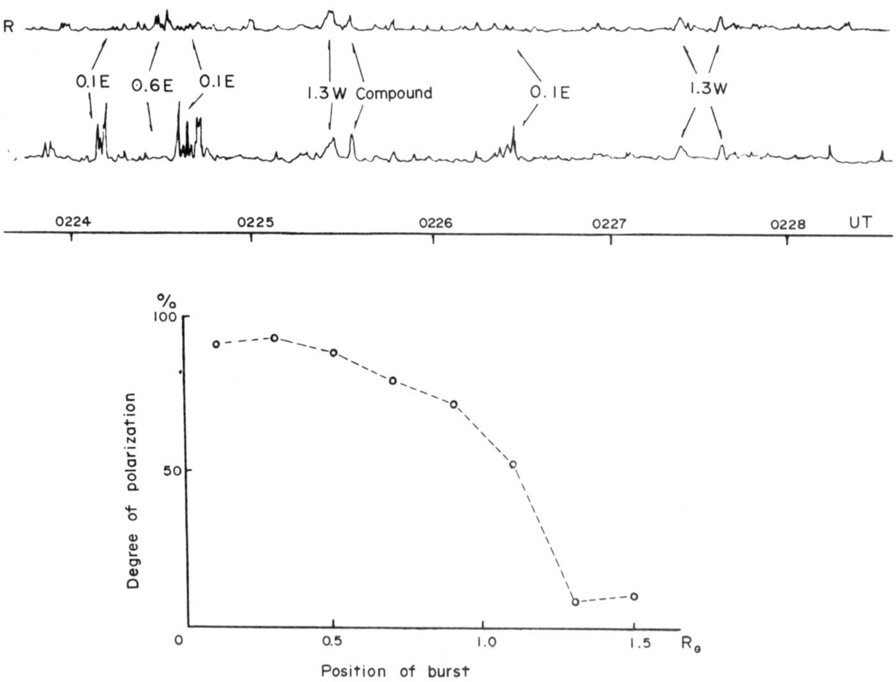

Fig. 3.3. *(Above)* an example of a limb event (March 23, 1959): the positions of the sources are indicated; *(below)* centre-limb variation in the average degree of polarization of type I bursts (Kai, 1962).

As a preliminary conclusion it may be said that the majority of noise storms are circulary polarized with a degree of polarization which is between 80 and 100%. Within limits or error the polarization is the same for bursts and continuum. But exceptions occur, and Fokker (1960) mentioned the following cases:

(1) Strong polarization of continuum and unpolarized bursts.

(2) Weak polarization of continuum and strongly polarized bursts.

(3) Steadily, unpolarized continuum.

(4) Continuum and/or bursts weakly or moderately polarized.

(5) Mixed polarization may occasionally occur from one source only.

(6) Bursts have identical sense but widely differing degrees of polarization.

It is important to note that linear polarization has not been detected in storm emission (Suzuki, 1961). Because of the Faraday effect in the corona this is in agreement with expectations.

3.3. POLARIZATION OBSERVATIONS WITH HIGH SPATIAL AND TIME RESOLUTION

For a full determination of storm polarization radio pictures of the sun at several frequencies, with high spatial resolution, with high time resolution and in two circular components are desirable. The radioheliograph at Culgoora (Wild, 1967b; Sheridan *et al.*, 1972) is the instrument which most closely meets these requirements, although not all of them are satisfied.

By means of observations made with the Culgoora radioheliograph, Kai (1970) investigated the spatial distribution of circular polarization for bursts of spectral type I and stationary type IV bursts at a frequency of 80 MHz.

Type I storms with narrow band bursts superimposed on a background continuum were similar in many of their properties to stationary type IV bursts. The latter are dominated by the continuum which covers a wide frequency range, and they are associated with large flares. From the heliograph observations it appeared that the two phenomena were of much the same physical nature.

In Fig. 3.4 contour plots of brightness distributions of type I sources observed in right-handed (RH) and left-handed (LH) circularly polarized emission are superimposed and show a small position difference between the two components. A list of data from the analysis is given in Table 3.2.

In general the separation between the two sources was about 3′ arc and their dimensions were roughly 6′ × 6′ arc. Both type I storms and stationary type IV bursts seemed to be polarized in the ordinary mode, in agreement with the results from the leading spot hypothesis (Payne-Scott and Little, 1951; Suzuki, 1961; Le Squeren, 1963; Kai, 1965a).

As has been described, the flux of noise storm radiation, continuum as well as bursts, is usually strongly circularly polarized. Since radioheliograph observations reveal that bipolar structure is quite common, it must be concluded that one source is usually more intense than the other. When the net degree of polarization is 100%, one source only is present.

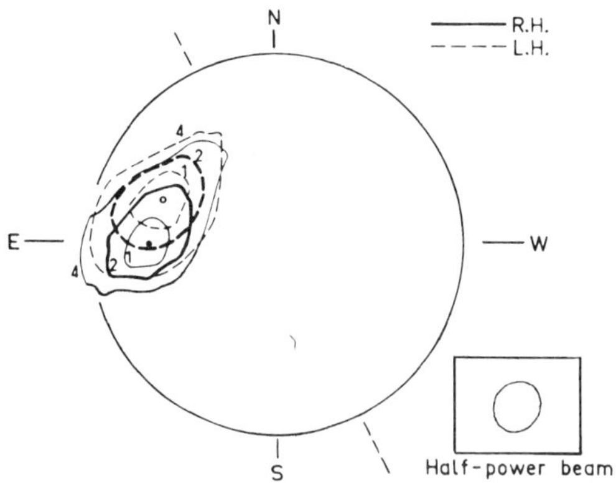

Fig. 3.4. Contour diagrams in RH and LH polarization. The brightest points are indicated by • for RH and o for LH polarization. The thicker contours represent half-power levels. The storm occurred on October 21, 1968. The net degree of polarization from the two sources was 60% RH. (Kai, 1970.)

TABLE 3.2

List of 30 MHz bipolar sources examined by Kai (1970)

Date (1968)	Time (UT)	Spectral type	Source configuration					Net polarization degree (%)	Sunspot	
			Preceding		Following		R-L separation (min of arc)		Magnetic classi- fication	Hemi- sphere
			polarity	size	polarity	size				
June 9	$\sim 01^h22^m$	stat. IV	R	8′×7′	L	8′×8′	3	50 R	$\beta f \rightarrow \beta p$	N
	$\sim 01^h01^m$	stat. IV?	R	12′×9′	L	12′×6′	2	50 R	$\beta \rightarrow \beta f$	N
June 16	$\sim 23^h10^m$	I	L	12′×8′	R	12′×9′	3	40 R	-	-
Sept. 4	$\sim 01^h00^m$	stat. IV	R	9′×6′	L	6′×6′	3∼4	60 R	$\beta p \rightarrow \beta\gamma$	N
Oct. 21	$\sim 03^h30^m$	I	L	8′×6′	R	9′×5′	∼3	50 R	$\beta\gamma$	N
Oct. 22	$\sim 02^h30^m$	I	R	9′×6′	L	6′×6′	2	60 L	αp?	S

From the 80 MHz observations, Kai was led to suggest that type I emission originates in strong magnetic fields of opposite polarity, possibly fields which close near sunspots, as shown in Fig. 3.5.

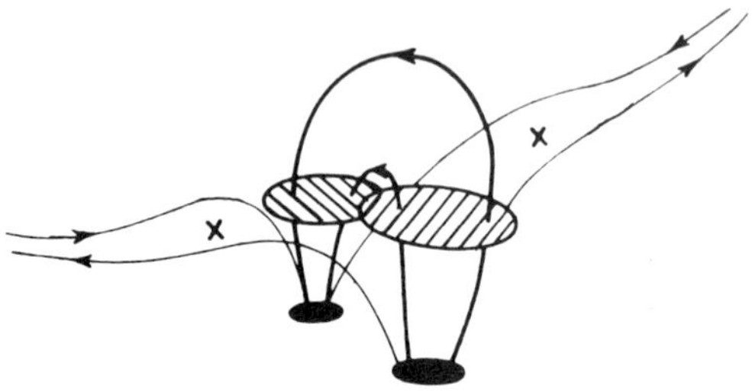

Fig. 3.5. Model for the location of type I sources (shaded areas). The regions marked x may be the location of bursts of spectral type III which originate in regions of weak magnetic fields. (Kai, 1970.)

Sunspots which are associated with noise storm sources may show complex magnetic patterns at the photospheric level, but high up in the corona the field may be of bipolar type.

After modification of the Culgoora radioheliograph, Dulk and Nelson (1973) were able to observe the two-dimensional positions, sizes and polarization of type I sources at two heights in the solar corona, i.e. near the 80 and 160 MHz plasma levels. Between August 31 and September 5, 1972, an unusually small, highly polarized type I source passed across the solar disk. The two-dimensional data combined with *H*-alpha pictures and magnetograms made it possible to determine where the sources were situated within the magnetic structure. Dulk and Nelson assumed heights of 1.2 $R_{\odot}$ and 1.45 $R_{\odot}$ for the 160 and 80 MHz sources respectively and projected the sources radially downward to the photosphere. The corrected source positions plotted on a magnetic map of the region is shown in Fig. 3.6. The corrected positions cluster strongly. The 160 MHz positions are about 0.5′ arc north - west of a sunspot of negative polarity, and the 80 MHz positions are about 2′ arc south of the sunspot. The position of the 160 MHz source was definitely in a region of

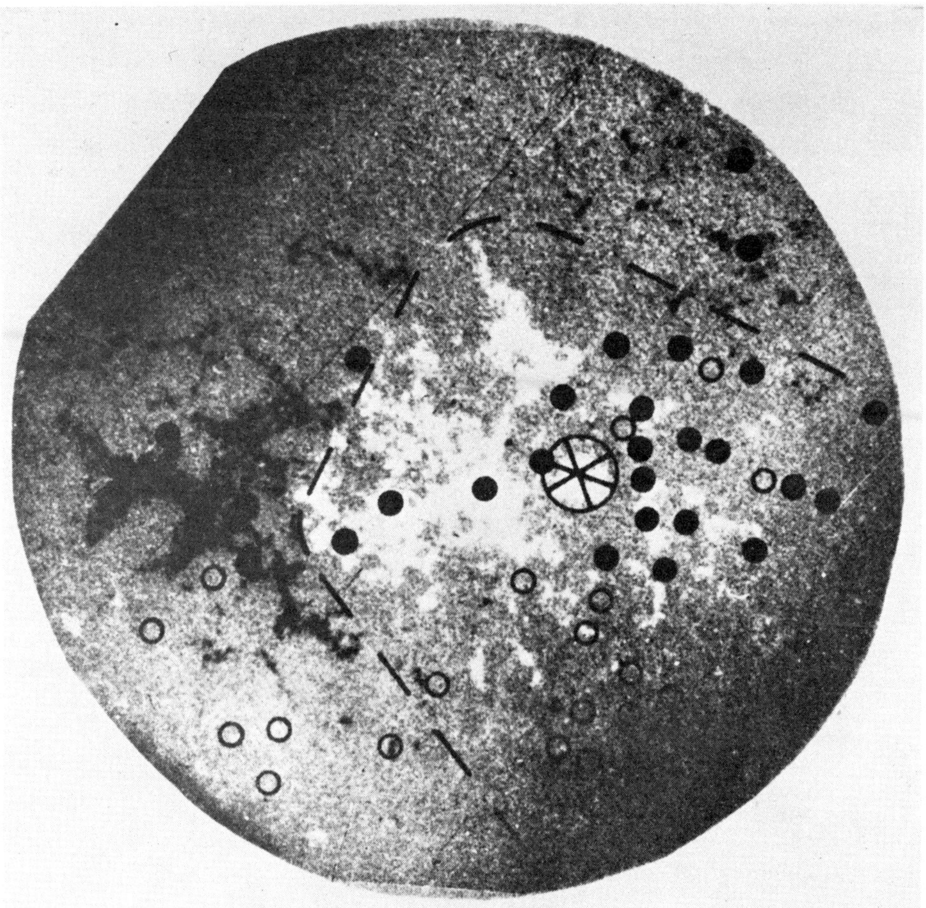

Fig. 3.6. Positions of type I storm sources at 80 MHz (o) and 160 MHz (•) after correction for height plotted on a magnetogram obtained at Culgoora; light regions show negative polarity, dark regions positive polarity (Dulk and Nelson, 1973).

negative polarity, and since the radiation was circularly polarized in the right-handed sense, this provides rather convincing evidence that the ordinary magneto-ionic mode was emitted. Also the 80 MHz radiation was RH polarized, and the southward displacement of 2′ arc suggests that the storm sources may have been contained in high, trans-equatorial magnetic loops.

In conclusion it may be said that all present evidence seems to show that type I storm radiation is emitted in the ordinary magneto-ionic mode. Of course, this is a result of high importance for the interpretation of the storm phenomenon.

3.4. THE EFFECT OF FARADAY ROTATION ON THE OBSERVED POLARIZATION

As mentioned in Chapter 3.2, linear polarization has not been found in storm emission. Since storm sources occur in regions of the corona where the magnetic field appears to be strong, the observation of linear polarization is unlikely. This follows from some simple considerations. Waves propagating in a plasma permeated by a magnetic field are subject to Faraday rotation, changing the orientation of the polarization ellipse. The amount of rotation is frequency-dependent and thus introduces difficulties for polarization observations. This problem was treated by Cohen (1958).

Consider waves propagating in a homogeneous plasma along a magnetic field B. It is then easy to show that the orientation of the polarization ellipse is changed by an amount Φ given by

$$\Phi = \frac{\omega}{2c} XYl, \qquad (3.4\text{-}1)$$

where $\omega = 2\pi f$, f is the observing frequency, $X = \omega_p^2/\omega^2$, $\omega_p = 2\pi \times$ electron plasma frequency, $Y = \omega_H/\omega$, $\omega_H = 2\pi \times$ electron gyro frequency, and l is the path length.

It is assumed that $\omega > \omega_p$ and ω_H. The formula (3.4-1) may be expressed in the following way:

$$\Phi = \frac{2.4 \times 10^{-3} N_e B_{\parallel}}{f^2} l \qquad (3.4\text{-}2)$$

where Φ is the rotation angle in radians, f is the observing frequency in megahertz, N_e electron density (number of electrons per cubic centimetre), $B_{\parallel}$ magnetic field in gauss (component in the direction of propagation), and l path length in kilometres.

When the radiation is received in a given frequency band there will be a difference in the rotation of the different frequencies. The dispersion is given by

$$\Delta\Phi = \frac{\partial\Phi}{\partial f}\Delta f = 2\Phi\frac{\Delta f}{f}, \qquad (3.4\text{-}3)$$

where $\Delta f = f_2 - f_1$ is the bandwidth of the receiver.

By integrating the Stokes parameters over the bandwidth, Hatanaka (1956) found the resultant degree of polarization (m) and the axial ratio r ($r = \text{tg}\beta$, $r = 0$, linear polarization, $r = 1$, circular polarization, $r > 0$ left-handed, $r < 0$ right-handed sense of rotation):

$$m = m_0\left(\frac{\sin^2\Delta\Phi}{\Delta\Phi^2} + \sin^2 2\beta_0\left(1 - \frac{\sin^2\Delta\Phi}{\Delta\Phi^2}\right)\right)^{\frac{1}{2}} \qquad (3.4\text{-}4)$$

$$\sin 2\beta = \left(\frac{\sin^2\Delta\Phi}{\Delta\Phi^2}\,\text{cotg}^2 2\beta_0 + 1\right)^{-\frac{1}{2}} \qquad (3.4\text{-}5)$$

Subscript $_0$ indicates values at the starting point. From (3.4-4) and (3.4-5) it is seen that when $\Delta\Phi \to \infty$, $m \to m_0 \sin 2\beta_0$ or $m \to m_0 \dfrac{2r_0}{1+r_0^2}$ and $\sin 2\beta \to 1$, i.e. $\beta = 45^\circ$, $r = 1$.

This shows that the effect of large Faraday dispersion is to make elliptically polarized radiation circularly polarized. However, the degree of polarization decreases. In the special case of strictly linear polarization ($r_0 = 0$) the signal may become completely depolarized.

In order to apply the theory on noise storms one should consider (a) Faraday effect in the source, and (b) Faraday effect during propagation through the solar corona.

If one, as an example, assumes the following values for the source region: $N_e = 5\times10^7$ cm^{-3}, $B_{//} = 1$ G, $l = 3\times10^4$ km and $f = 200$ MHz, the formula (3.4-2) gives $\Phi \simeq 10^5$ radians. The parameters may be changed somewhat, but in any case it seems plausible to suggest that the Faraday rotation angle in the source region is of the order of $10^4 - 10^5$ radians.

The rotation of the plane of polarization during propagation through the corona has been calculated by Hatanaka (1956) and by

Fokker (1971). Fokker assumed that the coronal electron density was equal to that given by the model of Newkirk (1961) multiplied by a factor of two. At distances larger than $2R_{\odot}$ from the solar centre the model was suitably extrapolated (Fokker, 1965b). The integral of $\int N_e B_{\parallel} dl$ was computed for a radial trajectory from a level near the critical electron density ($N_e = 5\times10^8$ cm^{-3} for 200 MHz) and upwards. For the magnetic field along the direction of propagation a value of 0.25 G was adopted at a height of 0.25 $R_{\odot}$ above the photosphere. The following assumptions as to the dependence on height were made: (i) $B_{\parallel} \propto (\frac{x}{R_{\odot}} - 0.5)^{-2}$ and (ii) $B_{\parallel} \propto (\frac{x}{R_{\odot}})^{-3}$. Here x is the distance from the solar centre. Under these assumptions the following values were found for the Faraday rotation: (i) $\Phi = 6.5\times10^5$ radians and (ii) $\Phi = 4.5\times10^5$ radians. Faraday rotation and dispersion angle for different receiver bandwidths are shown in Table 3.3. It is clearly seen that even with a small receiver bandwidth the dispersion angle $\Delta\Phi$ may amount to tens of radians. From the graphs in Figs. 3.7a and 3.7b it is seen that the observed polarization will be circular or nearly circular whatever the initial value of r_0 when $\Delta\Phi$ is about 5 radians or more. From Fig. 3.7b it is also seen that circular or quasi-circular waves will remain highly polarized, while a linear wave soon will become unpolarized. In the figure it was assumed that all radiation was completely polarized at the source ($m_0 = 1$).

TABLE 3.3

Faraday rotation and dispersion angle

Model	Rotation angle Φ (radians)	Dispersion angle $\Delta\Phi$ (radians) Bandwidth of receiver		
		1 MHz	0.1 MHz	0.01 MHz
Fokker (i)	6.5×10^5	6500	650	65
Fokker (ii)	4.5×10^5	4500	450	45
Source region	10^5	1000	100	10

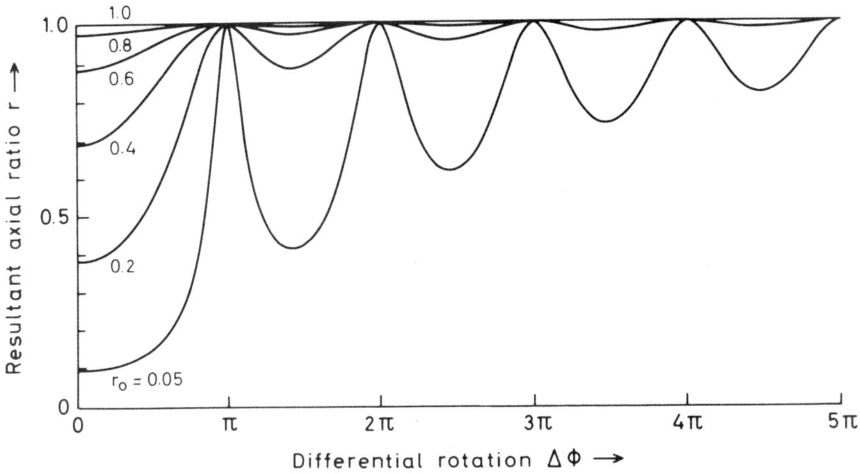

Fig. 3.7a. Change of axial ratio because of differential Faraday rotation. Original axial ratio r_0 as parameter.

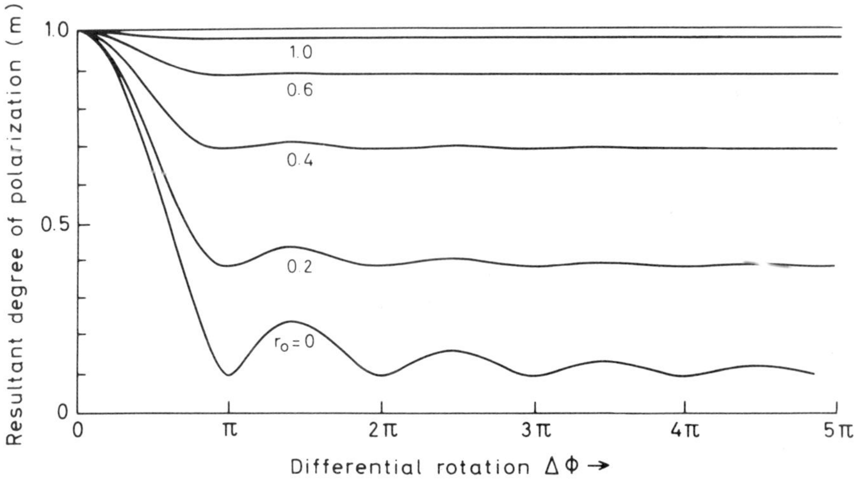

Fig. 3.7b. Depolarization because of differential Faraday rotation. Original degree of polarization, $m_0 = 1$. Original axial ratio r_0 as a parameter.

These results have large consequences for the polarization measurements of solar bursts. For receiver bandwidths commonly used, i.e. 0.1 - 1 MHz, the Faraday dispersion is a very important factor. The observed sense of circular polarization, whether right- or left-handed, may be correct, but if a linear component exists at the source it will not be observed. It is typical that the observations reflect just what should be expected from the effect of large differential Faraday rotation: circular polarization. In the case of noise storms the high degree of polarization shows that the signal has relatively large axial ratio at the origin.

3.5. MAGNETO-IONIC MODE COUPLING IN THE SOLAR CORONA

The Faraday rotation and dispersion discussed in the previous section takes place when there is no strong inhomogeneities in the medium. When the medium is changing more rapidly, the simple theory is not valid. Then the theory of coupling of the wave modes must be considered. Mode coupling may occur in the solar corona. This problem was treated by Cohen (1959), while magneto-ionic mode coupling in the ionosphere had been discussed by Försterling (1942), Rydbeck (1950), Budden (1952) and Gibbons and Neetney (1952).

The following assumptions were made in the theory of Cohen:

(1) The medium is slowly varying, i.e. the amount of change per wavelength in the medium is small.

(2) $X, Y, Z \ll 1$, ($X = \frac{\omega_p^2}{\omega^2}$, $Y = \frac{\omega_H}{\omega}$ and $Z = \frac{\nu}{\omega}$) (ν = collision frequency).

(3) The medium is stratified in planes parallel to the wave front (perpendicular to the direction of propagation). But it was assumed that the results were applicable also when this was not the case.

The volume polarization vector $\mathbf{P}$ is related to the electric wave field $\mathbf{E}$ by

$$\mathbf{P} = \sigma \mathbf{E} \qquad (3.5\text{-}1)$$

where σ is the susceptibility matrix. It was shown by Ratcliffe

(1959) that the wave equation then may be written

$$\nabla\times\nabla\times\mathbf{E} = k^2(\mathbf{E} + \frac{4\pi}{\varepsilon_0}\mathbf{P}) = k^2(I + \frac{4\pi}{\varepsilon_0}\sigma)\,\mathbf{E} \qquad (3.5\text{-}2)$$

where $k = 2\pi/\lambda$ is the wave number, I is the identity matrix and ε_0 is the permittivity of free space.

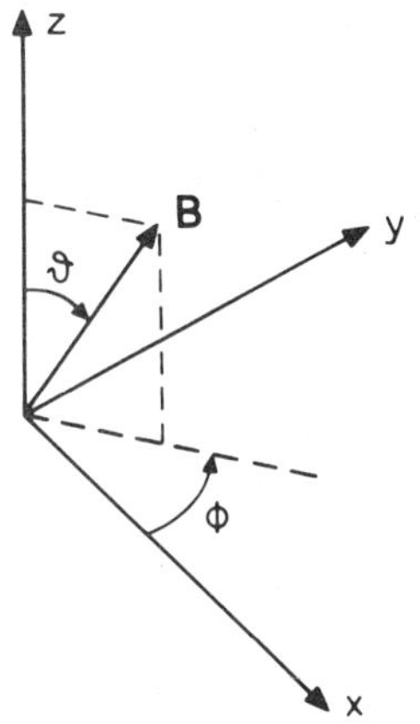

Fig. 3.8. Coordinate system.

Consider a coordinate system as in Fig. 3.8 and propagation in the z-direction. The x- and y-components of **E** can then be written as the sums of the components of two electric vectors

$$E_x = E_{x_o} + E_{x_e}, \qquad E_y = E_{y_o} + E_{y_e} \qquad (3.5\text{-}3)$$

where the subscripts o and e refer to the ordinary and extraordinary wave modes. Equations may now be derived for the components of the electric field vector. After some calculations (for details see Cohen, 1959) one arrives at the following coupled differential wave equations:

$$\begin{aligned} \Pi_o'' + n_o^2\Pi_o &= -\psi_o\psi_e\Pi_o + \psi_e'\Pi_e + 2\psi_e\Pi_e' \\ \Pi_e + n_e^2\Pi_e &= -\psi_e\psi_o\Pi_e + \psi_o'\Pi_o + 2\psi_o\Pi_o' \end{aligned} \qquad (3.5\text{-}4)$$

where

$$E_{y_o} = \Pi_o \exp\left(-\int_0^h \frac{R_o'}{R_o - R_e}\, dh\right), \quad E_{y_e} = \Pi_e \exp\left(-\int_0^h \frac{R_e'}{R_e - R_o}\, dh\right) \qquad (3.5\text{-}5)$$

and the coupling parameters are written

$$\psi_o = \frac{R_o'}{R_o - R_e} \exp\left(\int_0^h \frac{R_e + R_o'}{R_e - R_o}\, dh\right), \quad \psi_e = \frac{R_e'}{R_e - R_o} \exp\left(\int_0^h \frac{R_o' + R_e}{R_o - R_e}\, dh\right) \qquad (3.5\text{-}6)$$

The refractive indices of the ordinary and extraordinary waves are n_o and n_e respectively. The derivation is with respect to $h = kz$. The polarization ratios are

$$R_o = \frac{E_{x_o}}{E_{y_o}} \quad \text{and} \quad R_e = \frac{E_{x_e}}{E_{y_e}} .$$

The inhomogeneous wave equations (3.5-4) may be approximately solved following a method used by Budden (1952). One first solves the homogeneous equations

$$\Pi_o'' + n_o^2 \Pi_o = 0, \quad \Pi_e'' + n_e^2 \Pi_e = 0 \qquad (3.5\text{-}7)$$

An approximate solution of these is

$$\Pi_o = U_o \exp\left(-\frac{i}{2} \int_0^h (n_o + n_e)\, dh\right), \quad \Pi_e = U_e \exp\left(-\frac{i}{2} \int_0^h (n_o + n_e)\, dh\right) \qquad (3.5\text{-}8)$$

Substituting the approximate solutions of the homogeneous wave equation into the inhomogeneous equations and retaining first-order terms only, the following system of coupled differential equations is obtained:

$$\frac{dU_o}{dh} = \psi_e U_e - \frac{i}{2} \Delta n U_o , \quad \frac{dU_e}{dh} = \psi_o U_o + \frac{i}{2} \Delta n U_e \qquad (3.5\text{-}9)$$

where $\Delta n = n_o - n_e$.

Two special cases may now be considered; weak and strong coupling.

(a) *Weak coupling*

If $|\frac{i}{2}\,\Delta n| \gg |\psi_e|$ and $|\frac{i}{2}\,\Delta n| \gg |\psi_o|$

equations (3.5-9) may be written

$$\frac{dU_o}{dh} = -\frac{i}{2}\,\Delta n U_o, \quad \frac{dU_e}{dh} = \frac{i}{2}\,\Delta n U_e \tag{3.5-10}$$

The two functions U_o and U_e are now independent; the modes are uncoupled. Solving (3.5-10) one obtains

$$U_{\substack{o\\e}} = \exp(\mp\frac{i}{2}\int_0^h \Delta n dh) \tag{3.5-11}$$

Using this in (3.5-8) the wave functions may be written in the following form:

$$\Pi_o = \exp(-ik\int_0^z n_o dz), \quad \Pi_e = \exp(-ik\int_0^z n_e dz) \tag{3.5-12}$$

The wave travels as the sum of the ordinary and the extraordinary wave modes which are independent of each other. The propagation constants and polarization change continuously and are determined by the actual physical condition of the medium passed by the waves. This corresponds to the case of local homogeneity discussed by Booker (1936).

(b) *Strong coupling*

The condition for strong coupling is

$|\frac{i}{2}\,\Delta n| \ll |\psi_o|$ and $|\frac{i}{2}\,\Delta n| \ll |\psi_e|$.

Equations (3.5-9) are now approximately written

$$\frac{dU_o}{dh} = \psi_e U_e, \quad \frac{dU_e}{dh} = \psi_o U_o. \tag{3.5-13}$$

Using the definitions of $\psi_{\substack{o\\e}}$ in (3.5-6) it may be shown that one pair of solutions is

$$U_o = \exp\left(\int_0^h \frac{R'_e}{R_o+R_e}\,dh\right) \text{ and } U_e = -\exp\left(\int_0^h \frac{R'_o}{R_e-R_o}\,dh\right) \tag{3.5-14}$$

The other is

$$U_o = \frac{R_e}{R_e - R_o} \exp\left(\int_0^h \frac{R_o'}{R_o - R_e}\, dh\right) \quad \text{and} \quad U_e = \frac{R_o}{R_o - R_e} \exp\left(\int_0^h \frac{R_e'}{R_e - R_o}\, dh\right) \qquad (3.5\text{-}15)$$

The total electric field then has two solutions.

From (3.5-14), (3.5-3), (3.5-5) and (3.5-8) one gets

$$E_x = 0, \quad E_y = \exp(-ik \int_0^z \bar{n}\, dz). \qquad (3.5\text{-}16)$$

From (3.5-15) one has

$$E_x = \exp(-ik \int_0^z \bar{n}\, dz), \quad E_y = 0. \qquad (3.5\text{-}17)$$

The electromagnetic wave now propagates as a linear combination of (3.5-16) and (3.5-17). They have the same propagation constant $\bar{n} = \frac{1}{2}(n_o + n_e)$, and thus Faraday rotation and differential absorption of single wave modes are absent. Moreover, under the condition X, Y, $Z \ll 1$, the propagation occurs as if there was no magnetic field since $\bar{n}$ is identical with the refractive index for the medium without magnetic fields.

From (3.5-9) it appears feasible to define a coupling ratio Q determined by

$$Q = \left| \frac{\psi_o \psi_e}{(\frac{1}{2}\Delta n)^2} \right|^{\frac{1}{2}} \qquad (3.5\text{-}18)$$

When $Q \ll 1$ the two modes are weakly coupled. In the case $Q \gg 1$ the two magneto-ionic modes are strongly coupled, and the wave propagates as if the magnetic field was absent. In the case of arbitrary Q one may discuss quasi-longitudinal (QL) propagation and quasi-transverse (QT) wave propagation. For QL propagation the following expression was derived by Cohen:

$$Q = \frac{A}{f^2 XY \cos\Theta}, \qquad (3.5\text{-}19)$$

where A depends on the local geometry. One may regard the case Q=1 as the transition between the cases of weak and strong coupling and a transitional frequency f_t corresponding to this situation may be determined. When $f \ll f_t$ the modes are weakly coupled, when $f \gg f_t$ they are strongly coupled. For the solar corona, interplanetary medium and the ionosphere $f_t > 10^3$ GHz (Table 3.4) and hence in the QL regime the modes are weakly coupled for all actual radio frequencies.

TABLE 3.4

Values of f_t in different media

Medium	QL mode	QT mode
Interstellar space	10^5 GHz	50 kHz
Solar corona ($r = 2\,R_\odot$)	5×10^4 GHz	500 MHz
Interplanetary space	5×10^3 GHz	0.1 MHz
Ionosphere (F_2-layer)	10^3 GHz	50 MHz

In the QT case (Θ in the vicinity of $\Theta = \frac{1}{2}\pi$) the expression for Q takes the form

$$Q = \frac{A^* f^4}{N_e B^3 S} \qquad (3.5\text{-}20)$$

where S is the scale of the magnetic field ($\frac{1}{S} = \frac{d\Theta}{dz}$) and A^* is a constant. The value of this constant is equal to 10^{-17} when units of gauss, centimetre and hertz are used. When $Q = 1$, one finds for the transitional frequency

$$f_t^4 = 10^{-17} N_e S B^3 \qquad (3.5\text{-}21)$$

In the case of QT propagation (3.5-21) gives considerably smaller values for the transitional frequency than is found in the case of QL propagation. Some characteristic results are shown in Table 3.4 which was given by Krüger (1972).

When discussing polarization measurements of noise storms it is obviously important to consider mode coupling in the solar corona.

One of the assumptions made in deriving the expression (3.5-21) for f_t was that the observing frequency was much larger than the plasma frequency. The theory was extended by Melrose (1974a) to include the case when f is near f_p. He then arrived at the following expression for a QT region:

$$Q = (1-X)^{5/2} \frac{f^4}{f_t^4} \tag{3.5-22}$$

where f_t is given by (3.5-21). Thus for $f >> f_p$, $Q = 1$ at $f = f_t$, but for f close to f_p the factor $(1 - f_p^2/f^2)^{5/2}$ causes a suppression of the coupling factor.

Melrose (1974b) also noted that the spatial variations in the magnetic field assumed in existing coupling theories violate div $\mathbf{B} = 0$. For stratification along the z-axis div $\mathbf{B} = 0$ implies $B_z' = 0$ and B_z = constant, although B_x and B_y can be functions of z. Nevertheless, to treat coupling at a QT region it had been assumed that B_z was changing sign as a function of z. Melrose therefore developed a method for calculating the coupling coefficients for oblique incidence permitting treatment of coupling at a QT region without invoking non-physical magnetic structures. As a result of a detailed treatment of the problem it was found that the previous discussions based on the theory for vertical incidence were qualitatively and semiquantitatively correct despite the fact that non-physical magnetic field structures had been invoked.

Returning to (3.5-22) one may consider the coupling under conditions when the transition frequency is relatively near the plasma frequency, say equal to αf_p, where $\alpha > 0$. If L_B is the characteristic distance in kilometres over which the direction of the magnetic field varies and the frequencies are expressed in megahertz, the formula (3.5-21) derived by Cohen is equivalent to

$$f_t^4 \approx L_B f_p^2 f_H^3 \tag{3.5-23}$$

Using this together with (3.5-22), Melrose found that $Q = 1$ when

$$f_H^3 = g(\alpha) \frac{f_p^2}{L_B} \tag{3.5-24}$$

where

$$g(\alpha) = \frac{(\alpha^2-1)^{5/2}}{5\alpha} \tag{3.5-25}$$

In Fig. 3.9 $g(\alpha)$ is given as a function of f_t/f_p.

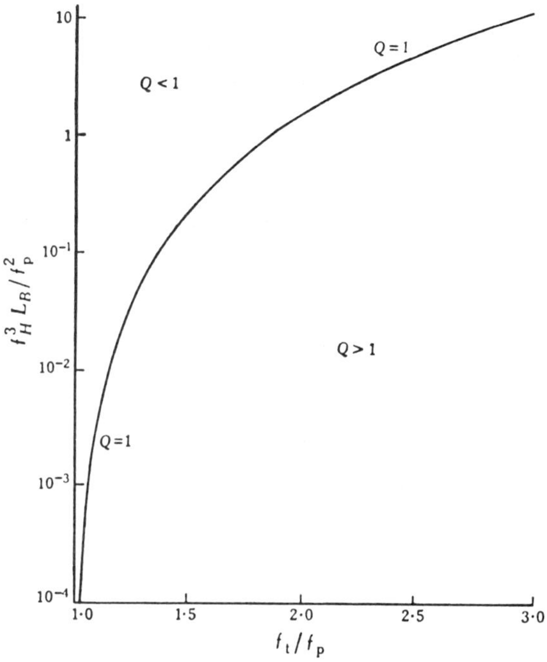

Fig. 3.9. Values of $g(\alpha) = (f_H^3 L_B)/f_p^2$ giving $Q = 1$ plotted as a function of the ratio between f_t and f_p: Regions of strong and weak coupling are indicated (Melrose, 1974a).

Mode coupling may also introduce a reduction in the degree of polarization of radiation passing a QT region. Suzuki (1961) suggested that this may explain the decrease in the degree of polarization for noise storms observed near the solar limb. Zheleznyakov and Zlotnik (1963) showed that radiation which was initially completely polarized in one mode became partially polarized when crossing a QT region. They derived a degree of linear polarization m_l given by

$$m_l = 2\left\{\exp(-2\delta)\,[1 - \exp(-2\delta)]\right\}^{\frac{1}{2}} \tag{3.5-26}$$

with $\delta = \frac{\pi}{2Q}$.

This component may then be depolarized by differential Faraday rotation.

3.6. MODE COUPLING AND POLARIZATION OF NOISE STORMS

Radio waves from solar storm sources may meet a QT region during propagation through the corona. The geometry only requires that the longitudinal component of the magnetic field be oppositely directed at the source and at a point further away along the ray trajectory.

During propagation through a QT region a circularly polarized wave changes sign (reverses direction of rotation) if $f \ll f_t$. If the frequency satisfies the condition $f \approx f_t$, the emerging wave becomes linearly polarized and if $f \gg f_t$, no change in the direction of rotation occurs. This is illustrated in Fig. 3.10.

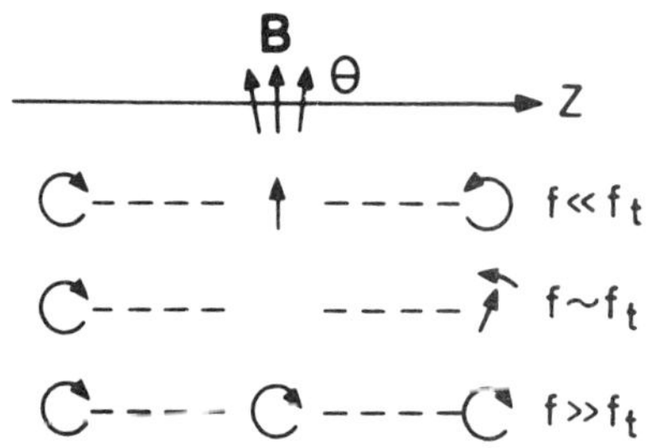

Fig. 3.10. Propagation through a QT region.

The identification of noise storm emission as ordinary mode from the magnetic field above the predominant sunspot is based on the assumption that no reversal of sense of rotation occurs during propagation. This assumption is strengthened by the fact that radiation from sources at the feet of bipolar structures are oppositely polarized. It gains further support from the finding that no polarization reversal occurs after central meridian passage of a noise storm (Payne-Scott and Little, 1951; Fokker, 1965a; Kai, 1970).

This behaviour is in fact contrary to expectations. If the sunspot magnetic field is approximated by a dipole field and if, moreover, $f \ll f_t$, one should find that both feet of the bipolar structure were similarly polarized: either both left- or both right-handed. At central meridian passage a reversal in the sense of rotation should take place. This behaviour was predicted by Piddington and Minnett (1951) who discussed propagation in a dipole field when there was no coupling between the magneto-ionic modes. The situation is illustrated in Fig. 3.11.

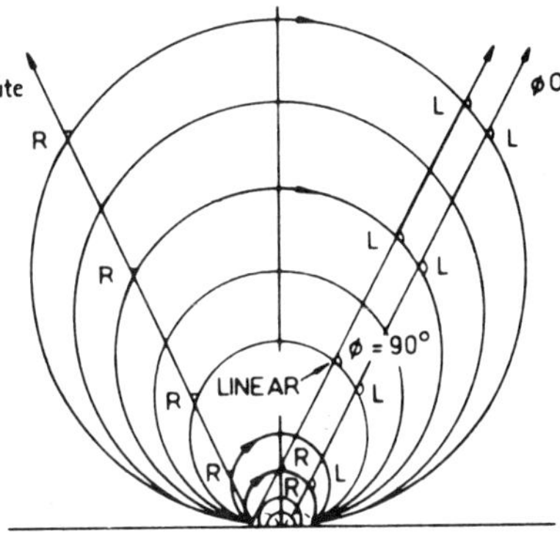

Fig. 3.11. Implications of magneto-ionic theory on the expected radiation from a bipolar region. The polarization corresponds to that of the extraordinary ray (Piddington and Minnett, 1951).

A problem now arises as was clearly pointed out by Melrose (1973b). The evidence from polarization measurements of noise storm radiation is only explained if QT regions are passed without polarization reversal ($f \gg f_t$) or if the magnetic field geometry is such that no QT region is encountered.

Consider the first-mentioned possibility. Noise storm emission is generated at heights near the plasma level ($f \approx f_p$). Figure 3.11 indicates that a QT region may be encountered quite close to the source. As an example take f = 100 MHz and assume that f_p = 50 MHz in the QT region. By reference to Fig. 3.9 it is found that

$$B^3 L_B < 300 \ (G^3 \ km) \tag{3.6-1}$$

for strong coupling to occur (Melrose, 1974a). A lower value of B in the QT region may be 1 G. L_B is of the order 3×10^5 km and (3.6-1) fails to be satisfied by a factor of more than 10^3. In the case under consideration a field of $B < 0.1$ G would serve to satisfy (3.6-1), but this value seems to be unlikely small.

Melrose (1973b) pointed out three possibilities for the magnetic field in the QT region to be weak: (1) The QT region could occur in a neutral plane as indicated in Fig. 3.12. Then neutral planes must occur down to heights comparable with the apparent height of type I sources. (2) If the magnetic field strength in loop structures in which the noise is generated is stronger than in the surrounding plasma, the field strength would decrease more rapidly along the ray path than would other-

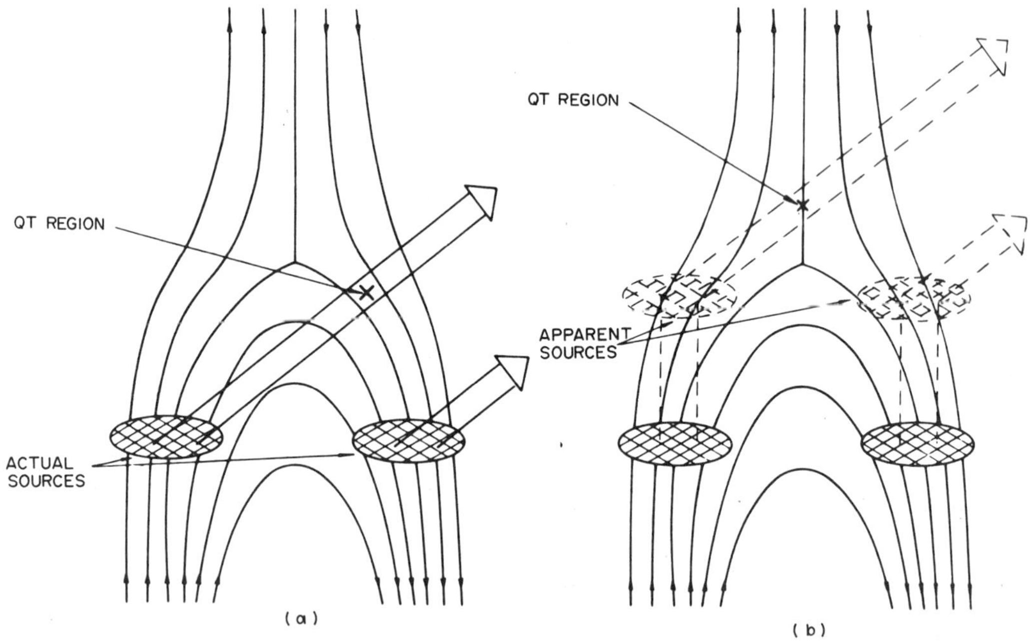

Fig. 3.12. Ray paths for sources in bipolar regions with overlying helmet structure. In (a) $f \gg f_p$ and the ray paths are almost straight lines. In the case of (b) $f \approx f_p$ and rays escape to the earth after refraction and scattering. A QT region is encountered in the neutral plane (Melrose, 1973b).

wise be estimated. (3) Radiation could be guided along field lines and away from possible QT regions by field-aligned density inhomogeneities (under-dense inhomogeneity acting as a waveguide).

As regards the first possibility it may be remarked that refraction followed by scattering from coronal inhomogeneities will increase the height of the first QT region. But then the radiation must be generated close to the plasma frequency. This would favour the hypothesis that noise storm radiation is generated at the fundamental plasma frequency.

Since there are large variations between different centres of activity, one should expect exceptional cases of polarization behaviour to occur, as is confirmed by existing observations.

Provided small inhomogeneities are present in the corona, the inconsistency between theory and observations can be resolved. In the treatment of mode coupling undertaken by Melrose (1974b) the theory of Cohen was generalized so as to permit the wave vector, the gradient in electron density and the magnetic field direction to be all different. Then the angle θ between the wave vector $\mathbf{k}$ and the magnetic field $\mathbf{B}$ may change along the ray path due to refraction when $\mathbf{k}$ and the gradient in electron density are non-parallel. Therefore a density inhomogeneity can cause effective coupling in a QT region. Since there is evidence for an inhomogeneous structure of the corona Melrose (1975b) attributed the coupling at QT regions to density inhomogeneities.

As has been shown, noise storm observations are compatible with $Q \gg 1$ in all QT regions. The expression for Q derived in the more general theory (Melrose, 1974b) may be re-written in the following form (Melrose, 1975b):

$$Q = (1-f_p^2/f^2)^{5/2} (f/f_{T1})^4 + (1-f_p^2)^{3/2} (f/f_{T2})^2 \qquad (3.6\text{-}2)$$

where

$$f_{T1}^2 \simeq 5L_B f_B^3 f_p^2, \qquad (3.6\text{-}3)$$

$$f_{T2}^2 \simeq 5L_N f_B^3. \qquad (3.6\text{-}4)$$

All frequencies are in megahertz and L_B and L_N are the respective characteristic distances in kilometers over which the direction of the magnetic field and the density changes.

The second term on the right-hand side of (3.6-2) is important because it implies that mode coupling in a QT region can occur as a result of a density inhomogeneity. The absence of a reversal in the polarization of noise storm emission requires that $f_{T1} \lesssim f_p$ or $f_{T2} \lesssim f_p$. The latter condition gives in combination with (3.6-4) the following inequality:

$$5L_N f_B{}^3 \lesssim f_p{}^2 \qquad (3.6\text{-}5)$$

or

$$L_N B^3 \lesssim 10^{-2} f_p{}^2 , \qquad (3.6\text{-}6)$$

where L_N is in kilometers, B in gauss and f_p in megahertz. If $B \approx 1$ G and $f_p = 100$ MHz, the inequality (3.6-6) leads to $L_N \lesssim 100$ km.

Noise storms are probably generated in closed magnetic field structures, but the first QT region may be encountered in a region where the field lines are open (Fig. 3.12). Here one may have sheets of plasma confined by their own magnetic fields. If it is assumed that the magnetic and gas pressures are roughly equal at a temperature of 2×10^6 K, $B^2/8\pi = 2N_e \kappa T$ leads to

$$B = 1.2 \times 10^{-2} f_p . \qquad (3.6\text{-}7)$$

Eliminating B by use of (3.6-6) and (3.6-7) gives

$$L_N \lesssim 10^4 f_p{}^{-1} \qquad (3.6\text{-}8)$$

Assuming $f_p = 400$ MHz and 10 MHz as limiting values in connection with noise storms, one obtains $L_N \lesssim 25$ km and $L_N \lesssim 1000$ km respectively. The existence of irregularities of this scale in the corona is not yet proved, but there are different kinds of evidence from X-ray, EUV and radio observations for inhomogeneities of small sizes, and the mechanism of strong mode - mode coupling explained in terms of small scale inhomogeneities proposed by Melrose appears quite plausible.

In sections 3.1 and 3.2 it was shown that limb storms tend to be unpolarized. Suzuki (1961) suggested that the effect might be explained by mode coupling in a QT region followed by depolarization due to differential Faraday rotation. The formula (3.5-25) shows that a circular component may be partially converted into linear polarization in a QT region, and thus substantiates the proposal made by Suzuki. The mechanism is particularly effective if there exists a region with $Q \approx 1$ near the plasma level (Melrose, 1974a). Then, amongst others, the QT approximation is valid over a wider range of angles than when $X < 1$.

3.7. CONCLUDING REMARKS ON POLARIZATION

Noise storms including continuum and bursts are generally almost completely circularly polarized. Linear polarization has not been detected. Even in type III bursts, in which case linear polarization has been reported by several authors, Grognard and McLean (1973) definitely claimed that linear polarization is non-existent. Considering the effect of depolarization by differential Faraday rotation it would in fact be surprising if linear polarization was present.

The circular polarization of storm radiation may result from the emission mechanism itself, but it may also be caused by the birefringent properties of the magnetized coronal plasma. The latter mechanism is highly probable.

When radio waves propagate in a plasma permeated by a magnetic field, the medium is birefringent. The refractive indices of the ordinary and extraordinary wave modes are given by (Ratcliffe, 1959)

$$n^2 = 1 - \frac{X}{1 - \frac{Y^2\sin^2\Theta}{2(1-X)} \pm \left\{\left[\frac{Y^2\sin^2\Theta}{2(1-X)}\right]^2 + Y^2\cos^2\Theta\right\}^{\frac{1}{2}}} \qquad (3.7\text{-}1)$$

The plus sign corresponds to the ordinary and the minus sign to the extraordinary mode. The angle between the direction of propagation and the magnetic field is Θ.

The strong polarization of noise storms in the ordinary mode indicates that the radiation is generated at a frequency close to the plasma frequency. Then the ordinary wave can escape freely, but the extraordinary wave meets a stop region above the level where $X = 1$. The stop region extends from the level $X = (1-Y^2)/(1-Y^2\cos^2\Theta)$ corresponding to $n = \infty$ and to $X = 1 - Y$ where $n = 0$. This is illustrated in Figs. 3.13 and 3.14.

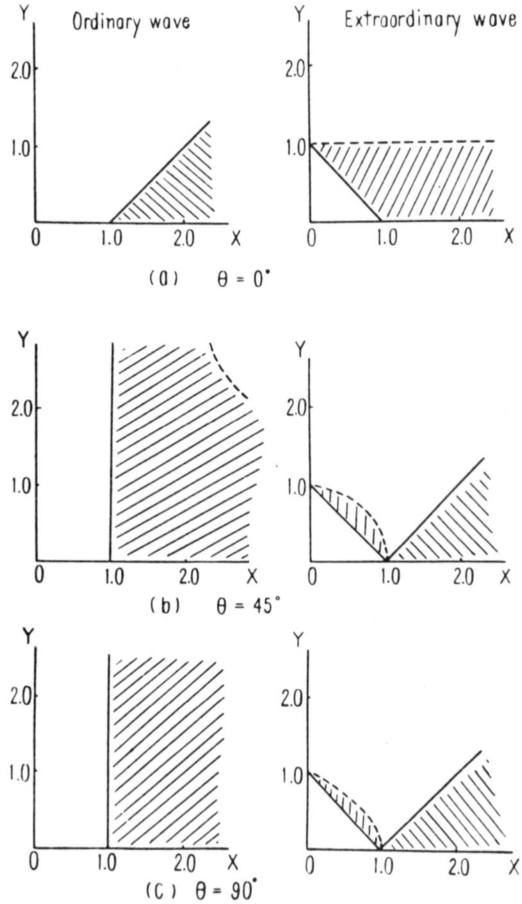

Fig. 3.13. Stop regions (shaded areas) for three values of Θ (Kai, 1962).

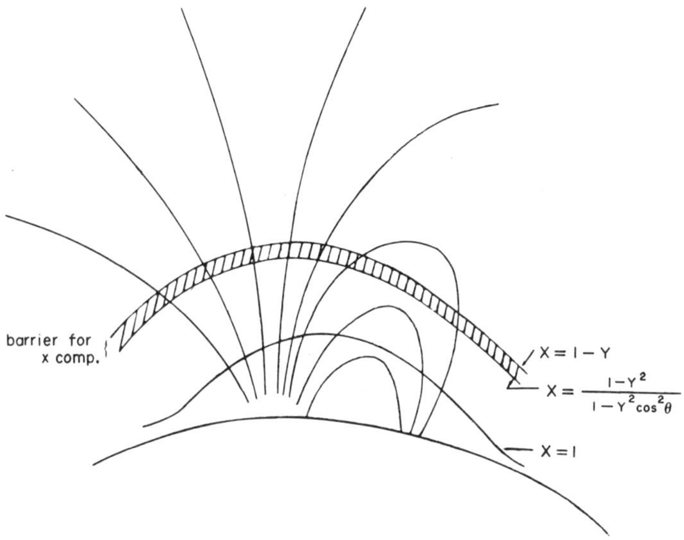

Fig. 3.14. Model indicating the levels of escape for the ordinary and extraordinary waves (Fokker, 1965a).

Observations show (Figs. 3.1 - 2 - 3) that unpolarized storms tend to occur near the limb. If a source is to be visible from the solar limb, it must also be situated sufficiently high in the corona, and observations show that this is the case (Le Squeren, 1963). Therefore the source may be situated above the level where $X = 1 - Y$ and both modes escape. Provided the differential absorption of the waves along the trajectory is very small, the received radiation will be unpolarized or weakly polarized. Near the limb the depolarization mechanism described in the previous section may also be effective. Thus unpolarized or weakly polarized storms are quite likely to occur at great distances from the central meridian of the sun.

Exceptions from the general behaviour of storm polarization are observed. Some unpolarized storms occur near the centre, strongly polarized storms are occasionally observed near the limb and unpolarized storms may evolve into strongly polarized ones. This indicates that the magneto-ionic conditions in the vicinity of the source are of great importance. The solar corona is quite inhomogeneous. This structure is particularly pronounced in active regions in which, amongst others, gigantic loop systems may be found. When this is taken into account, it appears fully possible to explain also the exceptional cases of polarization.

3.8. SOME MODELS OF CORONAL MAGNETIC FIELDS AS DEDUCED FROM STORM OBSERVATIONS

Noise storms are associated with active centres, and the location of the sources closely depends on the magnetic field structure. High resolution mappings of storm sources may be used to infer coronal magnetic fields. The technique is not limited to storms but has been applied in connection with various types of bursts. In some cases the observations have been compared with coronal magnetic fields calculated in the current free approximation from field measurements at the photospheric level (Newkirk, 1971; Dulk, Altschuler and Smerd, 1971).

In order to explain bipolar storm sources the model shown in Fig. 3.5 was proposed by Kai (1970). In this case a magnetic arch extends high into the corona and is rooted in or around sunspots of opposite polarity in a complex group.

Extensive observations of a long-lived type I storm were carried out by Kai and Nakajima (1974) with the Nobeyama 160 MHz interferometer. This instrument gives one-dimensional intensity distributions of the sun with a resolution of $2'$ arc every 0.2 s in LH and RH circular polarization. The storm and numerous transient bursts of spectral type III were observed between March 4 and 12, 1972, and were associated with one active region (McMath Plage Region No. 11769). The storm source was very stable and completely circularly polarized in the LH sense. It was associated with the leading sunspot which showed positive magnetic polarity, and the observations are therefore in agreement with the hypothesis of emission in the ordinary mode. No storm source was associated with the trailing spots which were less compact with correspondingly weaker field strengths.

During the period of storm activity a most striking feature was "correlated bursts". They occurred simultaneously in two sources separated by several minutes of arc. In a few cases transient bursts occurred simultaneously in three individual sources. Some of the correlated bursts were identified with bursts of spectral type III, which show fast frequency drift from high to low frequencies. They probably originated in regions of weak magnetic fields or open field configurations. The wave mode corresponding to the sense of polarization of type III bursts is not known, but Kai and Nakajima assumed that they were polarized in the ordinary mode. This assumption was supported by their observations. From the available data they then derived the model of the active region shown in Fig. 3.15.

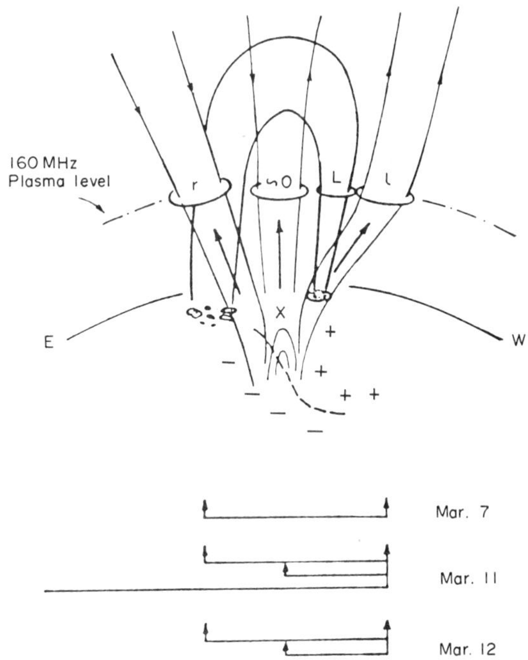

Fig. 3.15. Possible configuration of coronal magnetic fields. The storm source is in the strong field above the leading sunspot. Sources showing correlated bursts are indicated by arrows at the bottom. *L*, *l*, *r* and *O* indicate strong LH, partial LH, partial RH circular polarization and *O* almost no polarization. The dashed line is the neutral line dividing regions of opposite magnetic polarity. (Kai and Nakajima, 1974.)

The storm source is above the leading sunspot, in a region of strong magnetic fields and near the 160 MHz plasma level. The sources showing correlated bursts are associated with regions of weak magnetic fields which may exist in the region around the strong, closed magnetic fields. In most cases the correlated bursts occurred in sources of opposite polarity. Kai and Nakajima therefore suggested that part of the weak magnetic fields with opposite polarity merge in the lower atmosphere in a complex way and form neutral sheets above closed magnetic loops which cross over regions of opposite polarity. If fast particles are accelerated in the neutral sheet, they may travel along different paths and reach the 160 MHz plasma level at almost the same time in two or three different source regions, thus producing correlated bursts.

The models shown in Figs. 3.5 and 3.15 concern the magnetic field above and in the vicinity of a localized centre of activity. But there is now good evidence that field lines may also connect regions at the solar surface which are widely separated. An active region producing a noise storm lasting for several days was studied by Lantos-Jarry (1970). A systematic displacement of the continuum and burst sources was found, which could be explained by the model shown in Fig. 3.16. In this case an arch connects two regions of opposite polarity and the regions are separated by a large distance. In the narrow region with the strongest magnetic field the storm sources are situated. The arch spreads out, and the field lines come down in a region of opposite polarity delineated by quiescent filaments. From

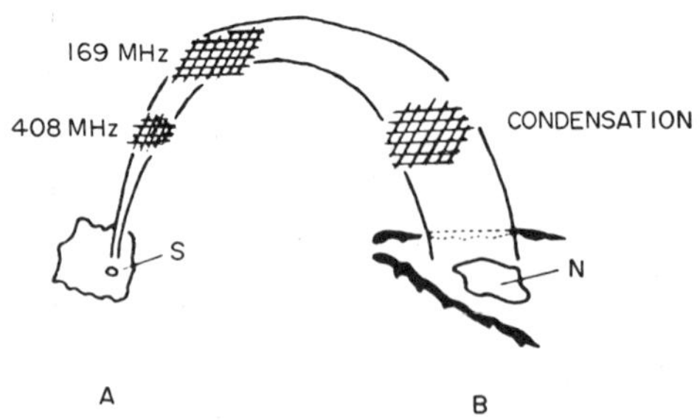

Fig. 3.16. Magnetic field structure as proposed by Lantos-Jarry (1970).

this part a faint thermal enhancement was observed at centimetric wavelenghts.

Further studies of selected, similar cases by Daigne *et al.* (1971) led to the following general results:

(1) The observed mean position of the continuum and type I bursts at 169 MHz is systematically displaced relative to the position of the optical region by about 10° to 20° in the east - west direction, the sense and magnitude depending on the storm centre observed. In each case the type I burst positions are widely spread.

(2) This displacement is always oriented toward a region of opposite polarity to that of the main spots of the eruptive structure giving rise to the noise storm.

(3) The region of opposite polarity is stable and most frequently delimited by a large quiescent filament. Filaments represent the limit of regions of opposite polarity. This region is an old calcium plage persisting for several solar rotations and is associated with a thermal enhancement observed at centimetric wavelengths. A thermal metric emission is observed inside the streamers associated with the presence of the filaments.

(4) A striking feature is the absence of any other plage between the active centre and the region which is at the foot of the inferred connecting magnetic arch. This property is visible on *H*-alpha spectroheliograms.

From this Daigne *et al.* concluded that there may exist large-scale coronal magnetic structures between regions of opposite polarity when they are at the limit of a large region free of any magnetic fields. On the other hand, the case of several neighbouring plages will lead to local magnetic field structures with magnetic field lines confining themselves to closed regions.

The findings quoted here are in excellent agreement with a case described by McLean (1969). Radioheliograph observations showed a storm at 80 MHz in which frequent changes of source position took place. The general features of the observations are illustrated in Fig. 3.17.

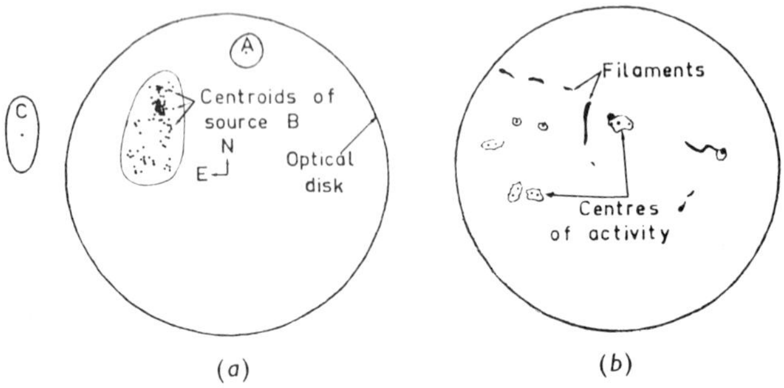

Fig. 3.17. (a) Map of the sun showing the position of sources *A*, *B* and *C*. Source *B* is represented by dots indicating successive positions of the centroid of this source during a sample period near 00 h UT, and an envelope drawn by eye around these points. (b) Sketch-map of the optical features of the sun on November 11, 1968, showing the positions of filaments and centres of activity. (McLean, 1969.)

At *A*, near the north point of the sun, a small source was very stable in position, but at *B* a second source was observed to move back and forth along a north - south line. Occasionally sources occurred at *C*. Fig. 3.17b is a sketch of the optical features of the sun while the storm occurred. Comparing Fig. 3.17a and b it is seen that source *A* is probably associated with the large centre of optical activity near the central meridian whereas source *B* appear to be connected with the dark filament to the east of this centre.

According to McLean there was evidence for a correlation between intensity variations in the sources *A* and *B*. He therefore suggested that field lines connected points along or near to the filament with points in the more compact centre via loops reaching high into the corona and intersecting the 80 MHz plasma level in the regions *A* and *B*. The field geometry may then be as depicted in Fig. 3.18.

Finally, a rather unconventional model of the magnetic field structure in the corona above active centres should be

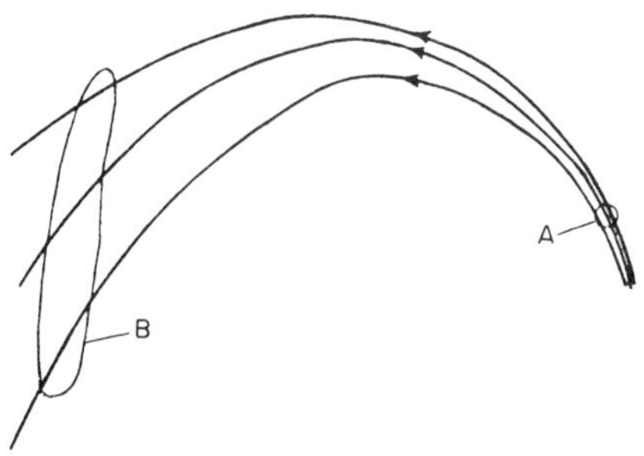

Fig. 3.18. Possible structure of field lines connecting sources *A* and *B* (McLean, 1969).

mentioned. McLean and Sheridan (1972) described 80 MHz radioheliograph observations which could be interpreted as evidence for a twist in the magnetic field. Dynamic spectra for November 19, 1970, showed a typical noise storm of moderate intensity between 200 and 100 MHz, but which was weak at 80 MHz. The activity on this day was exceptional so far as four sources were frequently simultaneously observed. Two of the sources, *A* and *B* in Fig. 3.19 were bipolar. The sources *C* and *D* were only detectable sporadically during the day and will not be discussed in the following. A very nice representation of the polarization and position observations is presented in Fig. 3.20. Black dots

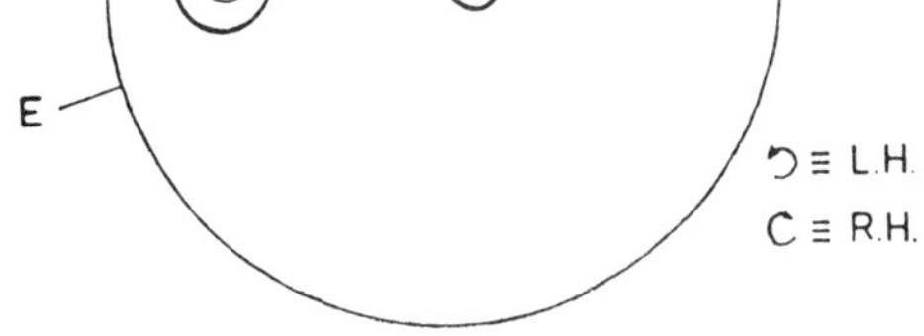

Fig. 3.19. Schematic diagram showing the positions of the four sources observed: two positions are shown for the bipolar sources *A* and *B* (McLean and Sheridan, 1972).

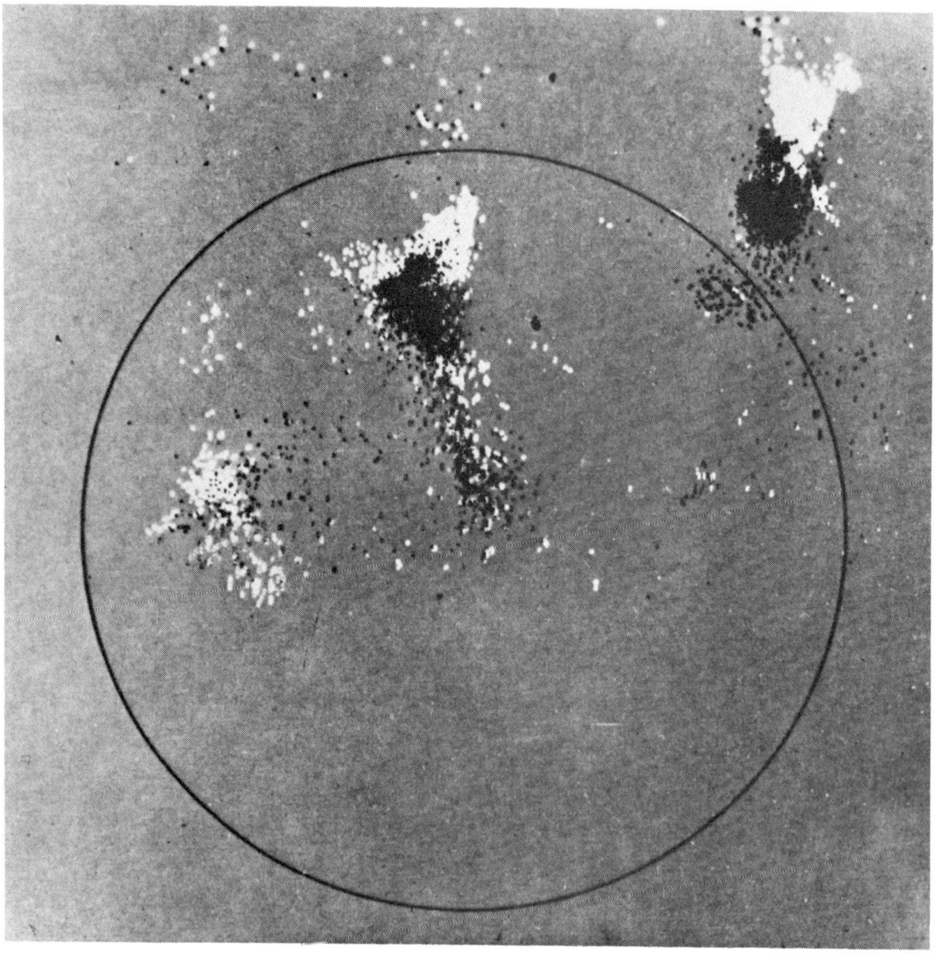

Fig. 3.20. Position data from all available heliograms on November 19, 1970: the centroid position of each source for each second of data has been marked by a small dot; black for RH and white for LH polarization. (McLean and Sheridan, 1972.)

indicate RH polarization and the white dots LH polarization. The bipolar structure is clearly seen and in both sources the RH component is systematically displaced to south and east as compared with the LH component. In Fig. 3.21 an *H*-alpha photograph of the sun taken at the same day and provided by the CSIRO Division of Physics is shown. The orientation of the sun is the same in all figures and it is then seen that source *A* can be

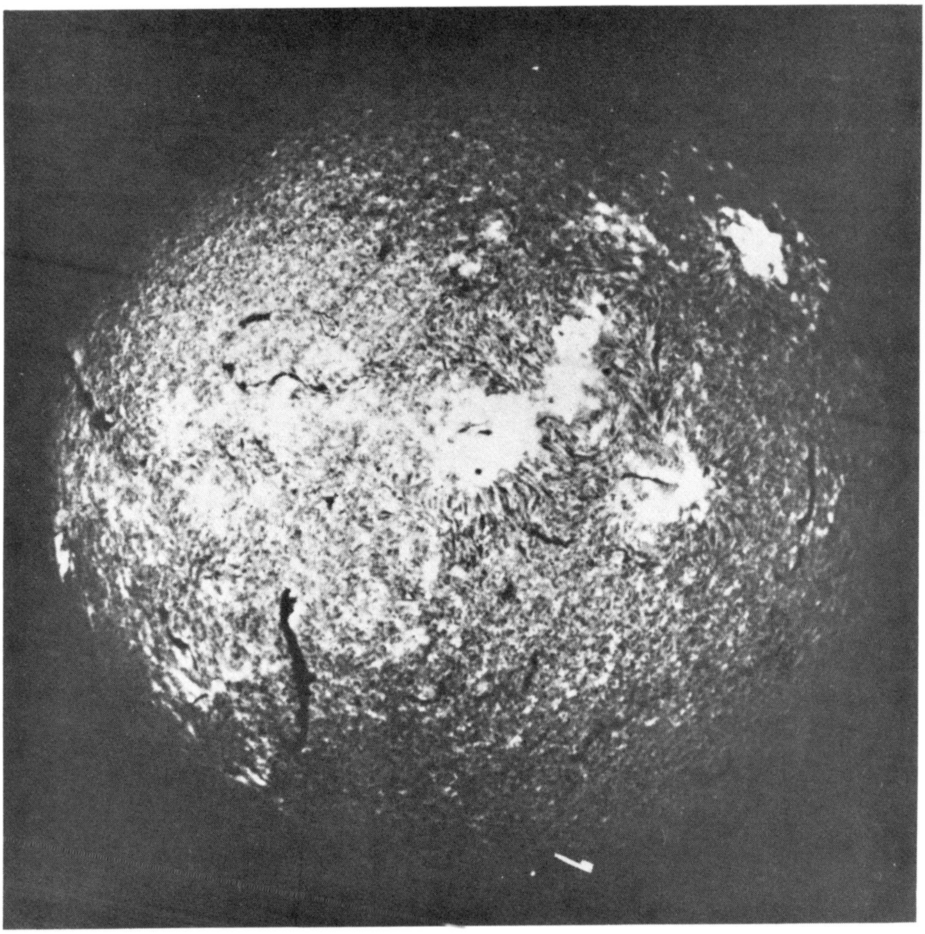

Fig. 3.21. Photograph of the sun in *H*-alpha for November 19, 1970 (McLean and Sheridan, 1972).

identified with the active centre near the north - west limb. But for centre *B* either the centre of activity near the centre of the disk, or the one a little to the north and west could have been associated with the source. In either case the radio source is displaced by about 8′ arc from the radial direction above the optical centre, but this is quite common and compatible with the results of Le Squeren (1973) when allowance is made for the height difference between the 169 and 80 MHz plasma levels.

It is rather firmly established that the dominating sense of polarization is determined by the field direction in the dominant sunspot in the associated centre of activity. This is in general the leading spot and storm radiation is emitted in the ordinary mode. The net polarization of sources *A* and *B* was LH circular. In the case of source *A*, the trailing spot in the active region was the larger and had the stronger magnetic field. For the source *B* the situation was less clear. Neither of the centres of activity considered showed a clearly dominant leading or trailing spot. But all three spot groups considered showed normal behaviour for sunspot groups at the northern hemisphere during the solar cycle in concern. The leading spots had south polarity, the trailing ones were of north polarity. For ordinary mode emission one should then find the RH components of sources *A* and *B* to lead, or to be to the west of the LH components. This situation is shown in Fig. 3.22a. Comparing with Fig. 3.20 it is seen that this does not correspond with the observations.

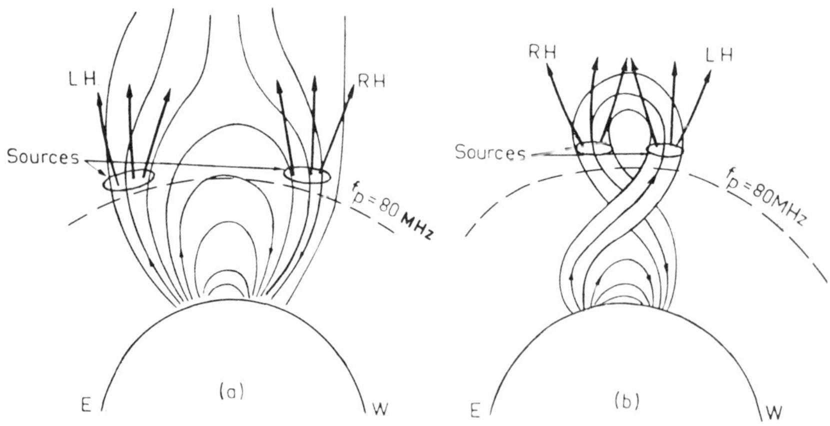

Fig. 3.22. Diagrams illustrating the expected positions and a possible interpretation of the reversal of positions of bipolar sources. (a) Simple model which gives wrong positions assuming emission in the ordinary mode. (b) Twisted field. The lines of force come out of the plane of the paper. (McLean and Sheridan, 1972.)

In the case of source *A* the orientation is almost the reverse of that predicted in Fig. 3.22a. McLean and Sheridan therefore found that the observations were best explained if a large-scale twist in the magnetic field was supposed, as illustrated in Fig. 3.22b. Depending on the amount of twist, the authors then could account for any orientation of bipolar sources. The situation described here is not exceptional, as is seen from the observations of bipolar sources by Kai (1970). The source orientations shown in Fig. 1.14 may well be explained when twisted fields are assumed. Twists in the field should be less detectable at higher frequencies since the source heights then are smaller. This is in agreement with results previously described.

A large-scale twist in the magnetic field above sunspots may occur as a result of the kink instability. A flux tube with fixed end points is kink unstable if the field lines encircle the flux tube more than once (Rosenberg, 1973).

4. DIRECTIVITY

4.1. INTRODUCTION

Storm radiation is generated in the solar corona where physical parameters such as electron density, magnetic field strength and temperature vary with altitude and heliographic position. The high intensity and the rapid flux variations observed in noise storms show that the emission is of non-thermal origin.

Several mechanisms which generate radio waves in a plasma emit energy in preferential directions. The directivity may be represented by a polar diagram. In addition, propagation effects may produce a beaming of the radiation in particular directions. It is well known that the beaming may be strong when the emission occurs near the plasma level. Fokker (1965a) pointed out that the visibility may also be governed by specific coronal structures that do or do not prevent the escape of radiation in certain directions (channelling).

In order to study the directivity of solar noise storms, analysis of the centre-limb variation of the storm intensity is carried out. Sometimes storms can be followed for many days, but frequently storms have shorter durations. In addition to this, storm emission is characterized by frequent variations in intensity. Therefore statistical investigations of the centre-limb variation of storm intensity have been undertaken.

A more direct approach to directivity determinations of solar bursts would be most valuable. Observations from the earth and a space vehicle have been undertaken in order to achieve this aim. Simultaneous signal reception from widely different directions is then possible, and better information on the directivity of the radiation mechanism can be obtained.

4.2. LONGITUDE EFFECTS

Early studies of storm radiation revealed that maximum intensity occurred near the central meridian passage of the associated sunspot (Hey, Parsons and Phillips, 1948). This was interpreted as evidence for beaming of the storm radiation approximately along the radial direction (normal to the spot). This discovery was verified by several later studies, and attempts have been made to determine average emission polar diagrams for storm radiation.

Machin and O'Brien (1954) calculated auto-correlation coefficients of daily flux values and found that the coefficient dropped to a small value for differences in time of a few days. It was suggested that this behaviour was caused by the polar emission diagram of the storm radiation. A trend with frequency seemed to exist indicating stronger beaming at lower frequencies. A similar statistical study was made by Jorand (1955) who interpreted the results in terms of the average height at which the emission takes place.

Using interferometer observations at 255 MHz, Fokker (1960) was able to associate storm sources with sunspot groups. He divided the solar disk in concentric circular zones (0° - 10°, 10° - 20°, etc.) and counted the total number of sunspots listed and the numbers of moderately and strongly noise active spots. The results are displayed in Fig. 4.1. It is possible to

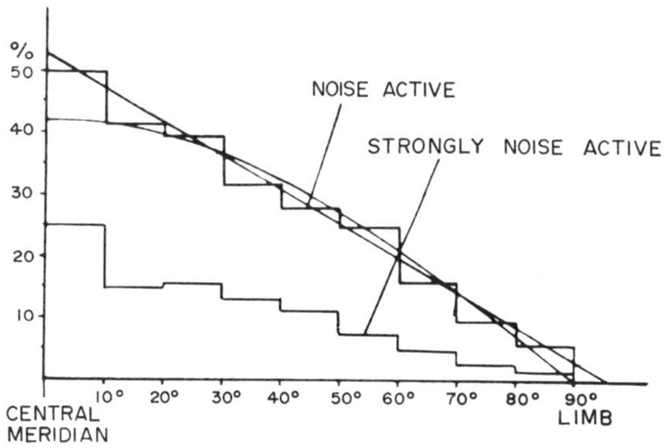

Fig. 4.1. Distribution of noise-active sunspots over the solar disk (Fokker, 1960).

represent the distribution either by a cosine law or a linear law. The emission is directive, but the directivity seems not to be frequency dependent because Fokker found a general agreement between the results of interferometer observations at 255 and 169 MHz.

Further studies of a possible frequency dependence of the width of the emission diagram were undertaken by Smerd (1964). Histograms of storm maximum in relation to central meridian passage of the associated spot group were constructed for several frequencies between 67 and 200 MHz and for 27 storms. No systematic change with frequency was found. Storm maxima seemed to occur within 2 days of central meridian passage. Particularly good data were available at 200 MHz, and here 18 storms showed a mean time difference between storm maximum and central meridian passage of the associated spot group of +9 h with a root mean square deviation of ±22 h. This was considered as evidence for a tilt in the dipole field of the associated spot group.

Four large storms could be followed for several days around storm maximum. The result is shown in Fig. 4.2. The low-frequency curve is based on data from one station only, and the small narrowing of the reception diagram at low frequencies may not be significant.

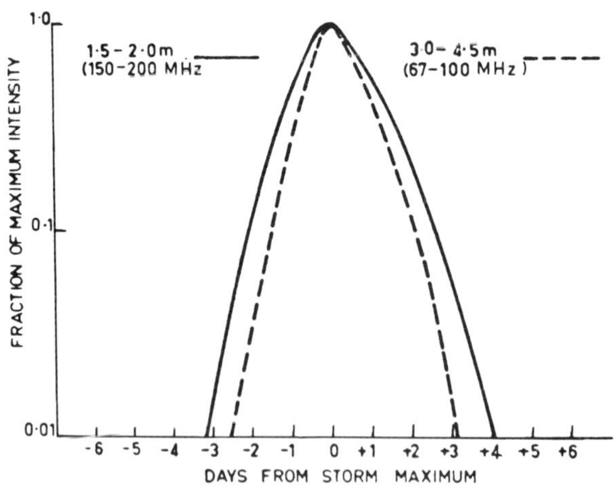

Fig. 4.2. Directivity of four storms (Smerd, 1964).

Suzuki (1961) investigated storms with high burst activity and storms with moderate burst activity. Figure 4.3 shows that the storms are concentrated near the centre of the solar disk and there is no significant difference between storms with high burst activity and those showing moderate burst activity. More storms were found on the western part of the sun than east of the central meridian.

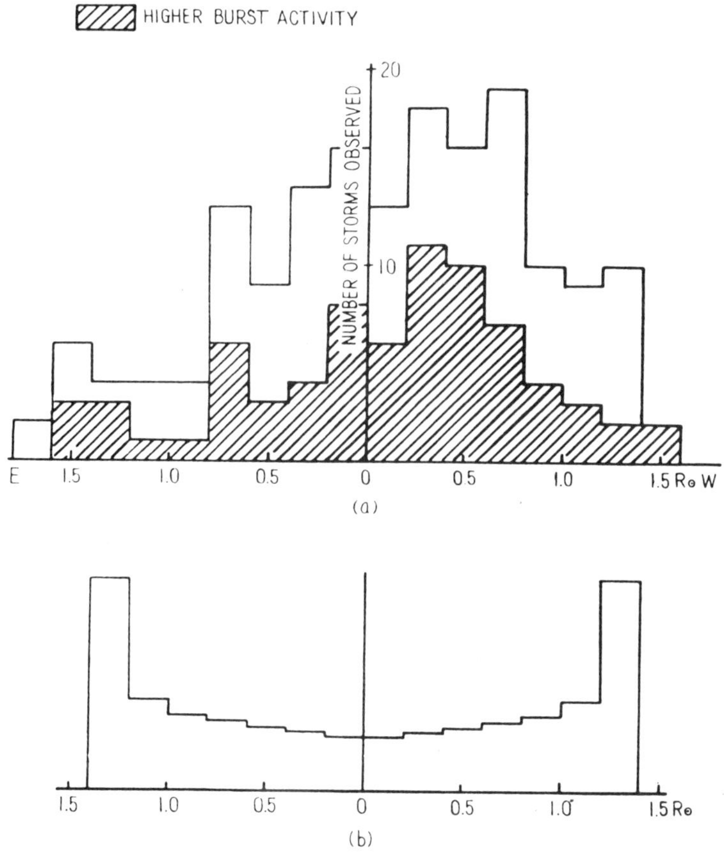

Fig. 4.3. (a) East - west distribution of noise storms observed at 201 MHz; (b) expected distribution with non-directive sources at 0.4 $R_{\odot}$ above the photosphere (Suzuki, 1961).

It is virtually impossible to deduce the true directivity of the storms from the histogram in Fig. 4.3a. Too many factors must then be taken into account. But Suzuki constructed a model histogram by assuming a source height of 0.4 $R_\odot$ above the photosphere and uniform directivity. The result is presented in Fig. 4.3b. A comparison between the results of storm observations and model calculations immediately shows that storms have a strong directivity.

A large observational material collected on 169 MHz was examined by Le Squeren (1963). Statistically 900 storm centres seemed to follow the relation

$$F = F_c(1-Kd) \tag{4.2.-1}$$

where F_c is the maximum flux at the centre of the disk, K is a constant, and d is the distance from the north - south axis as measured in units of $R_\odot$.

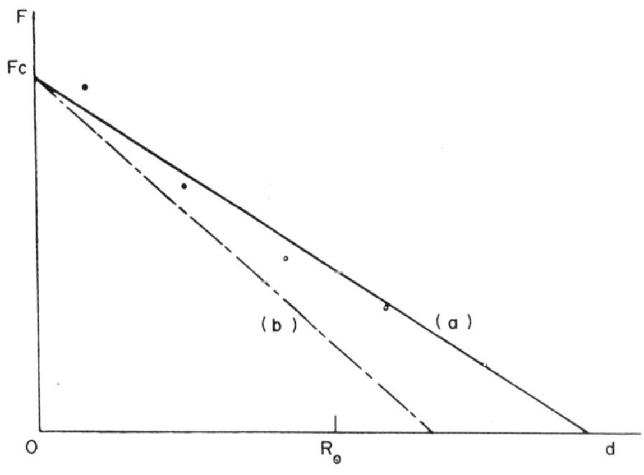

Fig. 4.4. Average decrease in flux of storms as a function of distance from the north - south axis: (a) storms with flux density less than 20×10^{-22} W m^{-2} Hz^{-1}; (b) more intense storms (Le Squeren, 1963).

In Fig. 4.4 the results are depicted. In the figure the material has been divided according to intensity. Curve (a) regards storms with $F < 20 \times 10^{-22}$ W m^{-2} Hz^{-1}, while curve (b) refers to more intense storms. For curve (a) the constant $K \approx 6/11$ and the intensity decreases to $\frac{1}{2}F_c$ when $d = (11/12)R_\odot$. Assuming that the sources are rather near to the equator, one may write $\sin L = d/\rho$, where L is the longitude and ρ is the distance of the source from the solar centre. One then gets

$$F = F_c[1-(6/11)\rho \sin L]. \qquad (4.2.-2)$$

Take $\rho = 1.4$ (in units of $R_\odot$). It is then found that the flux decreases to $\frac{1}{2}F_c$ when the longitude changes from 0° to 40°.

Le Squeren compared the above result with the directivity of the sources producing the slowly varying component in the dm and cm wavelength range and for which one had

$$F = F_c \cos L \qquad (4.2\text{-}3)$$

according to Gutmann and Steinberg (1959). In the latter case the flux is reduced to $\frac{1}{2}F_c$ when $L = 60^\circ$. Storm sources are therefore more directive than the bright plages responsible for the slowly varying component.

4.3. OBSERVATIONS OF THE DIRECTIVITY OF INDIVIDUAL BURSTS

In the preceding section information on the directivity of storm emission was derived from centre-to-limb analysis. This method has been used in statistical studies as well as in some cases when single storms were observed during many consecutive days. But storms are very variable and complicated in nature, and other factors than the directivity of the emission polar diagram may influence the results. The conclusions which regard the storm continuum are more qualitative than quantitative in nature, and the directivity of the short lasting storm bursts is in fact unknown. Steinberg and Caroubalos (1970) pointed out that the "intrinsic directivity" of short lasting solar bursts cannot be determined from centre-limb effects when it is assumed that their direction of maximum intensity is not radial. In order to obtain unambiguous evidence of burst directivity, simultaneous observations must be performed from the earth and a remote space probe.

Steinberg and Caroubalos reached their conclusion as follows. Assume that the events occur uniformly on the sun at constant height above the photosphere and in the plane of the ecliptic. All bursts are assumed to have the same emission polar diagram, $f(\phi)$, which is symmetric around an axis making an angle θ with the radial direction through the source. When the longitude of the source is L, the flux density received at the earth may be written

$$I(L) = Af(\phi) \tag{4.3-1}$$

where $\phi = L + \theta$. The situation is illustrated in Fig. 4.5.

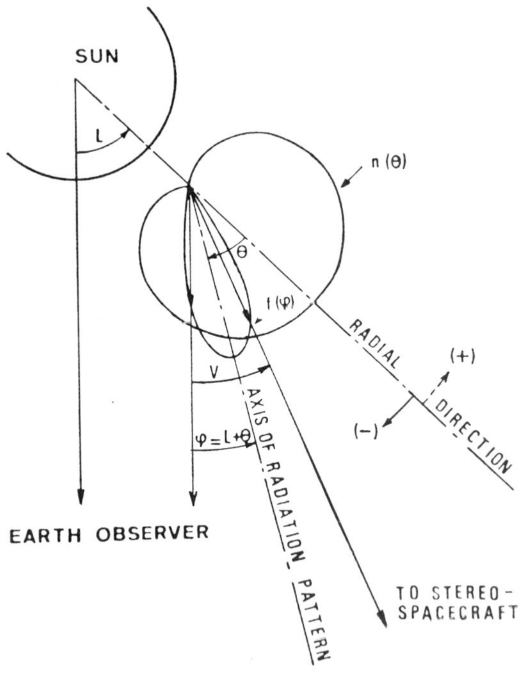

Fig. 4.5. Geometry of the observation of a directive burst when the axis of the directivity pattern is not assumed to be radial (Steinberg and Caroubalos, 1970).

A is the flux density (intensity) in the direction of the maximum of $f(\phi)$. If a burst is to be observed, the intensity must exceed the sensitivity treshold C of the receiving system

$$I(L) \geq C \tag{4.3-2}$$

or, using (4.3-1),

$$A \geq \frac{C}{f(L+\theta)}. \tag{4.3-3}$$

Then the probability to detect one event amongst $n(\theta)d\theta$ events at longitude L, the polar diagram of which is centred about the direction θ, is

$$P(L,\theta) = \int_{A_0}^{\infty} p(A)dA, \tag{4.3-4}$$

$$A_0 = \frac{C}{f(L+\theta)}$$

where $p(A)$ is the number distribution of A and $p(A)$ is assumed to be independent of θ. The relative number of bursts observed at the earth at a longitude L will be

$$N(L) = \int_{-\pi}^{+\pi} n(\theta) \int_{A_0}^{\infty} p(A)dAd\theta, \tag{4.3-5}$$

$$A_0 = \frac{C}{f(L+\theta)}$$

Focussing of the radiation because of propagation effects has not been taken into account in (4.3-5). This effect depends on the model of the solar corona and on the structure and the position of the source and is difficult to account for. Steinberg and Caroubalos simplified the problem by assuming that the effect of the propagation is that the rays leave the sun inside a cone, the aperture of which depending on the source position and the parameters of the medium. The focussing effect should be added to the source directivity to derive the apparent directivity as observed on the earth. This is taken into account by a factor $\alpha(L)$ in (4.3-1), thus

$$I(L) = A\alpha(L)f(L+\theta). \tag{4.3-6}$$

The expression (4.3-5) may then be written

$$N(L) = \int_{-\pi}^{+\pi} n(\theta) \int_{A_1}^{\infty} p(A)\,dA\,d\theta. \tag{4.3-7}$$

$$A_1 = \frac{C}{\alpha(L)f(L+\theta)}.$$

When $f(\phi)$ and $n(\theta)$ are narrow, the effect of the focussing function $\alpha(L)$ is less important. When $f(\phi)$ or $n(\theta)$ is wide, the focussing effect is important. In any case this discussion shows that one is unable to distinguish between the directivity due to the propagation and the directivity due to the emission itself from a single set of observations from the earth.

The functions $f(\phi)$ and $n(\theta)$ are unknown, but the intensity distribution function $p(A)$ has been determined by different authors (Fokker, 1960; Wild, 1951; Morimoto, 1963; Le Squeren, 1963; Hauge, 1958). Using different functions $f(\phi)$, $n(\theta)$ and $p(A)$, Steinberg and Caroubalos computed $N(L)$ for different combinations of the functions and for three values of the treshold sensitivity C. It was also attempted to take the effect of the focussing into account by estimating the function $\alpha(L)$. In Figs. 4.6 - 4.8 som examples of $f(\phi)$, $n(\theta)$ and $p(A)$ used in the computations are shown together with the assumed focussing function $\alpha(L)$. Some results from the calculations are shown in Fig. 4.9.

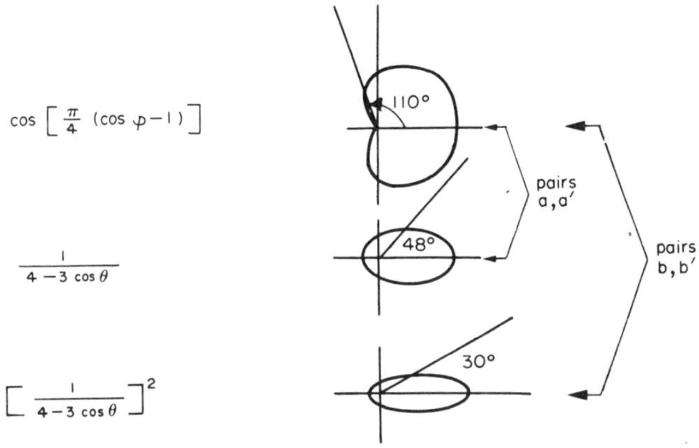

Fig. 4.6. Examples of functions $f(\phi)$ and $n(\theta)$ (Steinberg and Caroubalos, 1970).

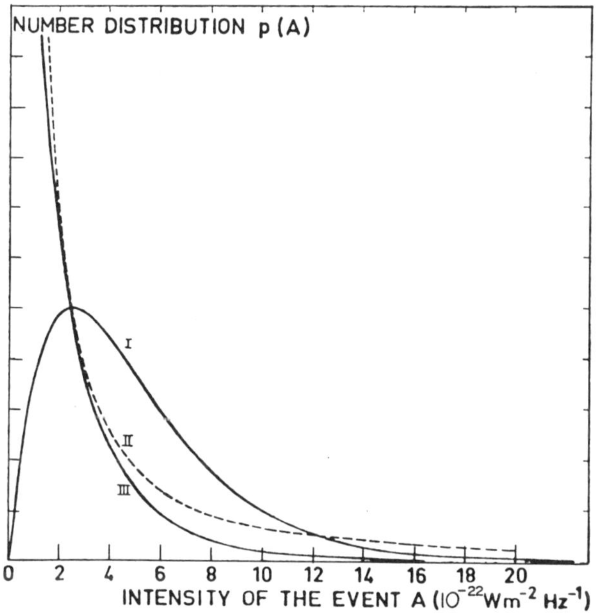

Fig. 4.7. Three models of number density distribution assumed for the bursts. Model I after Fokker (1960), model II after Wild (1951), Morimoto (1963) and Le Squeren (1963), model III after Hauge (1958). (Steinberg and Caroubalos, 1970.)

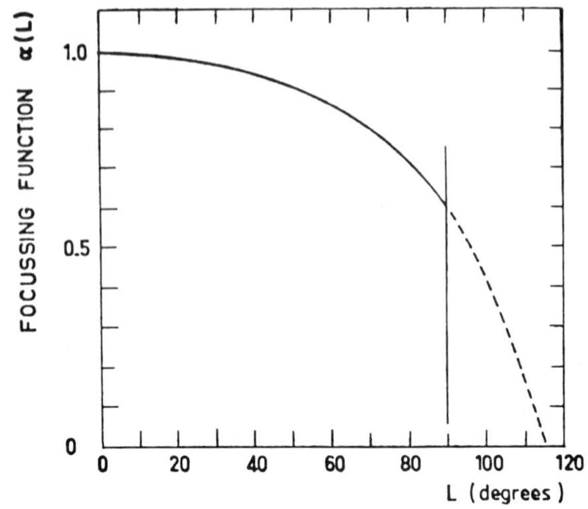

Fig. 4.8. The function used to take focussing into account (Steinberg and Caroubalos, 1970).

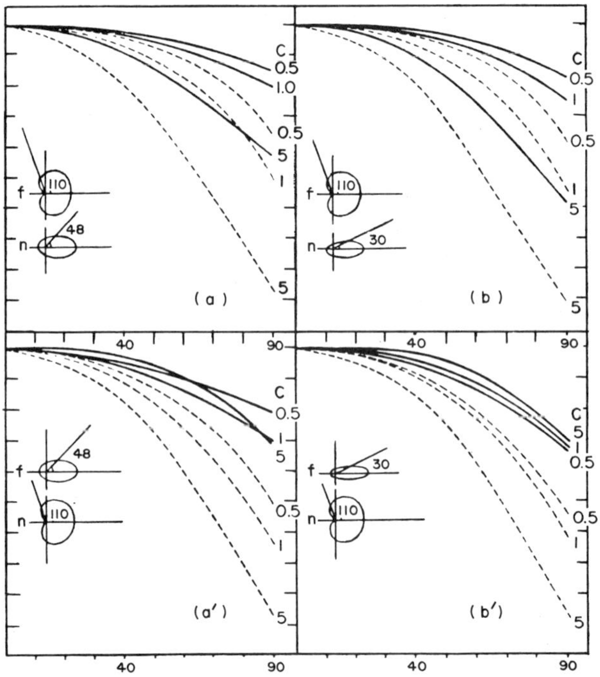

Fig. 4.9. Examples of computed centre-limb distributions $N(L)$ for various models of the intrinsic burst directivity (function f) and angular distributions of its main axis (function n). The function $p(A)$ is that of model III by Hauge. Continuous lines without focussing, dotted lines with $\alpha(L)$ taken into account. C is the treshold sensitivity of the equipment in units of 10^{-22} W m^{-2} Hz^{-1}. (Steinberg and Caroubalos, 1970.)

Different conclusions can be drawn from the model calculations, but the main outcome was that a given distribution $N(L)$ as determined from observations can be explained by a variety of models p, f, n and C. Therefore it is quite impossible to deduce any accurate information on the true directivity of a source from centre-limb distributions of bursts observed from the earth.

One is not much better off even if the observational material is so large as to permit determinations of the intensity distribution of the events at different longitudes. One may construct distributions $Q(L,I)$, but from (4.3-7) one can obtain

$$Q(L,I) = \int_{-\pi}^{+\pi} \frac{n(\theta)p[\frac{I}{f(L+\theta)\alpha(L)}]}{f(L+\theta)\alpha(L)} d\theta. \tag{4.3-8}$$

This relation demonstrates that it is impossible to obtain a proper directivity of the sources from observed distributions $Q(L,I)$ except when the emission polar diagram has an axis that points in the radial direction ($\theta=0$; see Fig. 4.5). Nothing is known about this condition from observations. In fact, for type I bursts it is most likely not accomplished.

In order to overcome the difficulties connected with analysis of centre-limb variations, Steinberg and Caroubalos suggested the use of space probes in heliocentric orbits to study the directivity of solar bursts. A receiver placed in the probe may then record bursts simultaneously with observations on the ground. It is essential for the experiment that the angle between the directions sun - space probe and sun - earth (angle V in Fig. 4.5, also called "stereo angle") can be made as large as possible. An experiment of this kind, called STEREO-1, was carried out between May 31, 1971, and February 24, 1972. The ground station was at Nancay, France, and the space terminal was on board the Soviet planetary probe Mars-3. During the experiment the angle V varied from 0° to about 80°. The observing frequency was 169 MHz. A post-detector bandwidth of 4 Hz was used.

The sun was observed for 185 h during which several noise storms and many hundreds of type III bursts were recorded.

Some preliminary results concerning type I bursts were published by Caroubalos and Steinberg (1974).

When the stereo angle V was small, i.e. less than 2°, all details of the type I burst emissions within the 4 Hz bandwidth as observed on the earth were also recorded on board the probe. However, in a few cases evidence of directivity effects were seen.

For angles between 10° and 20° the correlation between the probe recordings and those at the earth deteriorated, but many corresponding events could still easily be identified. But when the stereo angle was larger than about 20°, the correlation was no longer present and it was almost impossible to recognize corresponding type I bursts on the records. Two examples from the observational material are shown in Fig. 4.10.

An important parameter of the experiment is the intensity ratio I_{stereo}/I_{earth} which reflects the directivity of the bursts. For very small values of the stereo angle this ratio was about 1, but in the case when the stereo angle was 35° (Fig. 4-10b), the ratio varied between 0.1 and 30 in the cases when corresponding events could be identified. This was different from the results for type III bursts which showed less directivity.

The noise storm observed on November 18, 1971, was analysed in more detail by Steinberg et. al. (1974). The stereo angle was 35° and two sources were present on the sun: one close to the centre of the disk and another at the east limb. From the central source 300 bursts were analysed, whereas 80 bursts originating in the limb source were examined.

The investigation showed that the polar emission diagram of type I bursts has an approximately constant shape, but the orientation of the axis (angle θ in Fig. 4.5) changed from burst to burst. For the central source a symmetrical beam shape was found. Assuming a gaussian shape, the polar diagram in the ecliptic had a width between half-power points that was smaller than 25°. The limb source revealed an asymmetrical polar diagram, which may be explained by supposing that radiation reaching the space probe was subject to less scattering than radiation observed at the earth.

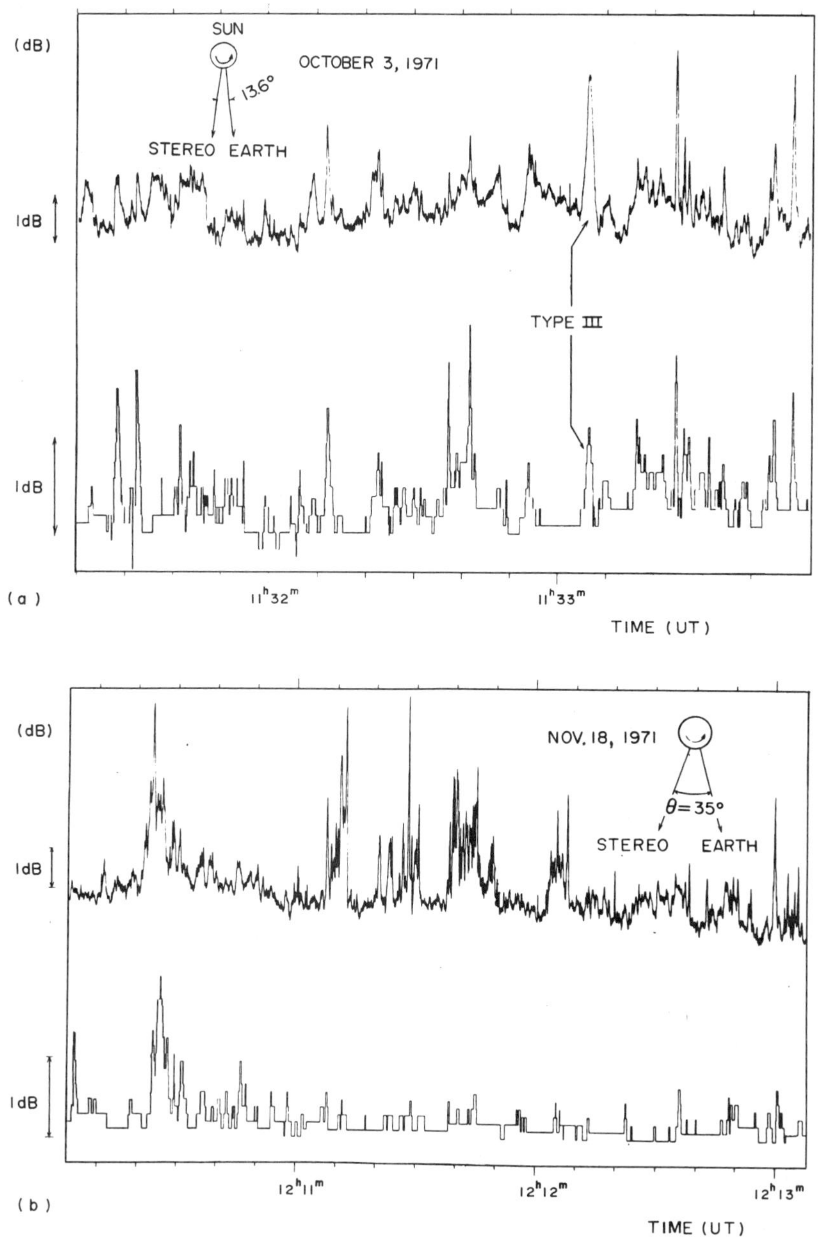

Fig. 4.10. (a) Time profiles of type I bursts on October 3, 1971. There is strong decorrelation, but identification of corresponding bursts is still possible. (b) Noise storm recorded on November 18, 1971. Some corresponding bursts can still be recognized. (Caroubalos and Steinberg, 1974.)

From the STEREO observations one can not exclude the possibility that the three-dimensional radiation pattern is hollow and symmetrical around the direction of minimum intensity. The directions of maximum intensity would then lie on a cone whose axis might be the direction of the local magnetic field. In this case the determined beamwidth would depend on the elevation of the axis of the cone relative to the ecliptic plane. But in any case a beamwidth smaller than that of a meridian cross-section of the hollow pattern can not be measured.

The beamwidth seems to be quite stable. Fluctuations in beamwidth are likely to be less than 20%.

The beam orientations determined for the central source showed a distribution that was roughly symmetrical and close to a gaussian with a standard deviation of about 4° and an orientation of about 20° east of the sun - earth line.

From the analysis it was further found that the maximum intensity I_m (intensity in the direction of the axis of the polar diagram) showed no dependence on the orientation of the beam.

When the effect of the random orientation of the beam was corrected for, the probability distribution of I_m of the sources themselves was determined to a constant factor. The distribution then showed, of course, a considerably smaller variation in burst intensities than the uncorrected distribution.

One may say that the data from the STEREO experiment confirm the ambiguous and qualitative results derived from centre-limb analysis. But in addition they revealed a considerably narrower beamwidth of type I bursts than had been expected. Even though more information on the polar diagram of type I bursts is needed, a lot has now been gained.

4.4. EAST - WEST ASYMMETRY OF NOISE STORMS

Noise storms do not occur with equal probabilities across the entire solar disk. As was briefly mentioned in section 4.1, more storms seemed to occur on the western hemisphere than on the eastern, and on the average, storm maximum occurred after central meridian passage.

The first evidence of an east-west asymmetry of storm emission was presented by Fokker (1960). Positions of noise sources were determined at 255 MHz with the interferometer at Nera Observatory in the period 1957 - 9. It was then found that sunspots which were associated with noise sources occurred more frequently on the western than on the eastern hemisphere. In accordance with this the interferometer position line was more often observed to lie on the western than on the eastern hemisphere. The same effect emerged when data from the 169 MHz multi-element interferometer at Nancay were used.

According to Maltby (1960) the probability to observe a noise storm from a given source was 1.4 times larger on the western hemisphere than on the eastern hemisphere. This is in agreement with the results of Suzuki (1961) who found that the number of sources on the western hemisphere was 1.35 times greater than on the eastern hemisphere as determined from his 201 MHz interferometer observations.

Investigating intense noise storms, i.e. noise centres giving a flux $F > 20 \times 10^{-22}$ W m^{-2} Hz^{-1}, Le Squeren (1963) obtained results which are depicted in Fig. 4.11. There is an

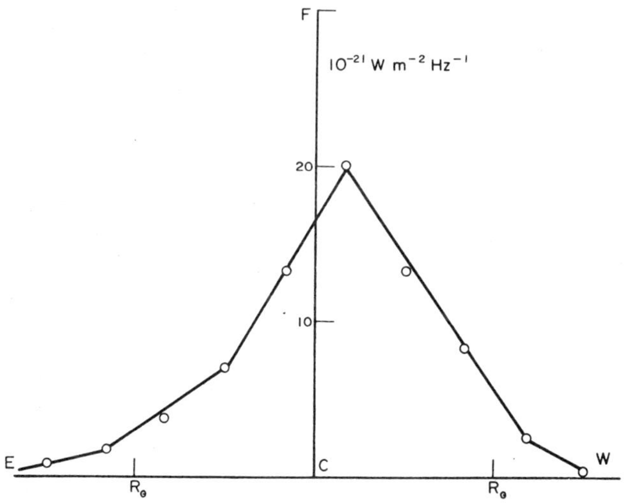

Fig. 4.11. Average variation in flux of 20 noise storms as a function of position (Le Squeren, 1963).

east - west asymmetry which indicates that the centres are best seen on the western part of the solar disk. This tendency gains support when one plots the intensity of different storms against the position of the associated optical centre, as can be seen from Fig. 4.12.

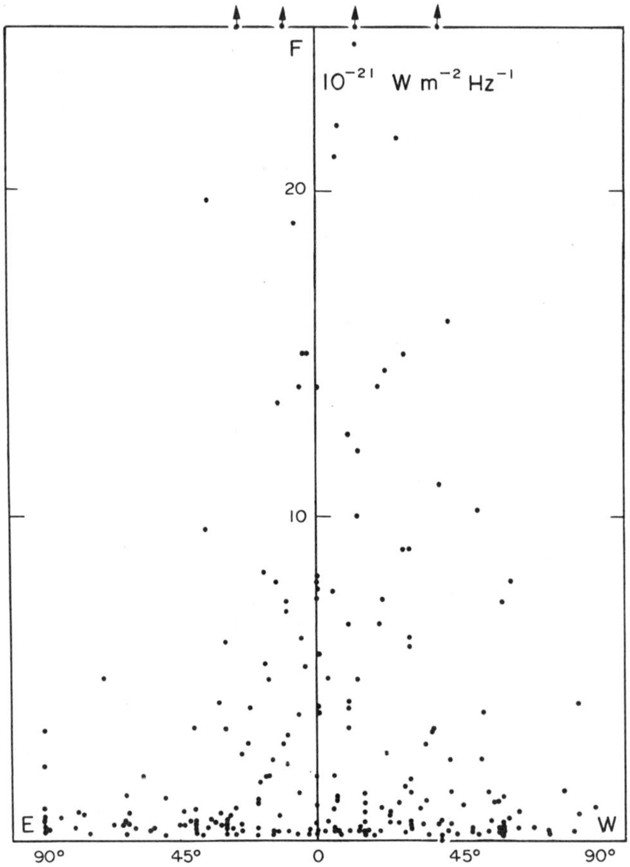

Fig. 4.12. Distribution of the intensity of active centres as a function of their position on the disk (Le Squeren, 1963).

Using observational material obtained with the 169 MHz multi-element interferometer at Nancay, Fokker (1963) made a new investigation of the east - west asymmetry. The observations covered the period from 1957 to 1962. Again a dominance of the western hemisphere was clearly present. "White regions" on the

Nancay diagrams (these are regions of high intensity) were not distributed symmetrically about the north - south diameter. The diagram was rather more nearly symmetrical about a line at 0.1 $R_{\odot}$ to the west.

On its way out the radiation may be guided by large-scale structural elements in the corona into specific directions (Fokker, 1960). Since such elements are likely to be aligned along magnetic field lines, the east - west asymmetry of noise storm radiation may be explained by a tilt in the magnetic structures. The excess of sources on the western hemisphere can be explained by a tilt in the mean emission diagram of 4°. The magnetic structures should then be lagging behind in the solar rotation (Fokker, 1965a). A tilt may result from an unequality of the fields that originate in sunspot regions of opposite polarity. A difference in magnetic field strength of 10% may result in an overall obliqueness of about 13°, and the tilt is in the right sense if the leading sunspot produces a stronger field than the following spot (Fokker, 1963). This is generally the case.

A somewhat different explanation was proposed by Le Squeren (1963). The east - west asymmetry was supposed to be due to a slight inclination of the storm centres with respect to the radial direction in such a sense that the centres were systematically to the east of the radial position, as demonstrated in Fig. 4.13. Then a smaller portion of the source volume would

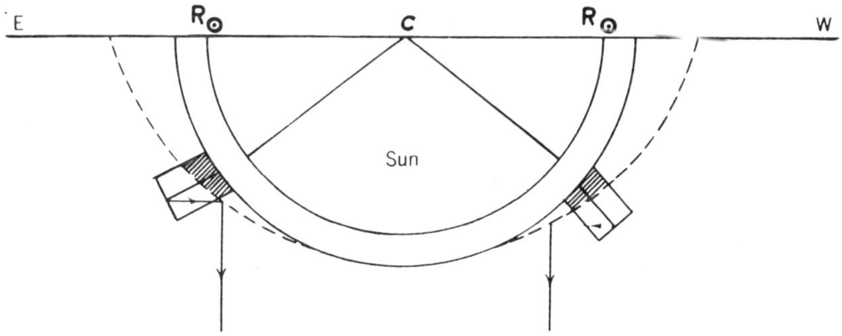

Fig. 4.13. Explanation of the east - west asymmetry of noise storms. The source is a column whose axis is inclined with respect to the radial position. Radiation generated in the hatched regions can not escape. (Le Squeren, 1963.)

be visible on the eastern than on the western hemisphere. Le Squeren found by comparing the distance of the noise sources with the distance of the associated sunspot from the central meridian for different positions on the sun, that an inclination of 10° with the radial position appeared likely for the noise sources.

Maltby (1960) pointed out that the east - west asymmetry he had detected could be explained if the angular emission diagram of the noise storm source had an eastward tilt from the normal to the the solar surface. This is just what was found by Steinberg *et al.* (1974). Bursts from a source near the centre of the solar disk showed an eastward tilt of the beam of about 20° from the sun - earth line.

The asymmetry of noise storm sources and stationary type IV bursts was discussed by Clavelier *et al.* (1968). They found that 7 out of 13 sources showed a double structure. The simple sources were situated to the east of the central meridian whereas the double sources occurred on the western part of the disk. The eastern component of the double sources was in general the least intense and the most transient. Clavelier *et al.* suggested that the western component was systematically more elevated than the eastern one. Then the leading source may occult the other source on the eastern part of the sun, whereas both sources will be observed on the western hemisphere. The situation is depicted in Fig. 4.14.

If the magnetic structure is somewhat tilted towards the east, the height of the western source may be decreased.

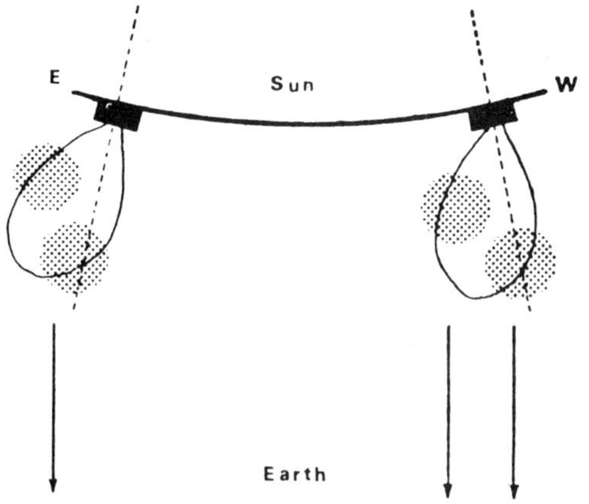

Fig. 4.14. Positions of two sources of a double structure. The leading source occults the trailing source for positions in the eastern hemisphere (Clavelier *et al.*, 1968).

5. CORONAL SCATTERING OF RADIATION FROM NOISE STORM SOURCES

5.1. THE CORONA AS AN INHOMOGENEOUS MEDIUM

When cosmic radio sources pass the sun they increase in angular size. Usually interferometers are used to measure the diameter of the radio sources as a function of the distance from the sun. The best-known example is the observations of the Crab nebula which passes close to the sun in June.

The observed increase in source size is explained as an effect caused by scattering of the radio waves on irregularities in the coronal electron density. It is now understood that the corona cannot be considered as spherically symmetrical and homogeneous. A minimum size of 10 km is indicated for the irregularities at a distance of 30 $R_{\odot}$ according to Hewish (1965). The maximum size may be some 10^6 km. The irregularities seem to be oriented along magnetic lines of force extending far out in the corona, and they probably have the form of aligned filaments. Hewish and Wyndham (1963) suggested that the scale of the irregularities at the solar surface may amount to 100 km, which roughly corresponds to that of the granulation and the chromospheric spicules. Thus the fibrillar structure of the spicules may be a general feature of the corona.

However, scattering measurements of occultations of radio sources by the solar corona are confined to distances which exceed 5 - 6 $R_{\odot}$. Extrapolation to low altitudes of results obtained at large heights is a dubious undertaking for which there is no real justification.

The complicated and irregular structure of the corona is also well known from optical eclipse observations and observations of prominences. Convincing evidence on the irregular structure has also been gained from observations undertaken outside the atmosphere of the earth.

In the great majority of papers on solar bursts one will

find that effects of scattering of the radio waves on coronal inhomogeneities have been neglected. This is permissible in many instances, but scattering cannot always be ignored. During the time after which computers became readily available, scattering effects have more frequently been taken into account.

Scattering calculations were started in an attempt to explain the size of the sources of type I bursts, and possibly their duration.

There are several difficulties connected with computations of scattering effects on solar bursts: the size and shape of the inhomogeneities are unknown; the variation of electron density is another parameter of which the knowledge is rather limited; and when it comes to the magnetic field, this is usually neglected. A spherically symmetrical corona is frequently assumed, although this may be a too rough approximation. But it is in order to mention that the models have shown an evolution with time so as to approach more and more realistic cases. In addition, quantitative data on coronal inhomogeneities are now emerging from X-ray and EUV observations taken from outside the atmosphere of the earth (Skylab). Thus the situation improves.

5.2. COMPUTATION OF CORONAL SCATTERING OF RADIO WAVES IN THE SMALL ANGLE APPROXIMATION

The first estimates of the role of scattering in connection with noise storms were made by Högbom (1959) and Fokker (1960). Further work on the scattering problem was next published by Fokker (1965b).

The nature of the scattering depends essentially on the scale of the inhomogeneities. Scattering on smooth large-scale inhomogeneities with a characteristic size which is considerably larger than the wavelength ($\ell >> \lambda$) and show a small variation in refractive index $[(\Delta n)^2 << 1]$ can be described in the approximation of geometrical optics except when $n \approx 0$. This means that during the scattering the wave type is not changed. An important effect consists in a broadening of the angular emission diagram caused by irregular changes in directions of the beams in the corona. If the scale of the inhomogeneities is of the same order of magnitude as the wavelength of the waves, transformation into other wave modes may take place. This problem will not be considered in the present chapter.

Electromagnetic waves propagating in the solar corona are subject to regular refraction due to the variation of electron density with height and irregular refraction by large-scale inhomogeneities.

At this point it should be stressed that the models which have been adopted in the scattering calculations which follow are arbitrary. The scale of the inhomogeneities, the relative variation of electron density, etc., were chosen so as to represent reasonable assumptions at the time of construction of the models. Moreover, some idealization is necessary in order to be able to carry out the computations with reasonable efforts.

Fokker (1965b) considered slabs of thickness d in which there were irregularities of a characteristic scale ℓ. A continuous field of fluctuations of the refractive index was assumed. The deviation of the dielectric constant in a specific irregularity is $\Delta\varepsilon$. Then, according to Chandrasekhar (1952), the r.m.s. deviation suffered by a ray travelling through the slab is given by

$$\psi \propto \left(\frac{d}{\ell}\right)^{\frac{1}{2}} \left[\overline{\Delta\varepsilon^{\;2}}\right]^{\frac{1}{2}}. \tag{5.2-1}$$

The treatment was restricted to one component of the angular deviations since most determinations of angular extent are one-dimensional. Assume a Gaussian shape for the distribution of the angular deviations of individual rays that pass through the slab. When a ray has passed through n slabs, the mean square of the angular deflection ψ becomes

$$\overline{\Psi^2} = \sum_{i=1}^{n} \psi_i^{\;2}, \tag{5.2-2}$$

where ψ_i is the r.m.s. deflection in slab number i. Regular refraction in the corona is now neglected.

Provided the direction of the refracted ray makes a small angle with the original ray direction, the extent of the scatter image may be determined relatively easily. Fokker divided the scattering medium into slabs of thickness d_i at distances s_i from the source level. In the slabs the ray suffers angular deflections ψ_i. This is illustrated in Fig. 5.1, where the rays are followed in the direction from the earth to the source.

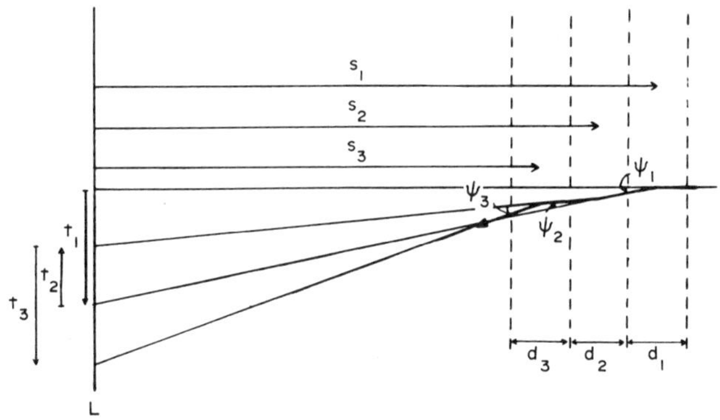

Fig. 5.1. Successive deflections of a ray in elementary slabs (Fokker, 1965b).

The total lateral deviation t' then becomes

$$t' = t_1 + t_2 + t_3 + \ldots . \qquad (5.2\text{-}3)$$

Because $\sum t_i t_j = 0$ and $t_i = \psi_i s_i$, one may write

$$\overline{(t')^2} = \int_0^\infty \psi^2(s) s^2 ds , \qquad (5.2\text{-}4)$$

where s is taken along the ray from the source and outwards. Assume that

$$\frac{\left[\overline{(\Delta N_e)^2} \right]^{\frac{1}{2}}}{N_e} = \text{constant}, \qquad (5.2\text{-}5)$$

where ΔN_e is the deviation of the electron density from the local average value N_e. One then has

$$\overline{(\Delta\varepsilon)^2} \propto \frac{N_e^2}{f^4} \tag{5.2-6}$$

If in addition the scale ℓ of the irregularities is proportional to the distance r from the centre of the sun, (5.2-4), (5.2-1) and (5.2-6) lead to

$$\overline{(t')^2} = c_0 \int_0^\infty \left(\frac{N_e^2}{r}\right) s^2 ds, \tag{5.2-7}$$

where c_0 = constant and $r = r_0 + s$ (r_0 distance of source level from the solar centre).

For the mean square of the total angular deflection of different rays from the source Fokker then got

$$\overline{\psi^2} = c_0 \int_0^\infty \frac{N_e^2}{r} ds. \tag{5.2-8}$$

In order to evaluate (5.2-7) and (5.2-8) it is necessary to have a model of the variation of electron density as a function of the distance from the solar centre. Fokker used the model shown in Fig. 5.2, which is constructed using data from several sources.

Assuming a height of 0.3 $R_\odot$ above the photosphere for the 200 MHz noise storm source and a ray propagating radially outwards, numerical calculation of $(N_e^2/r)s^2$ gave the curve shown in Fig. 5.3.

Fokker evaluated (5.2-7) numerically and used a factor c_0 which made $\overline{(t')^2} = 1$. Similar integrals were evaluated for other frequencies starting from appropriate heights. The relative angular extent of a source as a function of frequency as compared with the 200 MHz source is shown in Fig. 5.4.

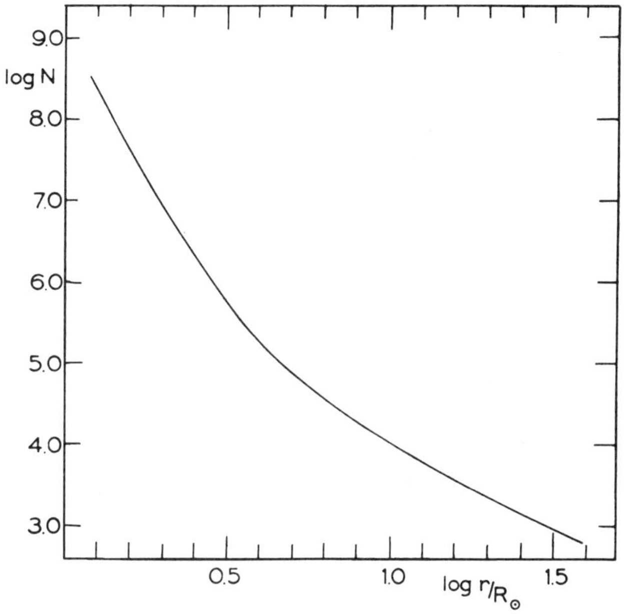

Fig. 5.2. Adopted model for the electron density N_e. Up to $\rho = r/R_{\odot} = 2$ the density model of Newkirk (1961) for the corona above an active region is used. For $\rho > 2$ the curve is extrapolated so as to run parallel to the model of the equatorial corona as given by van de Hulst (1950). The curve is connected with that given by Blackwell (1956) in the region $6 < \rho < 16$. Further out N_e falls off as the inverse square of the distance to the solar centre. (Fokker, 1965b.)

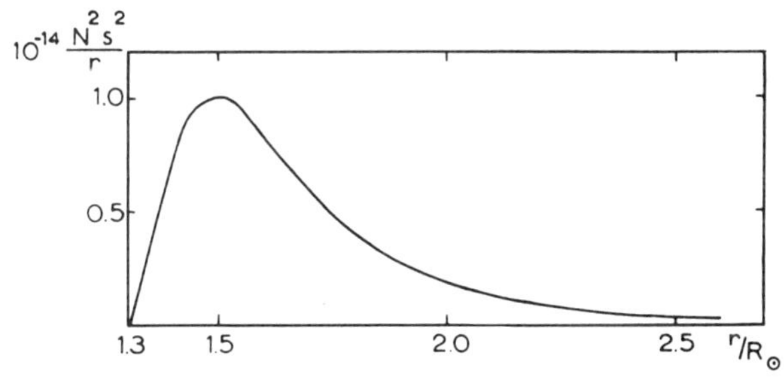

Fig. 5.3. Relative contribution to the extent of the scatter image in various parts of a radial ray starting at a height of 0.3 $R_{\odot}$ (Fokker, 1965b.)

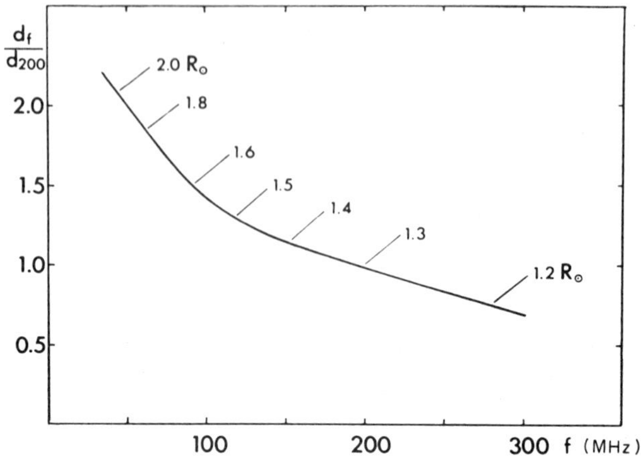

Fig. 5.4. Relative angular extent of the scatter image as a function of frequency (Fokker, 1965b).

The mean square total angular deflection may be derived. Observations indicate an angular extension of about 3' arc for type I bursts at 200 MHz (Fokker, 1960). The constant c_0 in (5.2-7) then becomes $c_0 = 1.0 \times 10^{-16}$ when N_e is expressed in cm^{-3}. Then (5.2-8) gives $\overline{\Psi^2} = 0.5$. This introduces difficulties because (5.2-4) and (5.2-7) are valid provided $\Psi << 1$. Because of this it may be necessary to follow a method of ray tracing in order to take scattering effects correctly into account. It is here assumed that the extent of the burst source is entirely due to scattering. There are arguments both in favour of and against this assumption.

5.3. SCATTERING AS DETERMINED BY RAY TRACING

When the deflections of the rays are large, it is very difficult to treat the scattering problem analytically. Therefore Fokker (1965b) continued the calculations using a statistical numerical treatment. This work was extended by Steinberg *et al.* (1971).

A ray theory can be used provided:

(a) The wave number k varies by a small amount over a wavelength,

$$\frac{1}{n^2}\frac{\partial n}{\partial s}\lambda \ll 2\pi \qquad \text{(W.K.B. approximation)}.$$

(b) The r.m.s. phase fluctuations $\delta\phi$ along the ray is much greater than one radian: $\delta\phi = \frac{2\pi}{\lambda}\,\delta n \Delta s >> 1$, where δn is the variation of the refractive index over a distance Δs along the ray path.

(c) The mean refractive index is constant.

The scattering can be introduced as a random small perturbation to the direction of the final wave propagation vector.

Steinberg *et al.* (1971) adopted the model of coronal electron density derived by Hansen *et al.* (1969) and which is close to that of Newkirk:

$$N_e = 4 \times 10^{4+4.32/\rho}$$

where N_e is in cm^{-3} and ρ is the distance from the solar centre as measured in units of $R_\odot$.

The electron temperature was assumed constant throughout the corona and was set equal to 1.9×10^6 K. Magnetic fields were neglected and the coronal inhomogeneities in refractive index were considered to be isotropic. The fluctuations were described by

$$\overline{\delta n_1(r_1)\,\delta n_2(r_2)} = \overline{(\delta n)^2}\,\exp\left[-\frac{(r_1-r_2)^2}{h^2}\right], \qquad (5.3\text{-}1)$$

where h is the radius of correlation and $\overline{(\delta n)^2}$ is the mean square fluctuation of n.

Furthermore, the scale of the inhomogeneities was supposed to increase with the distance from the sun as $h = 5 \times 10^{-5}\rho$. In the following the relative fluctuation in the electron density is denoted by $\eta = \left[\overline{(\Delta N_e/N_e)^2}\right]^{\frac{1}{2}}$

The scattering effects are determined by tracing a large number of rays which travel through different coronas. In the model coronas the average electron distribution and the statistics of the inhomogeneities are the same. But the inhomogeneities are microscopically different. An ensemble average of all these rays is made and the image parameters may be found.

It is convenient to divide the description of the computations in two parts: (a) direct rays, (b) reflected rays.

Direct Rays

The ray trajectory is divided into rectilinear steps of length Δs_i which are variable with the height. Then the refractive deviation θ_i due to the gradient in the average electron density is calculated. After this, two random angles $\Delta\psi$ and $\Delta\Omega$ in two perpendicular planes are added to θ_i. The angles $\Delta\psi$ and $\Delta\Omega$ are normally distributed around zero and they have the same standard deviation, but they are statistically independent.

The scattering angles are given by a theory due to Hollweg (1967) for the case in which $n \neq 1$. This is an extension of Chandrasekhar's (1952) theory:

$$\overline{(\delta\psi)^2} = \overline{(\delta\Omega)^2} = 2\pi^{\frac{1}{2}} \int_{\text{ray}} \frac{\overline{(\delta n)^2}}{n^2 h} ds. \qquad (5.3\text{-}2)$$

Since (5.3-2) is valid for small angles and a constant mean value of n, the ray path must be divided in linear steps Δs so that over each step $\Delta\psi < 0.1$ radian and $\Delta n/n < 0.1$. Then over each step one may write (5.3-2)

$$\overline{(\Delta\psi)^2} = \overline{(\Delta\Omega)^2} = 2\pi^{\frac{1}{2}} \frac{\overline{(\delta n)^2}}{n^2 h} \Delta s \qquad (5.3\text{-}3)$$

Using $n = [1 - (f_p/f)^2]^{\frac{1}{2}}$ results in

$$(\delta n)^2 = \frac{\eta^2}{4n^2} \left(\frac{e^2}{\pi m}\right)^2 \frac{N_e{}^2}{f^4} = \frac{\eta^2}{4n^2} \frac{f_p{}^4}{f^4}. \qquad (5.3\text{-}4)$$

Expressing f in MHz, h and Δs in $R_\odot$ and N_e in cm^{-3} gives

$$\overline{(\Delta\psi)^2} = \overline{(\Delta\Omega)^2} = 5.8 \times 10^{-9} \frac{1}{n^4} \frac{\eta^2}{hf^4} N_e{}^2 \Delta s. \qquad (5.3\text{-}5)$$

It is seen that the scattering properties of the medium depend on the factor η^2/h. Results are then valid for different values of η and h provided the ratio η^2/h does not change.

Reflected Rays

Each step Δs_i is extrapolated by an amount equal to half of its length. If the refractive index at this point is lower than a critical value ($n < 0.13$), specular reflection was assumed to occur at the end of the step Δs_i. At this point the scattering angles $\Delta\psi$ and $\Delta\Omega$ are added to the angle of reflection.

Steinberg *et al.* now assumed a point source radiating isotropically. Rays start from the source in all directions δ_i. To each direction δ_i there will be p different output directions Δ_{ij} ($j = 1, 2, \ldots.., p$) (Fig. 5.5).

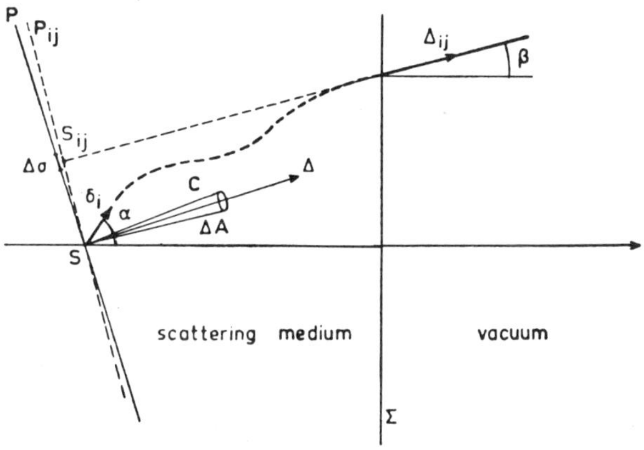

Fig. 5.5. Image formation from the computed rays Δ_{ij}. The cone C has a small solid angle ΔA about the observing direction Δ. (Steinberg *et al.*, 1971.)

To an observer looking in the direction from the earth towards the source the ray Δ_{ij} appears to come from the image point S_{ij} in the plane P_{ij} through the source S and perpendicular to Δ_{ij}. The apparent source brightness distribution when observed from a given direction Δ may be calculated by considering a cone of directions C of small solid angle ΔA centred on Δ. The brightness of the surface element $\Delta\sigma$ of the apparent source is proportional to $n_r/\Delta\sigma\Delta A$, where n_r is the number of rays within the cone C which have their image point S_{ij} within $\Delta\sigma$.

Losses were introduced in the computations by weighting each Δ_{ij} by the attenuation coefficient $\exp(-\tau_{ij})$, where τ_{ij} is the optical depth along the path i, j.

The propagation time of each ray was computed. In order to get a result which was independent of the boundary surface Σ, Steinberg *et al.* used the same method as was done by Wild *et al.* (1959). The travel time at the velocity of light from the image point S_{ij} to Σ was subtracted from the computed travel time from S to Σ.

Results of interest in connection with type I bursts emerging from ray-tracing technique in scattering calculations regard the size of the scattered image and the brightness distribution across it, the shift in position of the source, the widening in time of the pulse and the directivity of the radiation.

The computations of Steinberg *et al.* (1971) were carried out for a frequency of 169 MHz. Two source heights were chosen, $\rho = 1.12$ where $n = 0.38$ and $f_p = 0.92\, f$ and $\rho = 1.2$ where $n = 0.73$ and $f_p = 0.69\, f$. Since type I bursts seem to be generated near the plasma level, the results obtained at $\rho = 1.12$ may have the strongest bearing on the real situation. But limb sources are found to occur at larger heights than sources near the centre of the solar disk. Therefore the results obtained for $\rho = 1.2$ should also be taken into account.

Sizes of the scattered image between half-power points for different values of η and three selected longitudes are given in Table 5.1.

As is seen from the table, the source size increases with increasing values of η and for larger values of the longitude.

Extensive scattering calculations using ray-tracing were performed by Riddle (1972, 1974) for a frequency of 80 MHz. Riddle (1972) determined the effect of scattering on the source parameters for 80 MHz radiation emitted from the 78 MHz level ($f_p = 0.975\, f$). From occultation observations of cosmic radio sources it is known that the coronal inhomogeneities in electron density are elongated in the radial direction. This is important for the scattering of the waves from the occulted sources. Therefore Riddle allowed the inhomogeneities to be larger in the radial than in the transverse direction. But the results of the

TABLE 5.1

Size of scattered image at 169 MHz

f_p/f	η	η^2/h	Half-power width in minutes of arc		
			$L = 0^{\circ}$	$L = 45^{\circ}$	$L = 75^{\circ}$
0.92	1%	2	1.7	2	-
0.92	2%	8	3	3.5	4
0.92	3%	18	3.3	3.5	-
0.69	1%	2	1.3	2	2.5
0.69	2%	8	2.3	3	3.5
0.69	4%	32	3	4.5	5

calculations showed that the elongated shape of the inhomogeneities had only a minor influence on the scattering of radio waves from solar bursts. When the difference in geometry between the two situations is taken into account, this finding is not surprising. Riddle (1974) extended the studies to include the case when a source radiates from a position on the axis of a coronal streamer superimposed on a spherically symmetric coronal density distribution.

At frequencies of 60 MHz and 30 MHz scattering effects were determined by Leblanc (1973). Results of scattering calculations at a number of frequencies were reviewed by Leblanc (1972).

Some data which concern the size of the scattered image at different frequencies are given in Table 5.2.

TABLE 5.2

Size of scattered image at different frequencies

f (MHz)	f_p/f	η^2/h	Half-power width in minutes of arc		
			$L = 0^{\circ}$	$L = 45^{\circ}$	$L = 75^{\circ}$
169	0.92	8	3	3.5	4
80	0.975	5	1.8	2	3
60	0.92	8	4	4	5
30	0.92	8	4	4	5

The source sizes given for 80 MHz in Table 5.2 refer to the case of a spherically symmetric corona. When the source position was supposed to be on the axis of a streamer, somewhat smaller values were found for the half-power width.

At all frequencies listed in Table 5.2 the source size increases from the centre to the limb. Moreover, there is a tendency for the source size to be larger at lower frequencies. This is due to the fact that the scattering angle is proportional to $n^{-2}f^{-2}$ and that n decreases more slowly with height at low frequencies. Therefore the effective scattering takes place over a more extended layer for low frequencies than for the higher ones.

Another effect of scattering is that the brightness distribution over the scattered image becomes asymmetric for larger longitudes. A tail develops towards the solar centre. At 169 MHz this occurs already for $L > 20^{\circ}$ according to Steinberg *et al.* (1971). Riddle found the same effect for $L > 60^{\circ}$ at 80 MHz, and also at 60 and 30 MHz asymmetry occurs for large longitudes.

If the evolution of the burst in time is followed, the scattering calculations predict that the source should broaden and move towards the centre of the sun.

Scattering also introduces a shift in position of the maximum intensity of the image for increasing longitudes. The apparent position of the source is always further from the centre of the sun than the true position. The effect increases slightly when the anisotropy of the inhomogeneities is enlarged. In the isotropic case the shift amounts to some 3' - 4' arc at large longitudes for a frequency of 80 MHz (Riddle, 1974). The shift in position is larger at lower frequencies. However, position shifts of the scattered image are almost impossible to determine from observations of noise storm sources so the effect may be difficult to verify experimentally. It is of larger interest for the interpretation of the positions of the sources at the fundamental and second harmonic frequencies of type III bursts.

Radio rays in different parts of the scattered distribution have not travelled along paths of the same electrical length. If the source is assumed to radiate a pulse of extremely short duration, the intensity recorded as a function of time at the earth may be termed the "transient response" of the corona, as was done by Steinberg *et al.* (1971). In a large number of papers this effect has been neglected in the discussion of burst durations. This may not always be permitted.

As is seen from Fig. 5.6, the transient response of the corona has a shape in the intensity - time plane which is very similar to that of solar bursts of type I or type III.

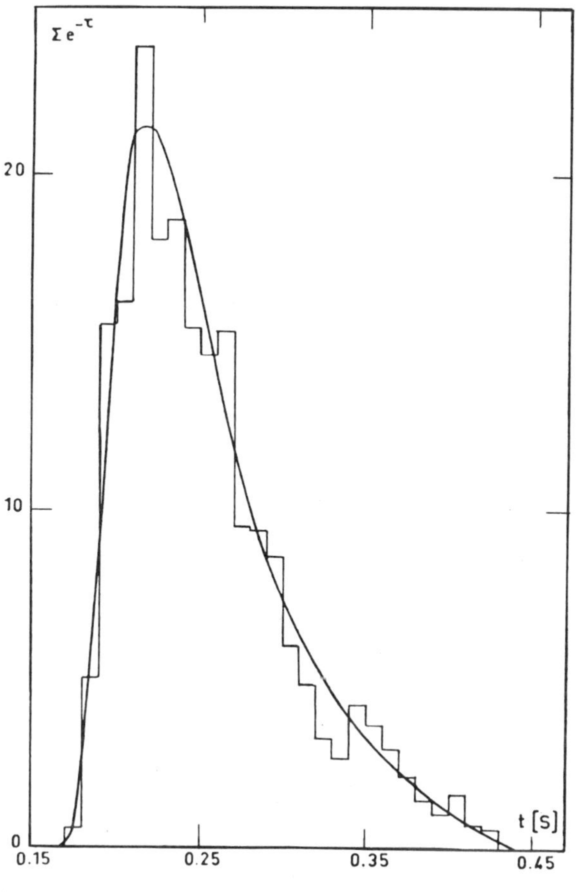

Fig. 5.6. Transient response of the corona for a source height $\rho = 1.2$, $\eta = 2\%$, $f = 169$ MHz and $L = 0^{\circ}$. An isotropic emitter is assumed. On the horizontal axis the time in seconds elapsed since the pulse. (Steinberg *et al.*, 1971.)

The width of the curve describing the response of the corona increases with decreasing height of the sources because scattering then becomes more important. The curve describing the transient response also gets a larger half-width when η is increased. In addition there is also an increase in the half-width when the source approaches the limb of the solar disk.

For the source at $\rho = 1.12$ the width amounts to 0.2 s at $L = 0^{\circ}$ and does not increase for larger longitudes when $\eta \geq 2\%$. For the source at $\rho = 1.2$ the saturation of the width occurs for $\eta = 4\%$ and the width is close to 0.2 s.

A summary of results for different frequencies is given in Table 5.3 after Leblanc (1972).

TABLE 5.3

Transient response of the corona at different frequencies.

f (MHz)	f_p/f	η^2/h	Half-power width in seconds of time		
			$L = 0^{\circ}$	$L = 45^{\circ}$	$L = 75^{\circ}$
169	0.92	8	0.2	0.2	0.2
80	0.975	5	0.4	0.4	0.6
60	0.92	8	0.2	0.4	-
30	0.92	8	0.25	0.45	0.6

It is important at this point to stress that the values in Table 5.3 were obtained for an isotropically radiating source. In the case of a strongly beamed source the widening of the pulse would be smaller, but in any case the influence of scattering should not be neglected in discussions of burst durations.

The directivity of a source as a function of heliographic longitude L was determined by Riddle (1974) by taking the ratio of the power received from the source when scattering, refraction and absorption were taken into account, to the power received from a source at the same position emitting isotropically the same power in the absence of any coronal effects. It was found by summing the rays in a group, each weighted by exp(-τ), and dividing by the number of rays that would fall in the same range of *L* from an isotropic source when the rays are unaffected by scattering or refraction. Directivity curves derived in this way are shown in Fig. 5.7. It is seen that the source is visible across the whole disk although limb sources will be relatively weak when the emission occurs near the plasma frequency.

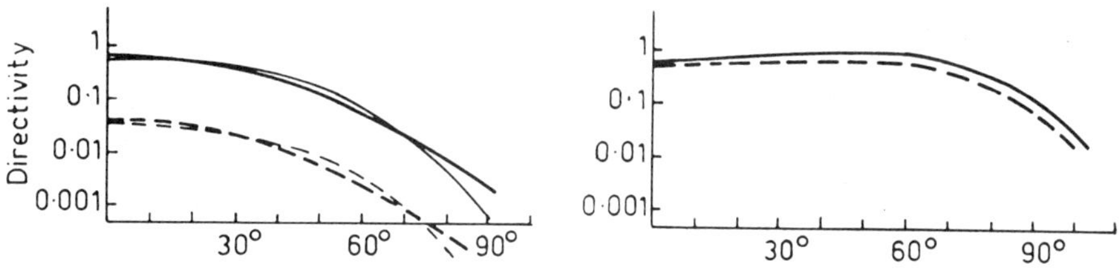

Fig. 5.7. Directivity at 80 and 156 MHz in a spherically symmetric corona with random inhomogeneities in electron density. Thick lines for isotropic inhomogeneities, light curves for anisotropic (elongated) inhomogeneities. $T = 2 \times 10^6$ K (full curves), $T = 10^6$ (dashed curves). At the source position $f_p = 0.975\ f$ (left) and $f_p = 0.5\ f$ (right). (Riddle, 1974.)

The transfer characteristics of the scattering corona were studied in more detail by Caroubalos *et al.* (1972). The same coronal model was used as in the investigations of Steinberg *et al.* (1971). The source was located at $\rho = 1.12$, corresponding to a plasma frequency $f_p = 115$ MHz. For the scattering parameter η^2/h a value of 8 was used, the temperature was 1.9×10^6 K in the whole corona and magnetic fields were neglected. The source was assumed to radiate a white spectrum. Computations were made for different primary polar diagrams, which were radially oriented and equal for all frequencies.

Consider first an isotropic source which is observed from different directions θ (θ is the zenith angle of the emerging radiation). A graphic representation of the results is given in Fig. 5.8. It appears from the figure that the cut-off frequency becomes independent of θ and equals the plasma frequency at the source. The spectrum of the emerging radiation depends to a considerable degree on θ, and the emission is beamed for a wide range of frequencies.

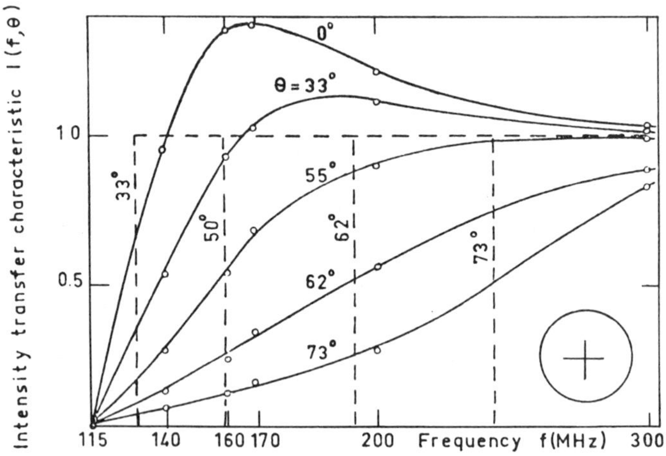

Fig. 5.8. Intensity transfer characteristics I (f, θ) in a scattering corona. Dotted lines indicate the cut-off frequencies for a homogeneous (η=0) corona. The source emits isotropically. (Caroubalos *et al.*, 1972.)

In addition to the isotropic emitter three models of primary emission patterns were studied. These had half-power beamwidths of 110°, 62° and 18°. They were chosen so as to be larger than, about equal to and smaller than the focussing cone of a source at ρ = 1.2 radiating at 169 MHz. This cone, which is determined by regular refraction in the corona, has a width of 57°. The directivity diagrams for selected frequencies and for the different primary patterns of the source are shown in Fig. 5.9.

The diagrams in Fig. 5.9 illustrate the competition between refraction and scattering. Consider first the source with the broad primary emission pattern in Fig. 5.9(a). It is seen that the focussing effect dominates over scattering at high frequencies, and the result is a smaller half-power width than that of the primary diagram. At frequencies lower than about 170 MHz the situation is reversed, scattering predominates and the received diagram moves towards a limiting pattern (represented by $J(\theta)$).

Turning next to the narrow primary pattern shown in Fig. 5.9(c), scattering is the most important factor at all frequencies and broadens the diagram of the received radiation. When $f \to f_p$ the diagram tends to the same limiting shape as in the case of (a).

Thirdly, Fig. 5.9(b) demonstrates the competition between focussing and scattering. When f decreases from ∞ to about 170 MHz, the diagram narrows towards a limiting shape; then

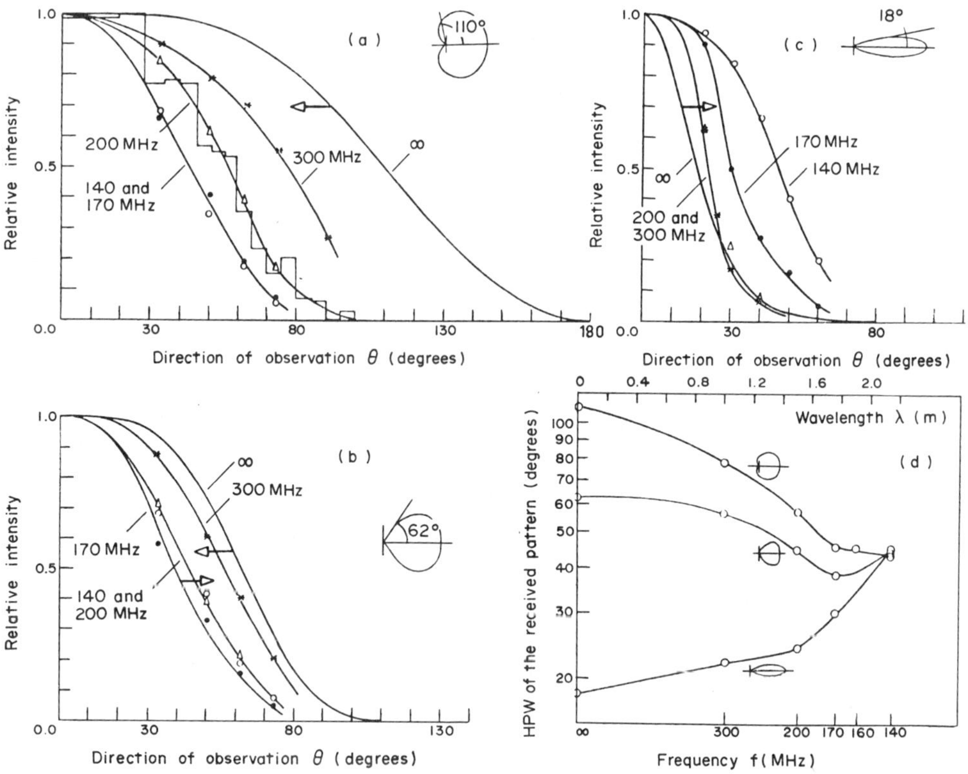

Fig. 5.9. (a), (b), (c), directivity patterns of the emerging energy from the scattering corona as a function of frequency for three different primary patterns. In (a) the histogram of the number of rays as a function of θ is given in order to indicate the errors involved in the statistics (f=200 MHz). In (d) it is shown how the half-power beamwidth approaches, in all cases, to about the same limiting cone imposed by scattering when $f \to f_p$. (Caroubalos *et al.*, 1972.)

scattering becomes the more important factor for lower values of f and broadens the diagram until the received pattern reaches the limiting form $J(\theta)$.

Figure 5.9(d) represents a summary of inferences drawn from the other figures. The important issue in this figure is that whatever the primary pattern of the source, the half-power beamwidth of the received radiation moves towards the same limiting cone — the cone imposed by scattering when $f \rightarrow f_p$.

One outcome of the investigations of Caroubalos *et al.*, (1972) is, amongst others, that it will be difficult to determine the primary pattern of the radiating source from the received directivity diagram.

5.4. OBSERVED FEATURES OF STORM EMISSION AS COMPARED WITH PREDICTIONS FROM SCATTERING THEORY

It is not quite easy to compare the results of scattering calculations with observational data. In several respects the theoretical models are idealized, and on the observational side the samples are frequently small and there are considerable variations between individual cases. The field is open to new theoretical developments and systematic observations with instruments of increased resolving power in space and time.

An attempt has been made in Table 5.4 to assemble available data on the angular sizes of the sources emitting the storm continuum and the bursts. The diameters quoted are measured between half-power points.

As regards the storm continuum there is an increase in the source size with decreasing frequency. In the 400 MHz range, which is around the upper limit where noise storms occur, the source size is less than 2' arc. In the cases when larger source sizes were found, the source seemed to be composed of several components each of which having sizes between 0.5' and 1' arc (Abrami, 1974).

In the intermediate frequency range from about 200 to 100 MHz different observations give values of the size of the continuum source which are around 3' arc. Le Squeren (1963) determined source sizes ranging from 9' arc to values which were smaller than 2' arc and could not be resolved. The average value amounted to 3.8' arc. Observations with the Culgoora radioheliograph have demonstrated that noise storm regions may be composed of several individual sources and that the two-dimensional pictures may be quite complex. A characteristic size of continuum sources at 160 MHz appears to be 3' arc.

TABLE 5.4

Diameters of type I continuum and burst sources

f (MHz)	Continuum source (min. of arc)	Burst source (min. of arc)	Reference
408	1-2		Clavelier (1967)
408	0.8-4		Abrami (1974)
200		3	Fokker (1965a)
200		1.5-3.5	Diemel (1969)
200	~3		Gnezdilov (1970)
169	3.8		Le Squeren (1963)
169		2.6	Bougeret (1973a)
169		3.5-5	Kerdraon (1973)
160	3		Dulk and Nelson (1973)
125		< 1.6	Goldstein (1959)
81		5	Högbom (1959)
80	6		Kai (1970)
65	6		Wild and Sheridan (1958)
45	10		Wild and Sheridan (1958)

At lower frequencies the source size at 80 MHz seems to be 6' arc (Kai, 1970), and it is still larger at the lowest frequencies where type I storms occur; probably of the order of 10' - 15' arc (Wild and Sheridan, 1958; Erickson, 1961).

The brightness distribution across the continuum sources was investgated by Fokker (1960) who found a stronger concentration in the centre than would correspond to a gaussian distribution. Observations during the May 20, 1966 eclipse (Gnezdilov, 1970) appear to be in accord with this result.

The trend outlined here — increasing source size with decreasing frequency — is the same as was found in the scattering calculations. It seems possible to construct models which would fit the observations. But there are some complicating factors. Le Squeren (1963) determined the diameter of the continuum sources as a function of the distance from the central meridian. It was found that more sources with small diameters occurred near the limb than in the centre of the disk. The average sizes in the centre and at the limb were 4.1' and 3.8' arc respectively. This is not in accordance with the behaviour to be expected when the effect of scattering is large; then the source size should increase from the centre towards the limb (Table 5.2). Even if the scattering elements were allowed to be elongated in the radial direction, the centre-limb effect mentioned should be present (Riddle, 1974). Further to this it is possible to determine quite persistent "quasi-point regions" within the continuum source from which bursts of definite space - time patterns are radiated (Bougeret, 1973b), and this finding is not in agreement with the suggestion that the size of the continuum source is determined by scattering.

Scattering introduces a position shift of the source away from the centre of the disk. This position shift has been verified by comparison of the positions of the fundamental and the harmonic components of type III bursts. But this argument in favour of scattering is weakened by the doubts which have been raised about the nature of double type III bursts. A fundamental together with a harmonic component may not exist (Mercier and Rosenberg, 1974; Daigne, 1975). Noise storm continuum sources are not radially aligned with the associated optical centre of activity, and Le Squeren found (Fig. 1.11) that the radio centre was displaced towards the pole and towards the north - south diameter of the sun. Studying Figs. 1.9 and 1.11 one finds some evidence for a bending away from the north - south diameter for limb sources. This might be interpreted as a scattering effect. But when the results concerning the position of noise storm sources which were described in section 3.8 are taken into account, one should not attach too much weight to this argument.

The directivity of the continuum sources is related to the

transfer characteristics of the scattering corona. Observationally, Fokker (1960) and Smerd (1964) found roughly the same directivity at different frequencies. The results presented in Figs. 4.1, 4.2 and 4.4 indicate a half-power beamwidth of the emerging radiation of the order of 40° - 50° (aperture angle 80° - 100°). It is tempting to conclude that the emission occurs near the plasma frequency and that the results are in rough agreement with the findings of Caroubalos *et al.* (1972). When the emission occurred near the plasma level, Caroubalos *et al.* found a half-power beamwidth of approximately 45° regardless of the primary emission diagram. In addition one would expect only small variations in the beamwidth in the frequency range where noise storms most frequently occur.

Theory and observations also agree in the respect that scattering makes the sources visible even at the limb of the solar disk, although with reduced intensity.

The above discussion shows that scattering effects are detectable in the observed properties of noise storm continuum sources, but they nead not account for all aspects. The magnetic field configuration is probably of importance for the size of the sources, as suggested by Kai and Sheridan (1974) and as indicated in Fig. 5.10.

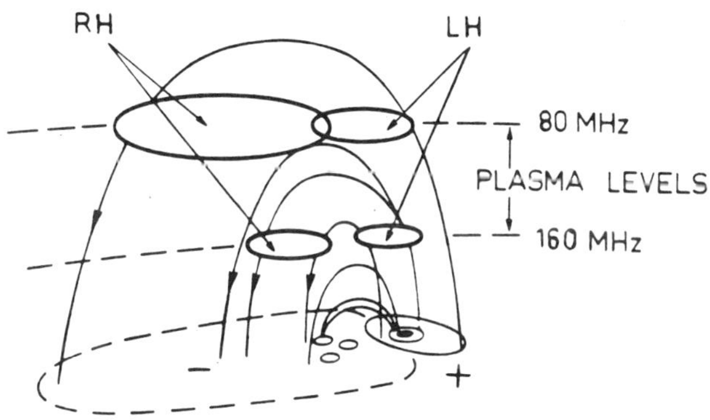

Fig. 5.10. A suggested model of magnetic field associated with bipolar sources observed at 80 and 160 MHz on October 30, 1972 (Kai and Sheridan, 1974.)

The large sizes of the RH polarized sources may be associated with the dispersed magnetic fields high in the corona over the trailing spot region.

Type I bursts have slightly smaller diameters than the continuum sources. In general the continuum source occupies a certain region, and type I bursts are emitted from different parts of this region. The angular size of the bursts is of the order of a few minutes of arc at frequencies between 100 and 200 MHz, and the duration of the bursts is only a few tenths of a second. Therefore Högbom (1959) and Fokker (1960) suggested that one may be observing the scattered image of a point source.

Kerdraon (1973) attempted to test the scattering hypothesis by performing centre-limb observations of the sources of type I bursts at a frequency of 169 MHz. Since a noise storm region may include several sources of type I bursts (Bougeret, 1973b) and each source shows characteristic shapes and diameters of the bursts, Kerdraon exercised great care to track the same source from the centre to the limb. Two periods were chosen: from May 6 to May 12, 1971, and from July 17 to July 22, 1971. It proved difficult to find noise storms lasting for a sufficiently long time to be followed from the centre to the limb.

Diameters of about 100 bursts were measured, but no systematic centre-limb effect was detected. The results are plotted in Fig. 5.11, which also shows the variations in diameters to be expected according to the scattering calculations of Steinberg *et al.* for a source height of $\rho = 1.2$.

From the figure Kerdraon concluded that the centre-limb variation of the diameter of type I bursts is too small to be explained by scattering on isotropic inhomogeneities when: $\eta = 1\%$, radius of correlation of the inhomogeneities, $h = 10^{-5}\,\rho$, the electron density distribution follows the model of Hansen (1969) and the altitude of the source is 0.2 $R_{\odot}$ above the photosphere.

The analysis showed that the angular size of bursts observed near the centre of the disk was not particularly small. Moreover, the size of bursts which are emitted during the same period of time from sources which are situated close to each other, may be quite different. This demonstrates that local effects exist.

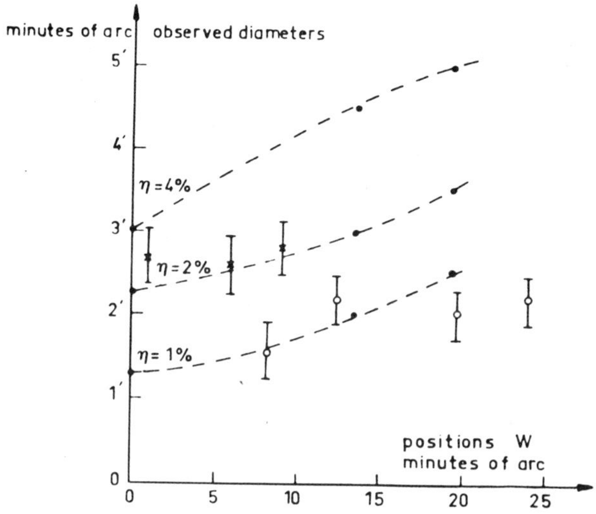

Fig. 5.11. Centre-limb effect on burst diameters. Dashed lines: broadening due to isotropic scattering (data from Table 5.1). Values measured in May 1971 are marked by o. The July 1971 data are given by x. (Kerdraon, 1973.)

The observations indicated an intrinsic size of the burst sources of the order of 20.000 km (or ≈ 0.5' arc).

The scattering calculations of Riddle (1974), performed for a frequency of 80 MHz, showed that the introduction of radially aligned anisotropic irregularities had little effect on the centre-limb variation of the source size. But it is conceivable that two effects may reduce the broadening near the limb: the irregularities may be anisotropic, but they are aligned in such a way that the propagation occurs almost along the larger axis of the irregularities. Next, as demonstrated by position measurements and suggested by polarization observations, the limb sources may be located in dense structures above the plasma level. Therefore the scattering effects may be reduced.

Kardraon (1973) concluded that the centre-limb observations of the diameter of type I bursts could partly be accounted for by a scattering process on radially elongated inhomogeneities. But the primary intrinsic source diameter was not negligible as compared to that of the scattered image. However, it seems small enough to be produced by local effects such as propagation in a strongly inhomogeneous medium. Zaitsev and Fomichev (1973) explained the duration of type I bursts in a similar way. The size of the inhomogeneities was 4000 km and the observing frequency 150 MHz. The emitting region would have about the same size, which is small enough to be considered as a point source. Local propagation effects would afterwards broaden this source. The mechanism is much the same as the one invoked by Bougeret (1973b) in an attempt to account for the space - time properties of type I bursts at 169 MHz.

Measurements of the directivity of type I bursts carried out during the STEREO experiment has consequences for the discussion of scattering in the solar corona. Steinberg *et al.* (1974) found that the beamwidth of type I bursts was smaller than 25°. If this is entirely due to scattering, the total angular deviation ψ would be 12.5°. Since some of the angular spread may be due to the source proper, it seems more likely that $\psi < 12.5°$, and the small angle approximation (section 5.2) may be applicable. One then has from (5.2-4)

$$t^2 = 4 \int_0^\infty \psi^2 s^2 ds, \qquad (5.4\text{-}1)$$

where t is the average diameter of the scattered image and ψ and s are as defined in section 5.2. The total angular deviation squared is given by

$$\Psi^2 = \int_0^\infty \psi^2 ds. \qquad (5.4\text{-}2)$$

At the frequency 169 MHz, t has been determined and is approximately 2' to 3' arc. It is now a matter of interest to see whether values of $\Psi \leq 12.5°$ and $t \approx 2' - 3'$ arc are compatible in the small angle scattering theory.

Steinberg *et al.* (1974) computed $\psi^2 \propto N_e{}^2\eta^2/h\, n^4$ (see 5.3-5) along a radial ray for various source heights using $N_e = 2.1 \times 10^{4+4.32/\rho}$, $h = 5 \times 10^{-5}\rho$, $\eta^2 = \overline{(\Delta N_e/N_e)^2}$ = constant and neglecting magnetic fields. For each altitude Ψ^2 was normalized to 12.5° or 5°. In the latter case the source proper is responsible for part of the angular width. Then the apparent size was calculated. The results are represented in Fig. 5.12 by the curves A_1 and A_2.

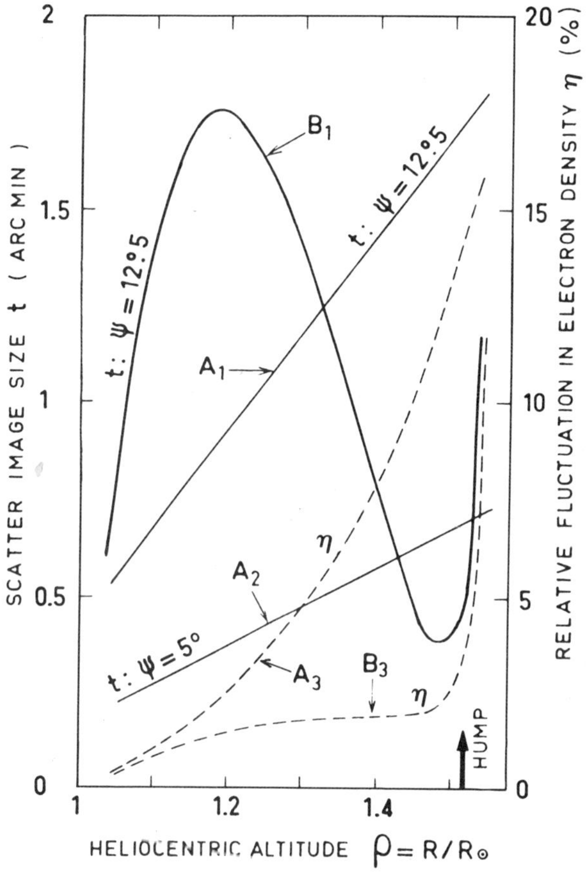

Fig. 5-12. Size of scattered image t versus height as given by ρ computed for different models. The r.m.s. angular deflection of the ray is normalized to 12.5° for curve A_1 and 5° for A_2. Curve B_1 corresponds to a distribution of scattering power with a large peak at $\rho = 1.5$. The dotted curves show for each source height the value of η which was necessary to produce $\Psi = 12.5^\circ$ in model A (curve A_3) and B (B_3). (Steinberg *et al.*, 1974.)

The results are not compatible with the observations. A source should be situated at $\rho = 1.65$ for the scattered image to be 2' arc when $\Psi = 12.5^\circ$. When $\Psi = 5^\circ$ is assumed, $t < 1'$ arc for all reasonable source heights.

Steinberg *et al.* then constructed a different model (*B*) in which the scattering power was strongly enhanced near $\rho = 1.5$. The region of strong scattering was assumed to have a gaussian shape and a width of 0.05 $R_\odot$ between half-power points. At maximum the scattering angles ψ^2 were multiplied by a factor of 100. The total angular deflection Ψ was normalized to 12.5° and the results appear as curve B_1 in Fig. 5.12. It now seems possible to obtain reasonable values for the source size for the observed beamwidth and at heights comparable with those which are deduced from observations.

For η constant, the multiplication of ψ^2 by 100 is equivalent to multiply N_e by a factor of 10. Perhaps coronal streamers of enhanced electron density may serve as "scattering curtains", but then considerable longitude effects should be expected on the size and shape of type I bursts.

It should be pointed out, as was done by Steinberg *et al.*, that the beamwidth determinations refer to one central source only (November 18, 1971) and that more material needs to be analysed before definite conclusions may be reached. Higher resolution in the observations of the source sizes of type I bursts is also a matter of importance.

The time profile effects of scattering may also be considered. According to Table 5.3 scattering may have som influence on the duration of type I bursts. The effect of the transient response of the corona increases towards the limb. Accordingly one may look for a centre-limb variation in the duration of type I bursts, although this is a difficult undertaking.

Suzuki (1961) found an average duration of 0.73 s for bursts observed at 201 MHz at a distance of 1.5 $R_\odot$ to the east of the central meridian on November 18, 1959. On November 21 the position was about 1.3 $R_\odot$ to the east of the central meridian and the average duration was 0.45 s. The observations were made with a receiver having a time constant of 0.25 s, which was not corrected for in the results. In the first case of November 18, Suzuki found that the associated sunspot group was behind the limb, and the observation may be rather exceptional. On this day the bursts were unpolarized, whereas on November 21 all bursts were highly polarized. Suzuki suggested that the difference in duration might be due to a difference in height of the burst sources at the two days. If the duration of the bursts is related to the collision frequency of electrons in the corona, the observations may possibly be explained, but one should not neglect the probability that scattering has had some influence on the burst duration.

Kai (1962) claimed that the average duration of 200 MHz storm bursts was longer at the limb than near the centre. This tendency, if present, is very weak and could not be confirmed by Peersen (1970) from observations made at 159 and 318 MHz with equipment of very short time constant. The material included more than 1500 bursts at 159 MHz and 1100 bursts recorded at 318 MHz during different storms.

Bougeret (1972) determined the duration of bursts of types (i), (d) and (s), examples of which are shown in Fig. 1.16, and found average durations of 0.38, 0.45 and 0.60 s respectively at a frequency of 169 MHz. Since (i) bursts show a concentration towards the centre whereas (d) and (s) bursts occur near the limb, Fig. 5.13, a centre-limb effect in duration may occur just because bursts of different space - time shapes are observed.

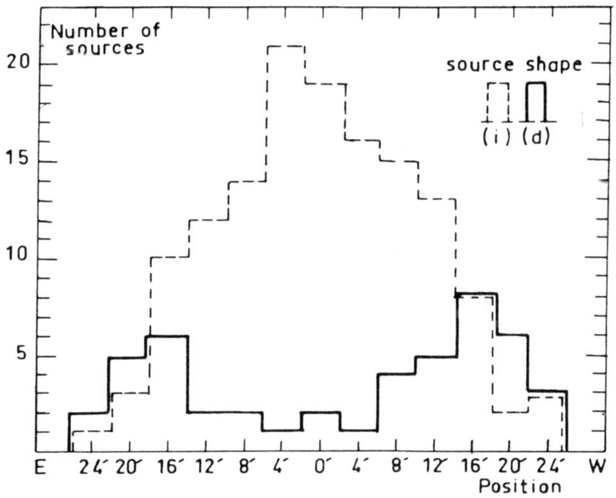

Fig. 5.13. East - west distribution of (i) and (d) bursts (Bougeret, 1973b).

It is possible that (d) bursts are strongly influenced by scattering, but further investigations are necessary to substantiate this suggestion.

One may conclude that as yet scattering effects have not been definitely detected in the duration of type I bursts. One should follow the same source of, for example, (i) bursts for several days in order to obtain some more convincing evidence.

The high directivity of type I bursts determined by Steinberg *et al.* (1974) for the storm of November 18, 1971, provides reasons to assume that centre-limb variations in the burst duration may be undetectable. The effect will be too small.

The theoretical and observational results which have been obtained concerning the scattering properties of the corona are most valuable. In particular the strong beaming of type I bursts calls for interest.

All evidence being taken into account it appears that there is some local scattering as opposed to the homogeneous models which were chosen to start with. The latter constituted a necessary step in the development. Furthermore the inhomogeneities are not weak and spherical but strong and highly elongated.

There is now a need for combination of different observational data of high precision concerning the same centres of activity. Scattering computations then can become less general and more "tailored" to realistic models of indvidual centra.

6. ORDERED BEHAVIOUR OF STORM BURSTS IN THE TIME-FREQUENCY PLANE

6.1. LARGE-SCALE FEATURES IN BURST ACTIVITY

A first view of high resolution spectral recordings showing storm burst activity gives the impression that the bursts occur at random in the time-frequency plane. However, a closer inspection of the records soon reveals that different types of ordering of the activity exist. Various aspects of these phenomena are examined in this chapter.

The most simple approach is to determine the frequency of occurrence of type I bursts as a function of time within a certain frequency band. In the frequency interval from 195 to 220 MHz Elgarøy (1965) found that more than 100 bursts might occur per minute during storms of high variability. But such large numbers of bursts were not generated without intermission. There was a typical tendency for the frequency of occurrence of the bursts to fluctuate, as may be seen from Fig. 6.1. Occasionally the burst activity showed a regular periodicity, but more common were irregular variations. One may describe the activity as bursts coming in "showers" which last some 5 min and in which the bursts seem to be independent of each other. This is in agreement with conclusions reached by Vitkevich and Gorelova (1961). Some evidence in the same direction was found by Elgarøy and Hauge (1958) from observations at 199 and 200.5 MHz. As appears from Fig. 6.2 the amplitude ratio of the bursts in the two channels undergoes changes. The points are very scattered, but if a smoothed curve is drawn it is seen that the average amplitude quotient shows variations during time intervals of the order of some minutes. This was interpreted by Elgarøy and Hauge as emission of bursts in bands which drifted with time.

The structure described here became more evident in an analysis of burst activity recorded simultaneously in the frequency bands 140-170 MHz and 300-330 MHz with high resolution radio spectrographs at Oslo Solar Observatory. Defining burst

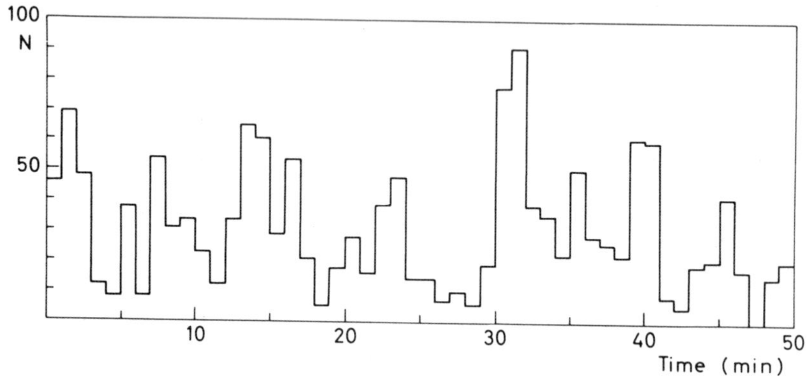

Fig. 6.1. The time variation of the number of bursts recorded per minute in the frequency range 195 - 220 MHz.

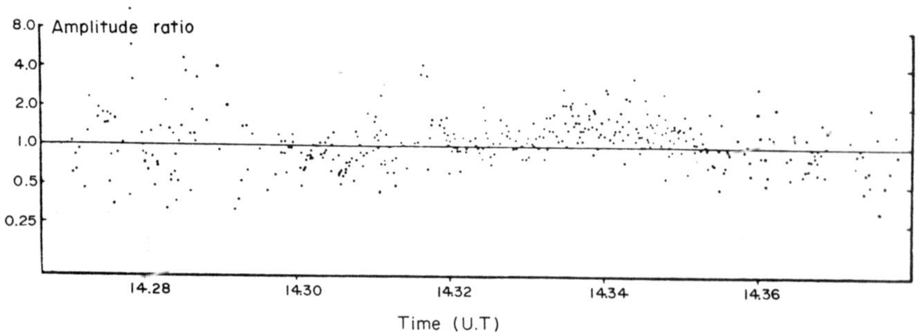

Fig. 6.2. Quotients for the ratio between maximum intensities of corresponding bursts on 199 and 200.5 MHz plotted versus time (Elgarøy and Hauge, 1958).

frequency as the number of bursts occurring on the records per minute, it was found from an autocorrelation analysis that the burst frequency showed a quasi-periodic variation in time. The characteristic period was about 3 - 5 min.

Correlating the burst activity in the two observed frequency bands, a maximum in the cross-correlation coefficient occurred for a time lag of 6 min between the 300 MHz and the 150 MHz bands. This shows that the burst activity sometimes comes in showers drifting at a rate of about 0.4 MHz s^{-1} and with a separation in time of roughly 3 - 5 min. It is noteworthy that the drift rate of the showers is about the same as occurs for burst chains.

The features described above are not found in all noise storms. This is not surprising since frequently emission from more than one centre may be recorded simultaneously. It is tempting to suggest that the large-scale ordering of the burst activity points towards a mechanism in which an exciting agency travels upwards (sometimes downwards) in the corona producing large numbers of type I bursts. This is more clearly demonstrated by the occurrence of chains of bursts which possibly may be regarded as showers in their "cleanest" form.

6.2. CHAINS OF TYPE I SOLAR RADIO BURSTS

Sometimes type I bursts are grouped together in tens or hundreds to form chains of narrow bandwidth and with durations from some seconds to several minutes. One example of a chain was reported by Wild (1957) and the phenomenon was also observed by Elgarøy (1961) and by Vitkevitch and Gorelova (1961). The characteristics of type I burst chains are illustrated in Fig. 6.3. Some examples from recordings in the 150 MHz range are reproduced in Fig. 6.4.

The equipment used by Wild and Tlamicha (1965) which was the same as that used by Hanasz (1966), permitted investigations of chains of duration longer than approximately 10 s. Elgarøy and Ugland (1970) made high resolution observations of chains, and in their case the records were suitable for investigations of short-lasting chains. In order to make the reading of the records simple and reliable, Elgarøy and Ugland adopted the following rules of guidance for the reduction of the material:

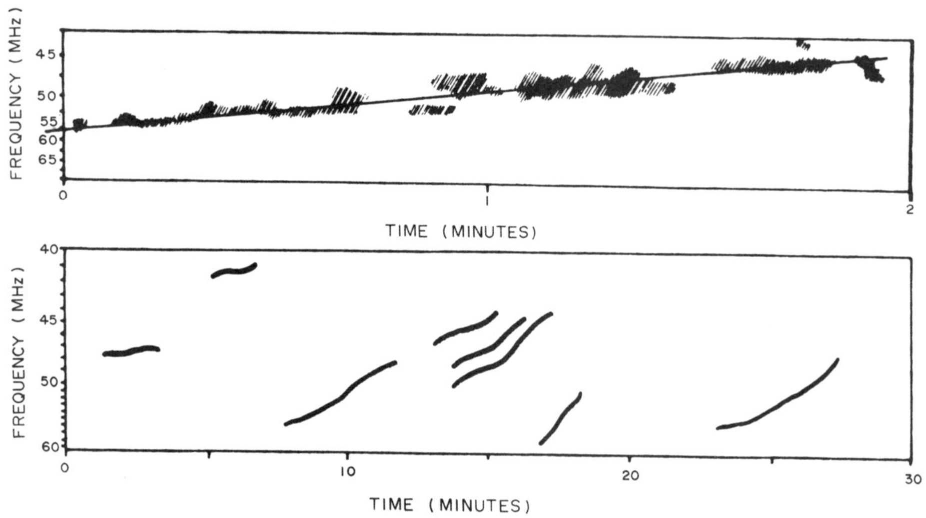

Fig. 6.3. Tracing of dynamic spectra. *(Above)* individual burst chain; the hatched areas indicate concentrations of type I bursts. *(Below)* several chains occurring in succession. (Wild and Tlamicha, 1965.)

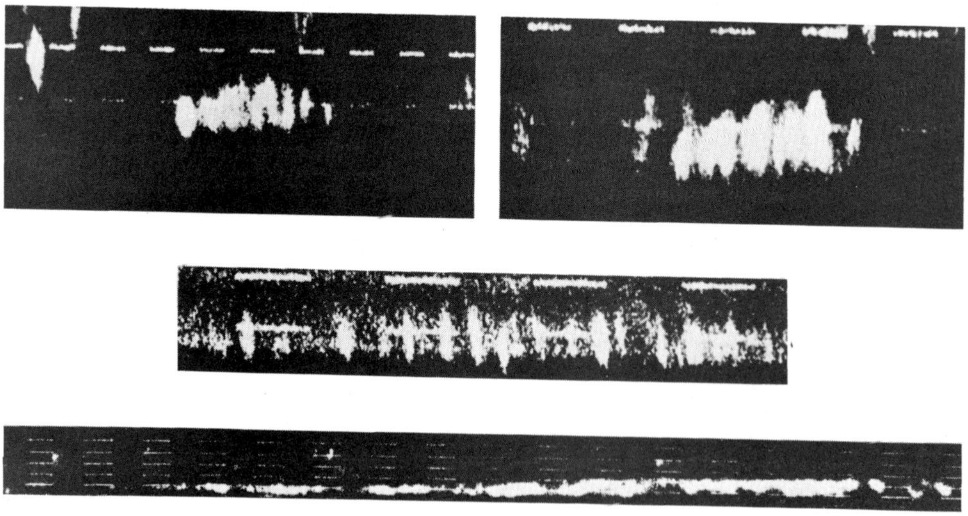

Fig. 6.4. Chains of type I bursts. Time increasing towards the right. Frequency decreases in the upward direction. The markers have a length of 1 s and a frequency spacing of 5 MHz. (Elgarøy and Ugland, 1970.)

(1) a chain must contain at least four bursts or have a duration of 2 seconds or more;

(2) the time separation between successive bursts in a chain must be less than 1 second;

(3) bursts following each other must have roughly the same centre frequency.

The same selection rules were adopted by Chernov *et al.* (1972), whereas Loonen (1971) allowed short interruptions in the sequence of bursts. Interruptions may occur because the conditions for generation of radiation cease to be satisfied. Another possibility is that propagation effects for a time prevent the radiation from reaching the observer.

Distribution of Chains with Frequency

Hanasz (1966) determined the distribution of chains with frequency in a material covering 1000 chains observed between 25 and 210 MHz with the Australian radio spectrograph at Dapto. The measurements were made at the mid frequency of the chains, and the results obtained are plotted in Fig. 6.5. Chains seem to occur in the range from 30 to 200 MHz with a maximum in the distribution around 100 MHz. But the upper frequency limit is probably influenced by decreasing sensitivity of the spectrograph. High resolution spectrograph observations at Oslo Solar Observatory has revealed many chains at frequencies between 200 and 225 MHz, but above 300 MHz normal type I chains are very infrequent. This is in accord with the results of Loonen (1971) who noted a marked drop-off at a frequency between 200 and 260 MHz. Thus it seems that chains almost cease to occur somewhere between 225 and 300 MHz. This may be a significant observation because it shows that chains are limited to a smaller part of the radio spectrum than type I bursts. The circumstances favouring chain production are probably easily disturbed, or not satisfied at all, low in the corona.

Assuming that chains are generated at the plasma frequency, one can convert frequencies into heights in the corona. Adopting electron densities equal to two times the values given by the Newkirk (1961) model for the axis of a coronal streamer, Hanasz found that the distribution shown in Fig. 6.5 corresponds to a height range extending from about 0.2 to 1.2 $R_\odot$ with a maximum occurrence of chains at a height of about 0.3 - 0.4 $R_\odot$ above the photosphere.

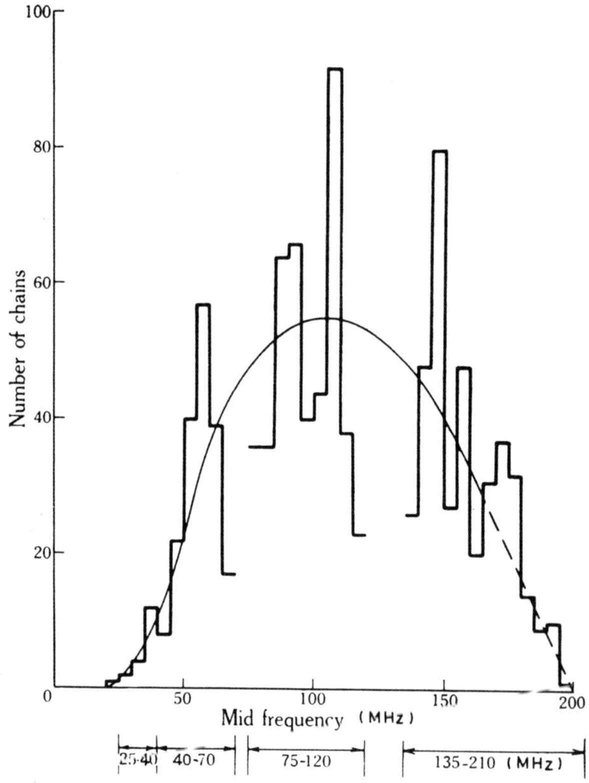

Fig. 6.5. Distribution of chains with mid frequency. The frequency coverage of receivers and gaps between them are shown under the frequency axis. The smooth curve represents the mean distribution of chains with frequency; the high frequency part of the distribution (dashed) may be distorted by the decrease in receiver sensitivity. (Hanasz, 1966.)

Frequency Drift Rate

A prominent feature of most chains is a systematic drift in frequency. Elgarøy and Ugland found a distribution of chains with frequency drift velocity as depicted in Fig. 6.6.

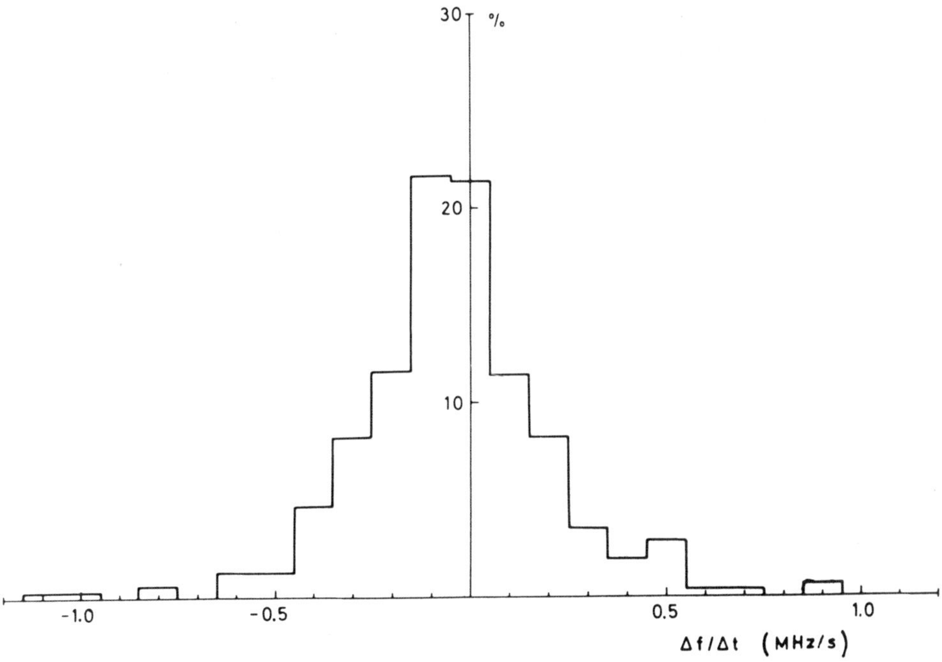

Fig. 6.6. Distribution of chains with frequency drift velocity (Elgarøy and Ugland, 1970).

The observations refer to the frequency range from 140 to 170 MHz. Similar results were obtained by Loonen (1971). It is seen that both senses of drift direction occur, i.e. towards lower and towards higher frequencies. This is different from type II bursts which only show drifts from higher to lower frequencies. About 50% of the chains had a negative drift, 30% showed a drift towards higher frequencies and the remaining 20% revealed no detectable frequency change. These results are in accord with those of Hanasz (1966) who found that 26% of the chains had positive drift rates and that the frequency drift rates showed a considerable spread. Values of $(1/f)(\Delta f/\Delta t)$ were mainly found between $-2.5 \times 10^{-3}\,s^{-1}$ and $+1.7 \times 10^{-3}\ s^{-1}$. In Fig. 6.6 this corresponds to limits of -0.4 MHz s^{-1} and $+0.25$ MHz s^{-1} respectively, and there is evidently no difference between long-lived and short-lived chains as concerns frequency drift.

Wild and Tlamicha (1965) suggested that the observed frequency drift of chains could be due to the motion of localized disturbances passing through the corona. The preponderance of frequency drift in the negative sense was taken as evidence in favour of this interpretation.

Assuming that the emission frequency is equal to the electron plasma frequency at the source position, the observed frequency drift velocities may be converted to radial velocities of the exciting agency of the chains. Adopting again en electron density equal to that given by Newkirk (1961) for the axis of a coronal streamer multiplied by a factor of two, the radial velocity v_r is found using the following expressions:

$$f_p = 9 \times 10^{-3}\sqrt{N_e}\ , \qquad (6.2\text{-}1)$$

where f_p is in MHz and N_e in cm^{-3}. The adopted electron density model is given by

$$N_e = 1.65 \times 10^5 \times 10^{4.32/\rho}\ , \qquad (6.2\text{-}2)$$

where ρ is the distance from the centre of the sun in units of $R_\odot$. From (6.2-1) and (6.2-2) one obtains

$$f_p = 3.65 \times 10^{2.16/\rho} \qquad (6.2\text{-}3)$$

Since it was assumed that the emission frequency was equal to the plasma frequency, the radial velocity v_r in km s^{-1} is found from

$$v_r = \frac{d\rho}{df} \cdot \frac{df}{dt} \cdot R_\odot = -\frac{3.5 \times 10^6}{(\ln f - 1.3)^2}\,\frac{1}{f}\,\frac{\Delta f}{\Delta t} \qquad (6.2\text{-}4)$$

where $\Delta f/\Delta t$ is the observed frequency drift velocity.

A histogram showing the distribution of radial velocities is shown in Fig. 6.7.

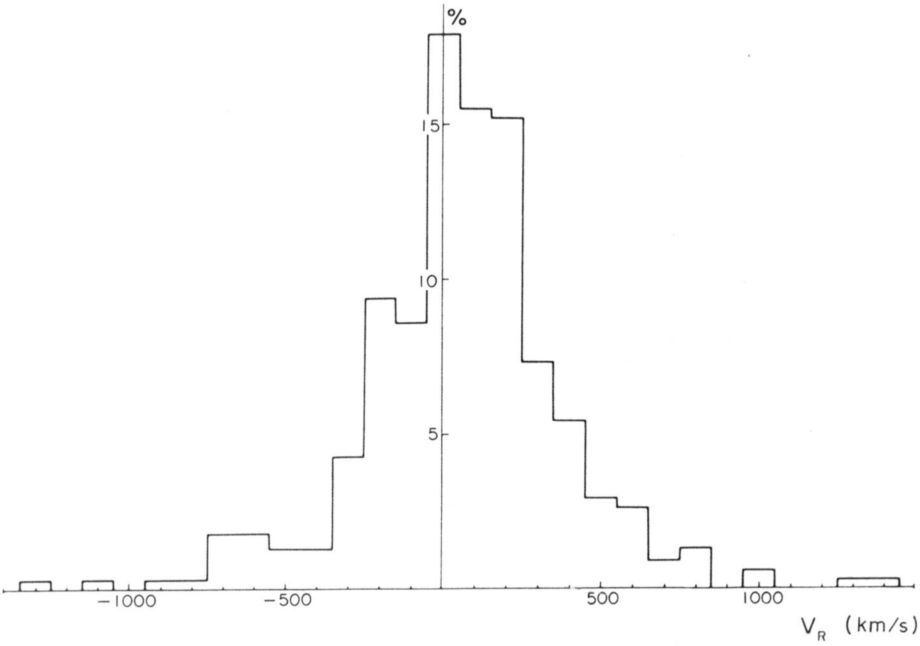

Fig. 6.7. Distribution of chains with radial velocity (Elgarøy and Ugland, 1970).

The radial velocity of agencies responsible for the generation of chains ranges from +800 km s^{-1} in the outward direction to -700 km s^{-1} in the inward direction. There is a clear distinction between the propagation velocities found for chains and for type II bursts. Comparison with the distribution of radial velocities of type II bursts as determined by Weiss (1965) shows that the radial velocities of type II bursts are 2 - 4 times larger than those of type I chains.

Lifetime of Chains

The lifetimes of chains have been determined by Hanasz (1966), Elgarøy and Ugland (1970), Loonen (1971) and Chernov *et al.* (1972).

The distribution of lifetimes derived by Hanasz is reproduced in Fig. 6.8 and is reliable for lifetimes of 20 s upwards.

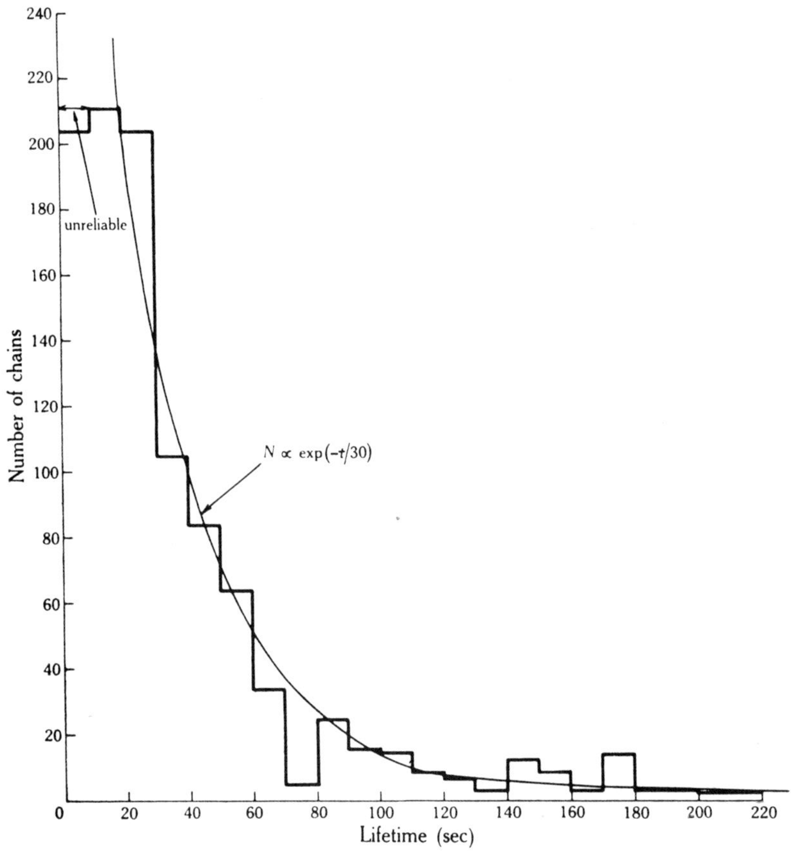

Fig. 6.8. Distribution of chains with lifetime (Hanasz, 1966).

In the range from 20 to 200 s the distribution followed quite well an exponential law:

$$N(t) \propto \exp(-t/30) . \qquad (6.2\text{-}5)$$

The sample contained 1000 chains observed in 1959 and 1960. For chains occurring at somewhat higher frequencies than the majority of those in the material of Hanasz, Loonen found an exponential decay of lifetimes with a time constant of about 20 s. About 450 chains occurring during an active period in

February 1970 were analysed.

The distribution of chain lifetimes derived by Elgarøy and Ugland is given in Fig. 6.9 and differs from that of Hanasz. Totally 316 chains which were observed during different noise storms in 1966-8 were included. It was found that about 60% of the chains had lifetimes less than 6 s. The average lifetime amounted to 5.1 s. The longest lifetime observed amounted to 26 s. This is in good agreement with the results of Chernov *et al.* who observed in the frequency band from 190 to 220 MHz and selected 77 chains during a storm in May 1969. They found lifetimes varying between 2 and 30 s. For the majority it was no longer than 10 s and the average lifetime was equal to 7.2 s.

One should not regard the histogram in Fig. 6.9 as representing the true distribution of the lifetime of burst chains. The histogram serves to give information about the "short-lasting end" of the full distribution, which could not be analysed on the records studied by Hanasz.

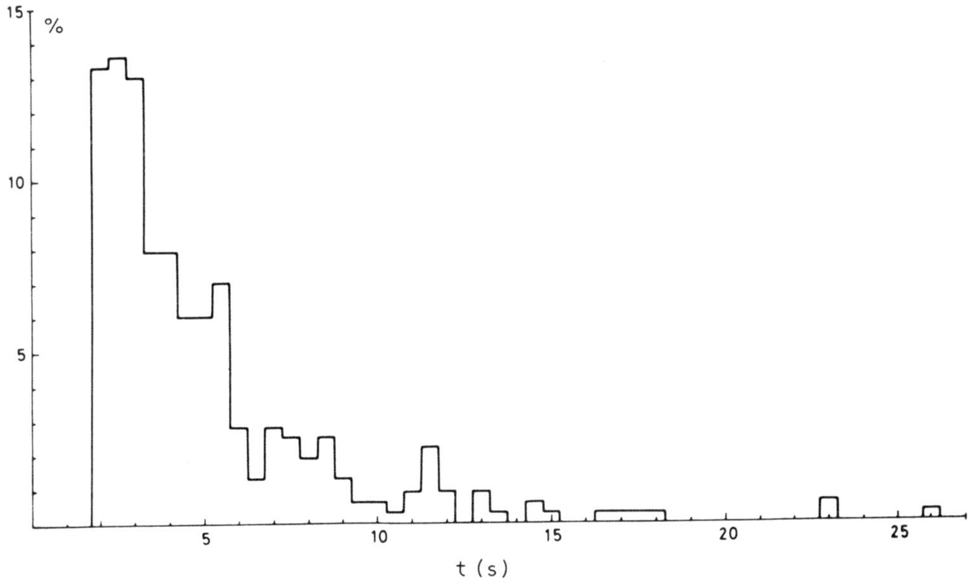

Fig. 6.9. Distribution of the number of chains with lifetime (Elgarøy and Ugland, 1970).

It is easy to explain the obvious difference between the distribution curves in Figs. 6.8 and 6.9. Firstly, the requirements set by Elgarøy and Ugland for a succession of bursts to be recognized as a chain were rather strict and would reduce the number of chains with large lifetimes. Secondly, the observing range was rather small and tended to descriminate against events with long lifetime when these possessed at a frequency drift.

Present evidence thus shows that the lifetimes of chains range from a few seconds to about 3 minutes; the majority having a duration less than 1 minute.

Positions of Chains

The positions of noise storms with and without chains were determined by Hanasz. It was found that the positions of chains did not differ from the position of the background continuum and the type I bursts, and both types of storms showed the same distribution across the solar disk.

A search for systematic movements of the sources of chains was made by Hanasz using 200 MHz interferometer recordings from the Tokyo Astronomical Observatory in combination with the spectral records. The "burst clusters" referred to by Suzuki (1961) were found to correspond to chains on the spectral recordings. Five chains were identified, but the evidence of systematic movement was inconclusive. However, in the paper by Suzuki (1961) it is mentioned that the majority of burst clusters show systematic movement.

The most important outcome of the position determinations of chains is that type I bursts, background continuum and chains are emitted from a common source region.

Polarization

Chains showed the same polarization properties as type I bursts on the recordings of a swept-frequency polarimeter (Hanasz, 1966). This is in accordance with the properties of moving burst clusters for which Suzuki (1961) found that the sense of polarization was the same as the general polarization of the noise storm. Chernov *et al.* (1972) studied the polarization of chains occurring during two noise storm periods in 1969. In both cases the development in polarization during several successive days was the same for type I bursts, storm continuum and chains.

Bandwidths

Elgarøy and Ugland (1970) measured the bandwidth of the first and last bursts in the observed chains. When it was difficult to separate individual bursts, the bandwidth was determined at time points situated 0.2 - 0.3 s away from the ends of the chain. The results of the bandwidth determinations are given in the histograms in Fig. 6.10. It is seen from the figure that the bandwidth is larger in the starting phase than near the end of a chain. The average bandwidth at the start of a chain was 5.1 MHz, whereas the bandwidth at the end was 4.6 MHz. In 62% of all cases the bandwidth decreased with time, whereas an increase was found in 22% of the cases. Results in agreement with those described here were obtained by Chernov *et al.* (1972) for chains occurring during the noise storm in May 1969. Chernov *et al.* found an average bandwidth of 4.8 MHz at the start of the chains and a value of 4.4 MHz at the end. In 55% of the cases the starting bandwidth was the larger. An increase in the bandwidth was found in 18% of the cases. The same result was not found in the storm recorded in June 1969. In that case the bandwidth decreased in 28% of the cases whereas an increase was found in 45% of the chains. This demonstrates that there may be differences between individual storms, as was also pointed out by Elgarøy and Ugland.

Intensity

Elgarøy and Ugland selected 25 chains which were examined more closely with a photometer. Photometer scans along the axis of the chains in the direction of increasing time in most cases showed a moderate decrease in intensity and then a sudden drop to zero. This is an interesting property. The intensity of the emission does not decay exponentially or according to some power law, but is rather quickly cut out at some stage.

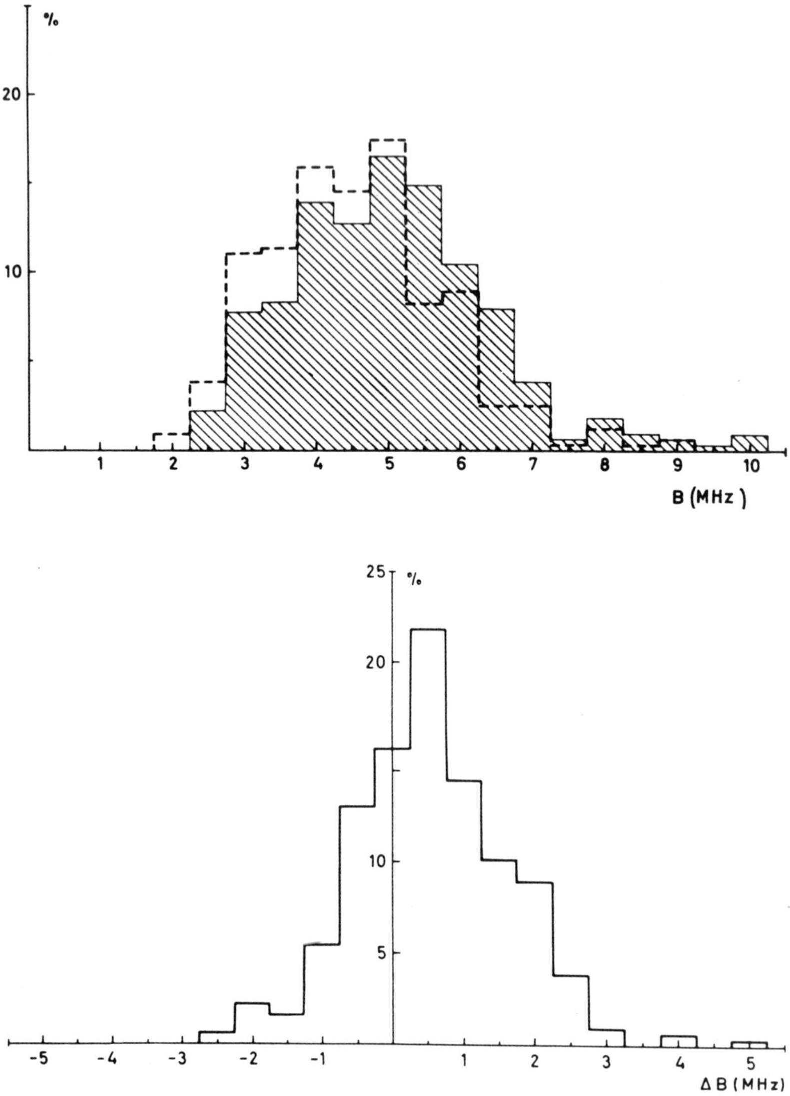

Fig. 6.10. *(Above)* distribution of chains with bandwidth. The full curve (hatched area) represents bandwidths measured at the start of the chains. The broken curve refers to values measured at the end of the chains. *(Below)* distribution of the number of chains of different bandwidth changes $\Delta B = B_s - B_e$.
(Elgarøy and Ugland, 1970.)

Fine Structure of Chains

The duration of bursts in chains and of bursts occurring outside chains on the same records were studied by Elgarøy and Ugland. It was found that histograms of the number of bursts of different duration in the two cases were quite equal. The average duration of bursts in chains was 0.40 s, and the average duration of bursts outside chains 0.42 s. Similarly no difference was found in the bandwidth of bursts occurring in chains and those outside.

Several other features may be of interest for the discussion of the origin of chains. The following may be the more important:

(a) Strong bursts may occur at any time in a chain, but there is a statistical tendency for bursts at the end of chains to be less intense than those near the beginning.

(b) Individual bursts in a chain may show different frequency drifts. The frequency drift velocity is usually not the same as that of the chain. Bursts of opposite drift directions may occur in the same chain. In some cases all bursts in a chain have no detectable frequency drift.

(c) In several cases individual bursts are bridged by narrow band weaker emission.

(d) Some chains are very complex; two or three bursts may be observed simultaneously on slightly different frequencies.

(e) Chains may have a "wavy" appearance on dynamic records. The emission frequency oscillates around that expected on the assumption of a linear frequency variation with time.

(f) Individual bursts in chains reflect the same general properties as those occurring outside. For instance, on a day when a storm was characterized by many reverse drift, short lasting type I bursts, the same type of bursts was predominant in chains.

Chains in Pairs

Multiple chains, and in particular pairs of chains running almost in parallel, are frequently present on spectrograph records. This is the same phenomenon of splitting as described by Elgarøy (1961). Very often, however, the two components in a pair do not start simultaneously. Differences in time up to 1 min were observed by Hanasz (1966). Further to this, measurements of drift rates showed that in most cases the upper and the lower bands were inclined to each other. Frequently there was a complete lack of detailed correspondence between two chains in a pair, but in some cases the intensity variations in one chain were repeated in the other. During periods lasting for hours, chains might occur almost only in pairs. It seems therefore that sometimes the conditions in the corona are particularly favourable for pair production.

The histogram in Fig. 6.11(a) shows the distribution of frequency separations between chains in pairs. Most pairs have

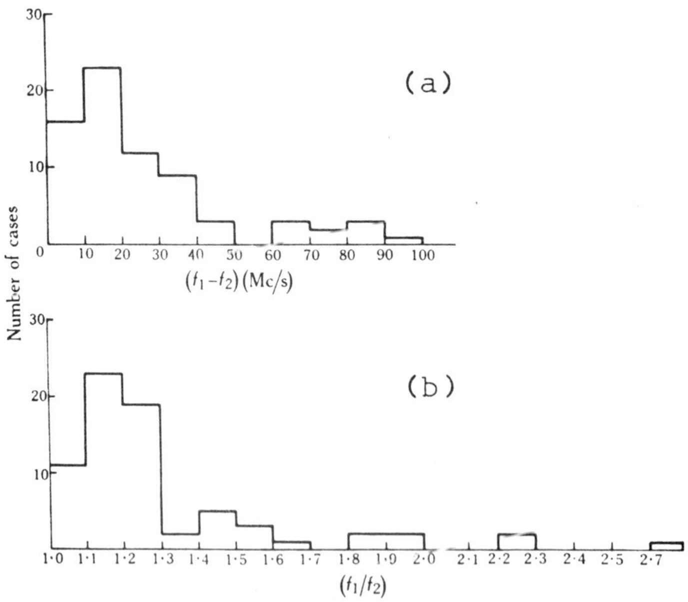

Fig. 6.11. Distributions of (a) frequency separations of pairs of chains, and (b) frequency ratios of pairs of chains (Hanasz, 1966).

a frequency separation between the components in the range from a few to 40 MHz. Type II bursts show a correlation between frequency of emission and frequency separation. This property is not found for chains. On the average the frequency separation between the components in pairs of chains is about twice as large as found for band splitting in type II bursts (Smerd *et al.* 1975).

It is evident from Fig. 6.11(b) that harmonic structure is not present in chains.

Association of Chains with Other Activity

Chains are usually accompanied by type I bursts and type III bursts. Sometimes weak type III bursts seem to grow out of type I chains (Fig. 6.12). An attempt to detect flares and type II bursts related to the appearance of type I chains in noise storms failed (Hanasz, 1966). No chains were observed on days with type II bursts. Thus strong disturbances may destroy coronal structures favourable for the generation of chains. In connection with Fig. 6.12 one should note that the type III bursts occur on the low-frequency side of the chains. This

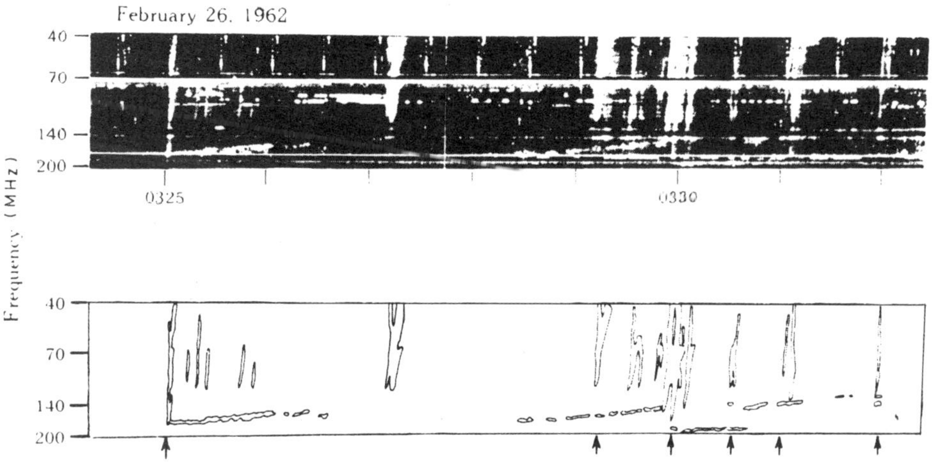

Fig. 6.12. Weak type III bursts which appear to "grow" out of type I chains (Hanasz, 1966).

behaviour is probably related to the result obtained by Malville (1962) who found that the starting frequencies of type III bursts usually were lower than the upper frequencies covered by noise storm emission.

Clustering of Chains

The distribution of chains with frequency varies from day to day, as can be seen from Fig. 6.13. In general chains are found to occupy the same frequency range as the noise storm

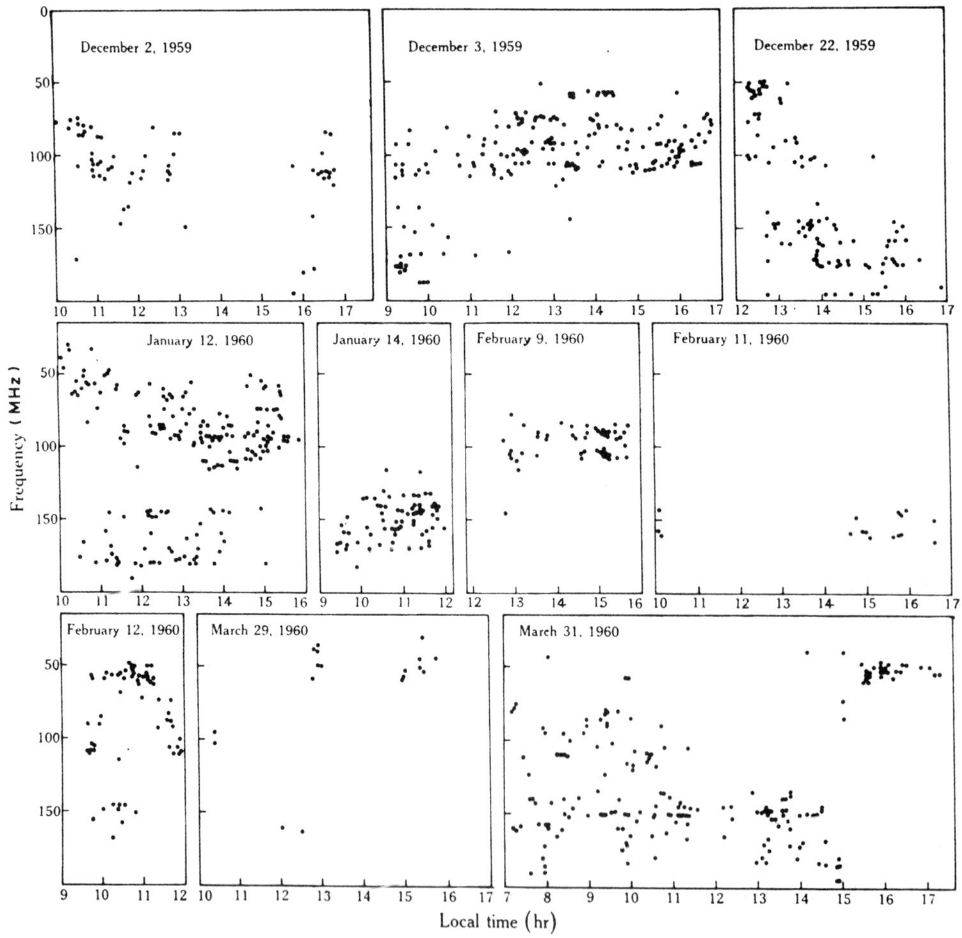

Fig. 6.13. Positions of chains in the time-frequency plane (Hanasz, 1966).

radiation. On some days chains are concentrated in relatively narrow frequency bands, and, as already noted, chains are less likely to occur at the highest frequencies where noise storms occur. Therefore conditions suitable for chain production may exist only in limited regions of the corona.

Centre-limb Variations

The most interesting property of burst chains is their frequency drift. Elgarøy and Ugland looked for a possible centre-limb variation in this parameter and found the results given in Table 6.1.

TABLE 6.1

Distribution of frequency drift velocities for chain positions near the centre and near the limb of the solar disk

Source position	Per cent		
	$\Delta f/\Delta t > 0$	$\Delta f/\Delta t = 0$	$\Delta f/\Delta t < 0$
Near centre	28	23	49
Near limb	29	15	56

For central sources the average drift velocity amounted to 0.18 MHz s^{-1} and it had increased to 0.23 MHz s^{-1} for limb sources. The difference may be explained by reference to Table 6.1 which shows that chains with no frequency drift occur less frequently near the limb than near the centre of the solar disk.

The average lifetime of chains near the disk centre seemed to be somewhat longer than for limb sources.

These results may be connected with escape conditions, beaming of emission and the geometrical configuration of the source region. However, more comprehensive observational material should be analysed before definite conclusions are drawn.

Conclusions

Chains represent the most clear-cut and ordered phenomenon encountered during noise storms. The observations show that bursts in chains have the same properties as type I bursts occurring outside chains. The basic emission mechanism is thus likely to be the same in the two cases, and chains may offer a very valuable opportunity to arrive at a better theoretical understanding of the noise storm phenomenon.

The frequency drift of chains suggests that they are generated by moving disturbances. Radial velocities inferred from the frequency drift velocity are about an order of magnitude smaller than the thermal velocity of electrons in the corona. It is therefore likely that acceleration of electrons to super-thermal velocities takes place in the slowly moving disturbance. This suggestion is substantiated by the association in several cases between chains and type III bursts.

Different theories involving weak shock waves and magneto-hydrodynamic waves have been proposed to explain the generation of chains, but the discussion of possible mechanisms for chain production is deferred to the part dealing with theories of noise storm emission.

6.3. TIME SPLITTING OF BURSTS

Short-lasting solar bursts occur at times in pairs: two bursts are observed, the second being a repetition of the first after a time delay varying from less than a second to several seconds.

Roberts (1958) identified a type of solar burst which he called reverse drift pair. He suggested that a pair was composed of a direct and a reflected ray. But in order to achieve agreement between theory and observation, several assumptions were necessary. The more important requirements were that the emission occurred at the second harmonic of the plasma frequency, the sources were observed only near the centre of the solar disk and the gradient of electron density was steeper than that given by the Baumbach - Allen model by a factor 1.5.

High-resolution observations in the 200 MHz range (Elgarøy, 1961) showed that type I bursts also may occur in pairs (Fig. 6.14). The time interval between the two bursts was measured in 80 cases and in about 50% it was between 0.5 and 0.7 s.

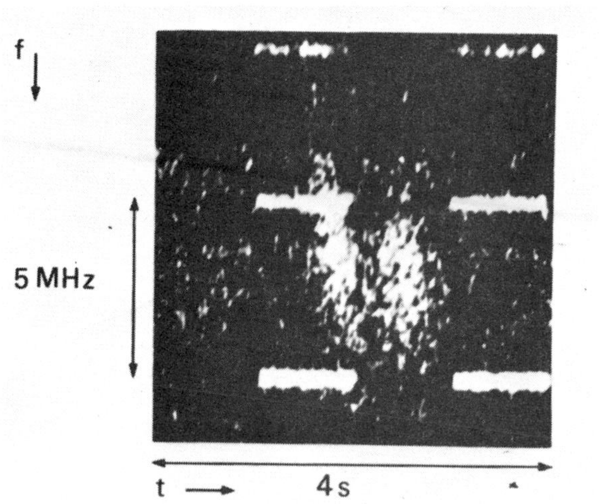

Fig. 6.14. Pair of type I bursts observed in the 200 MHz range (Elgarøy, 1961).

The split bursts observed by Bougeret (1973b) with the Nancay east - west radioheliograph operating at 169 MHz appear as a juxtaposition of a main burst and a second burst, which is always weaker and shifted in position in the direction of the centre of the disk. On November 18, 1968, the split bursts showed the following average characteristics:
first component: diameter 2.5′ arc, duration 0.50 s;
second component: diameter 2′ arc, duration 0.55 s,
angular shift 3.5′ arc, time separation 0.17 s,
amplitude ratio 5.

Drifting bursts (in the space - time plane) could be produced when the components of split bursts were not resolved, but they may also be explained by a real continuous drift effect.

It is possible that the split bursts observed by Bougeret correspond to the double storm bursts observed by Elgarøy, but a connection of this type has not yet been proved.

Ellis (1969) observed the fine structure of solar bursts in the frequency range 25 - 100 MHz. He detected bursts with narrow bandwidth (~0.03 MHz) and a frequency drift velocity in the range of 1 - 2 Mhz s^{-1}. They occurred during noise storms, and in many of the records they appeared to be double: two similar bursts occurred successively. The second component was found at a somewhat higher frequency than the first one. In Fig. 6.15 the distributions of time and frequency intervals between the two components are shown. The time and frequency intervals varied between 0.5 - 2 s and 0.1 - 2 MHz for bursts at about 30 MHz.

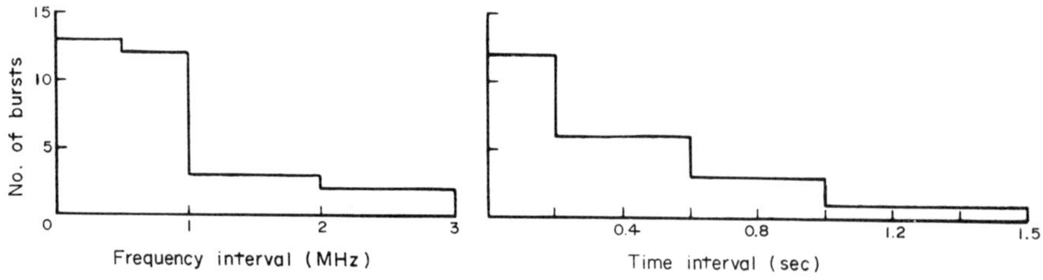

Fig. 6.15. Distribution of frequency intervals and time intervals between pairs of fast drift storm bursts (Ellis, 1969).

Ellis suggested that the bursts were similar to those denoted type I(fd) by Elgarøy (1961). He also observed another type of bursts which he called drift pairs. These showed in general the same properties as the reverse drift pairs described by Roberts (1959), but the direction of frequency drift was from higher to lower frequencies. Drift pairs had larger bandwidths and slower frequency drifts than the fast drift storm bursts. A remarkable feature of the drift pairs was that the second component occurred at a higher frequency and at a later time than the first, i.e. the second component was displaced diagonally in the frequency - time plane. The average frequency displacement at frequencies around 30 MHz was 0.8 MHz, and the average time difference amounted to 1.2 s. Drift pair bursts were usually not significantly polarized.

Yoh and James (1967) observed with a multi-channel receiver at frequencies around 38 MHz. The sensitivity of the receiver was particularly good. They found several storms of narrow band bursts with short duration. Occasionally two bursts with similar characteristics occurred 1 s apart. At times the two bursts drifted together in frequency, at other times the second component appeared stronger in a different frequency interval than the first. These results are in good agreement with those obtained by Ellis (1969).

Further observations were carried out at 25 MHz by Sastry (1969). A radiometer was operated during periods of enhanced solar emission. Sastry found that the enhanced radiation was composed of compound bursts and single bursts of durations ranging from about 0.5 s to several seconds. The bursts appeared to be superposed on a slowly varying continuum. Several double bursts were recorded. On the average the delay time between the two bursts was about 1.7 s.

The 25 MHz observations were continued by Sastry (1971) with a receiver operating in two or three neighbouring channels. Observations were performed during times of enhanced solar radio emission, and three types of noise storms were recorded: enhanced background continuum with only irregular variations, enhanced background continuum with short-lasting and narrow band burst superposed and, finally, short-lasting and narrow band bursts without continuum enhancement. Double bursts occurred with and without accompanying enhancement of the continuum emission, and in most cases the time distance between the two components in a pair was between 1 and 2 s. Measurements of the polarization of about 200 double bursts (Sastry, 1972) showed that 30% of them were polarized. It should be noted that in such cases both components showed the same sense and equal degrees of polarization.

Extensive observations with high sensitivity were performed by Baselyan *et al.* (1974a) at frequencies between 12.5 and 25 MHz. They found time-split bursts with an average value of the time distance between the components of 2.5 s. In 50% of the cases the second component was shifted in frequency by 10 - 30 kHz towards lower frequencies with respect to the first component. In 40% of the cases an equally large shift occurred towards higher frequencies, and in 10% of the cases there was no frequency shift to be detected. In some cases a splitting into three components was found.

All evidence presented above leaves no doubt that time splitting occurs, but the explanation of the phenomenon is not obvious. Several mechanisms may produce time-split bursts, but they cannot be applied on the observations without difficulties.

Roberts (1958) suggested that reverse drift pairs could be explained by assuming that the burst energy had propagated via two paths of different lengths in the regular solar corona. Provided the frequency is higher than the plasma frequency at the source level, Jaeger and Westfold (1950) had shown that echoes should occur, and time delays of several seconds between the two components were predicted. In general there are two rays to the earth from any accessible point in the corona, as illustrated in Fig. 6.16. The time delay between the reception of a disturbance via the direct and the reflected paths depends on the height of the source, its position on the disk and the frequency. When the source is at the plasma level, the delay is zero because the two paths are identical.

When scattering is taken into account, the problem becomes more complicated. Scattering has little influence on the direct ray when the source radiates at the harmonic of the plasma frequency, but the reflected ray is strongly affected (Riddle, 1974). The arrival of the reflected radiation is spread out over a period of time which is long compared to that of the forward ray. Since the correspondence between properties such as bandwidth, duration and intensity of the members of a pair is quite good, the echo hypothesis is difficult to maintain.

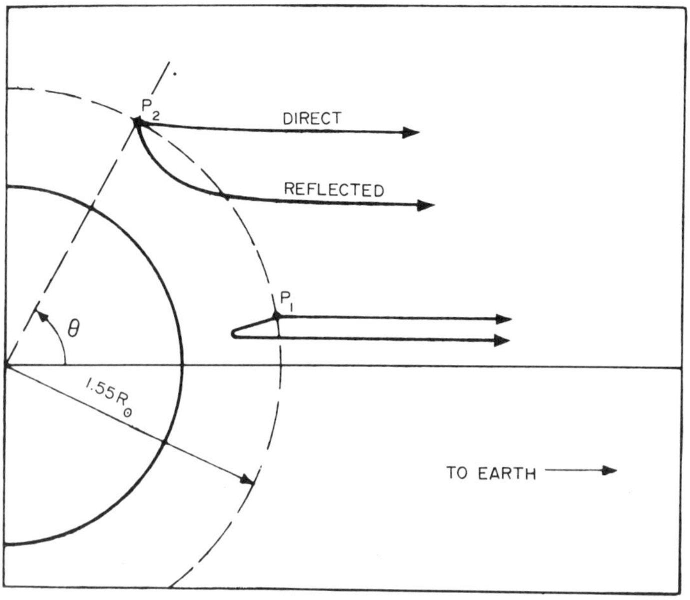

Fig. 6.16. Paths of the direct and the reflected rays in the corona for a source located at the 30 MHz plasma level and radiating at 60 MHz (Roberts, 1958).

Elgarøy (1968) proposed that the time splitting might occur because energy is propagated via two magneto-ionic modes. Magneto-ionic splitting is well known from ionospheric physics. Using plausible models of electron density and magnetic field distribution in the corona, Elgarøy derived delay times which might reach some seconds. According to this suggestion both components should come from the same position and they should show opposite polarizations. It was shown by Sastry (1972) that in a sample of about 200 bursts showing time splitting in the 25 MHz range, about 30% were polarized. The two components were always polarized in the same sense and to the same degree. It is therefore not likely that the observed time splitting can be regarded as an effect of different velocities of the two magneto-ionic modes in a magnetized plasma.

Ellis (1969) pointed out that his observations of drift pairs indicated an emission mechanism in which backward radiation from the source occurred at a slightly higher frequency than radiation emitted in the same direction as the source motion. This is also an echo-mechanism, and if scattering is important, the correspondence between the two components should not be as good as actually seen in the observations. Baselyan *et al.* (1974a) found that the delayed burst could be shifted both to higher and to lower frequencies as compared to the first. In some cases, as was also noted by Ellis, even triple bursts were observed. Perhaps scattering is less important than assumed under conditions when bursts split in time are generated, and that the delayed component is reflected from moving, sharply defined disturbances in the corona. Then the frequency displacement of the second component might be interpreted as a Doppler effect.

Finally, one should mention the possibility that two distant sources may be triggered almost simultaneously by a disturbance initiated by an instability high in the corona about midway between the two centres (Wild, 1969a). But it is difficult to explain the close correspondence which frequently exists between split pairs. Another possibility is that the same source is excited two or even three times by a "train" of disturbances. In this connection one should note that type III bursts may occur in pairs and that the phenomenon may be explained as a result of a plasma pinch effect rather than as an echo effect (Caroubalos *et al.*, 1974).

It is impossible at present to give a satisfactory explanation of the time splitting of solar bursts. One needs better observations, including positional data, polarization and spectrum with high resolution.

6.4. FREQUENCY SPLITTING

At times type I bursts occur in pairs at slightly different frequencies. Several cases of such frequency splitting were noted by Elgarøy (1961) from high resolution spectral observations in the 200 MHz range (Fig. 6.17). In some cases several

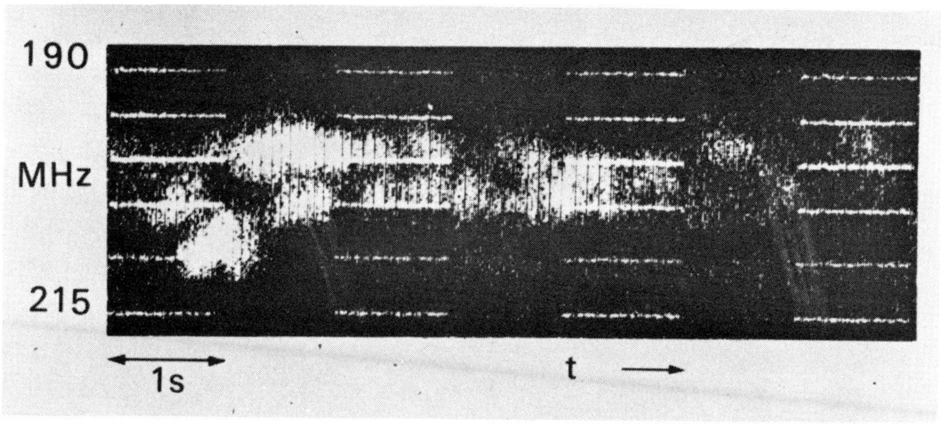

Fig. 6.17. Type I bursts showing frequency splitting (Elgarøy, 1961).

successive bursts showed frequency splitting, and it happened that three components could be recognized on the records. In such cases the distance between the central component and the upper component equalled the frequency distance between the central component and the low-frequency component. The average distance between the components of split bursts found in the 200 MHz range by Elgarøy was 8 MHz. The largest and the smallest values were 12 and 5 MHz respectively.

Frequency splitting of chains of type I bursts was reported by Wild and Tlamicha (1965) and by Hanasz (1966). This phenomenon may be different from the splitting observed in single bursts because pairs of chains show no detailed correspondence between the components and a tendency for the components to start at different times. But there are some arguments in favour of a common origin of the two components, notably the tendency for split chains to occur on particular days and the fact that the frequency distance between simultaneous components is rather small.

Frequency splitting in slow drift type II bursts was detected by Wild (1950), and occur both in the radiation at the fundamental frequency and in the harmonic component. In the harmonic the frequency separation is twice as large as in the fundamental. In many cases the split bands of each harmonic component are identical, but it also happens that they differ in bandwidth and intensity. The frequency split increases with frequency according to the relation

$$\Delta f = 0.27f - 3.0 \tag{6.4-1}$$

where Δf is the frequency split and f the frequency in MHz (Smerd *et al.*, 1975).

Very nice examples of frequency splitting during noise storms were observed at frequencies below 60 MHz by Ellis and McCulloch (1967), who detected large numbers of short duration bursts with frequency splitting. Frequency splitting of narrow-band short-duration bursts appear to be quite common at decametric wavelengths.

The frequency interval Δf between the components increases with frequency from an average value of 0.1 MHz at 25 MHz to 1 MHz at 60 MHz. Observational results due to Ellis and McCulloch are given in Fig. 6.18. In particular at higher frequencies there is a considerable spread in the values of the frequency splitting. At a given wave frequency there is also a variation of the frequency splitting with time, as is demonstrated in Fig. 6.19. But the average value of Δf does not vary much during a particular event. This shows that the explanation of the phenomenon must permit rapid changes either with time in a certain source region or with position if the bursts are generated at different places or with a property of the bursts as, for example, frequency drift velocity.

In approximately 10% of the cases Ellis and McCulloch detected bursts with three components. In these the central element used to be the stronger, and the frequency interval between the central component and the upper one was about equal to the separation between the central component and the one at the lower frequency. Some examples of split bursts recorded by Baselyan *et al.* (1974a) are given in Fig. 6.20. The results of Baselyan *et al.* are in excellent agreement with those of Ellis and McCulloch. Baselyan *et al.* also mentions the possibility of

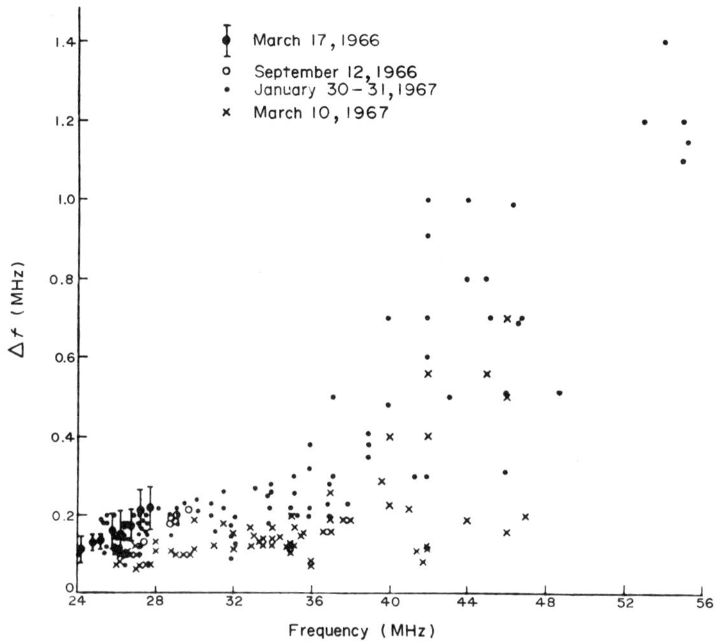

Fig. 6.18. Variation of the frequency interval between the burst components with wave frequency (Ellis and McCulloch, 1967).

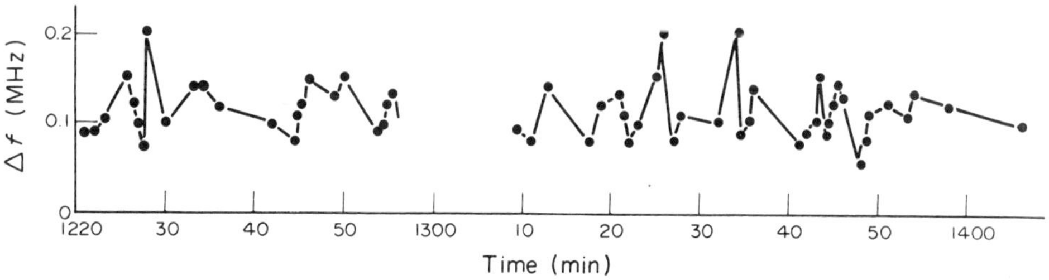

Fig. 6.19. Variation in the frequency interval for successive bursts recorded on March 17, 1966, at about the same frequency (25.75 - 26.25 MHz) (Ellis and McCulloch, 1967).

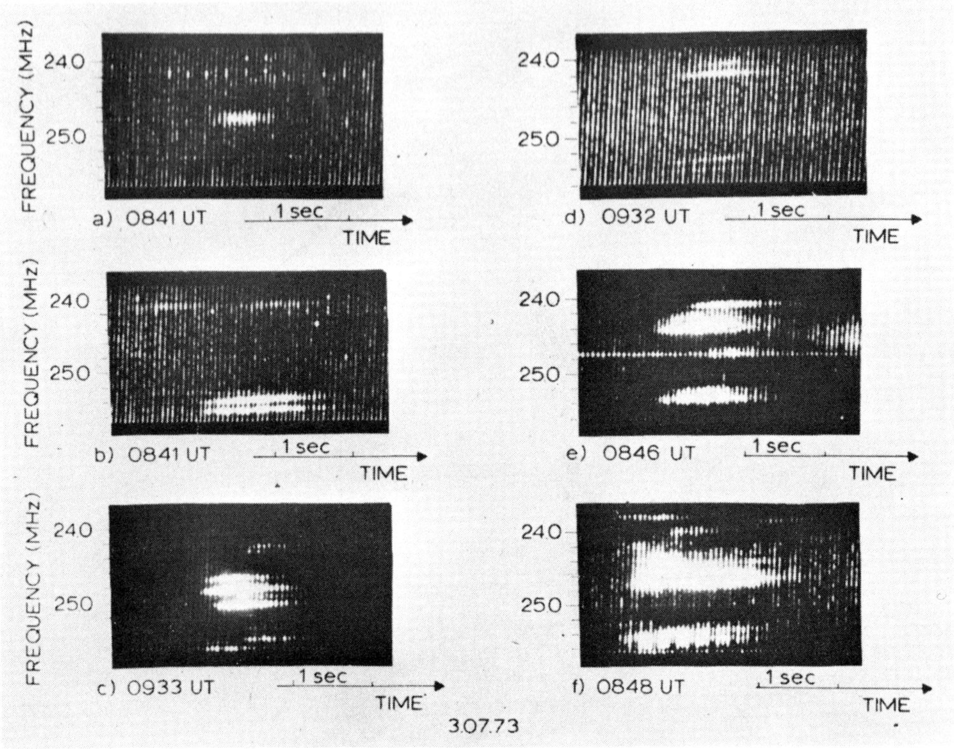

Fig. 6.20. Dynamic spectra of (a) single burst, (b) split pair and (c) triplet. The figures (d), (e) and (f) show the presence of two different scales of the frequency splitting. (Baselyan *et al.*, 1974.)

two different scales in the frequency splitting. The first scale characterizes the splitting inside each pair and amounts to 0.1 - 0.2 MHz. The second scale is the frequency interval between two pairs and is approximately 0.8 - 1.3 MHz (Fig. 6.20 (d)-(f)). But one should bear in mind that the double pairs may represent two independent elements in a row of bursts.

The components in pairs or triplets showing frequency splitting usually did not begin at the same time. Very often the low-frequency component began before the one at the higher frequency.

The duration of the components in frequency split pairs is similar to the duration of normal type I bursts at the same frequencies. Each component had a bandwidth of about 0.05 MHz s^{-1} (Ellis and McCulloch, 1967; Baselyan *et al.*, 1974a). The frequency drift velocity and the frequency distance between the components in a particular frequency interval were not correlated, as is convincingly demonstrated by Fig. 6.21.

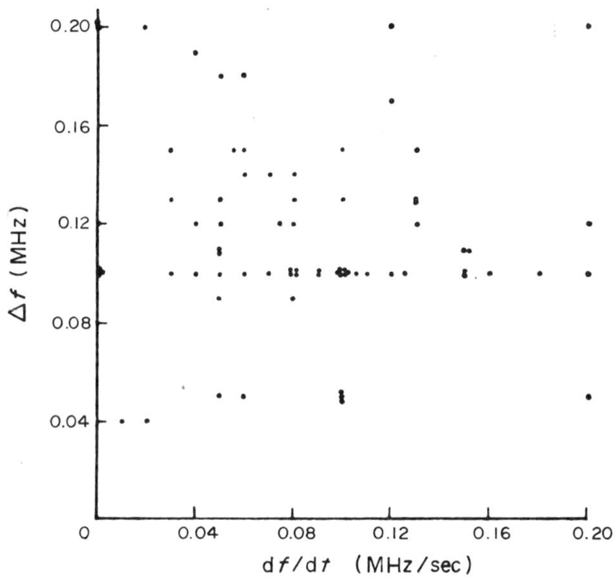

Fig. 6.21. Scatter diagram of the variation of the frequency intervals between bursts against their frequency drift velocity in the frequency range 26 - 27 MHz determined from records obtained on March 17, 1966 (Ellis and McCulloch, 1967).

Baselyan *et al.* (1974a) also detected more diffusive bursts in the 25 MHz range. These had longer durations, larger bandwidths and showed a frequency splitting with a distance between the components which was about 1.5 - 2 times larger than for ordinary split decametric bursts.

The burst types described above have a tendency to form chains that have several properties in common with type III bursts (Ellis and McCulloch, 1967; Ellis, 1969; de la Noë and

Boischot, 1972; Baselyan *et al.*, 1974b). Further discussion of this property is undertaken in connection with problems concerning the relation between metric and decametric storm activity.

Split pairs and triplets, as well as single bursts of the same type, are generally much more polarized than type III bursts. In some cases they have a degree of polarization of almost 100%. The bursts are circularly polarized, and the components in a pair or a triplet are always polarized in the same sense (de la Noë and Boischot, 1972).

Different attempts have been made to explain the frequency splitting of bursts. In connection with type II bursts, Wild (1950) and Roberts (1959) suggested that some kind of Zeeman splitting occurred, but they did not discuss the physical mechanism behind the splitting in detail. Zaitsev and Fomichev (1973) explained the splitting as the result of Doppler effect in weak shock waves propagating in a plasma with a strong magnetic field in the case of type I chains. The splitting may also be connected with electromagnetic or electrostatic resonances in a magnetized plasma (Ellis and McCulloch, 1967). Baselyan *et al.* (1974) suggested that the splitting was associated with some peculiarities of the transformation of plasma waves into electromagnetic waves near the plasma frequency and the harmonic of the plasma frequency. It was furthermore supposed that the peculiarities leading to fine structure begin to arise at stream velocities twice the velocities of streams generating ordinary type III bursts. The latter condition was deduced from the frequency drift velocity of chains of decametric bursts. Further discussion of frequency splitting is given in section 9.6.

7. PERIODIC AND QUASI-PERIODIC PHENOMENA

7.1. INTENSITY VARIATIONS IN THE BACKGROUND CONTINUUM

Inspection of radiometer records of noise storms in the metre wavelength region frequently gives the impression that the storm continuum undergoes quite regular variations. Accordingly a search for periodic intensity variations was undertaken by Hauge (1958) and continued by Sveen (1971).

Sveen analysed records from several noise storms observed in the 200 MHz range with the radiometer at the Oslo Solar Observatory during the time period from 1957 to 1970. Two different types of enhanced radio emission were investigated:

(a) normal type I storms showing enhanced background emission and with short-lived bursts superposed;

(b) continuum storms with strongly enhanced emission almost free from short-lived bursts, usually following a type IV burst.

Records from several days of enhanced radio noise emission observed during the last two sunspot maxima were selected for analysis. Different methods were used: periodograms were constructed, autocorrelation calculations were performed and power spectra were derived. The methods were tested on data with known periodicity and they were applied on sets of random numbers in order to obtain information on the significance of the observational results.

It was found that some noise storms show periodic intensity variations, but the fluctuation rate varied from day to day and from hour to hour. Thus short periodic intensity variations do not remain uniform throughout long time intervals. Usually type I storms showed fairly constant fluctuation periods for time intervals lasting between 1 and 2 h. One example of periodic intensity variations in the background continuum emission is illustrated in Fig. 7.1. The period of the intensity variations is about 265 s. Short periodic variations mostly occurred at high flux densities. No correlation was found between the burst variability factor and the length of the fluctuation period.

Different type I storms showed characteristic periods in the range from 3 to 7 min, the average value being about 5 min.

When the data samples covered long time intervals, say 12 h of continuous observation, a smoothed spectrum was usually found.

Some continuum storms of the second category mentioned above also showed periodic intensity variations. In these cases the periods which were found were in the range from 15 to 20 min.

There was no correlation between the amplitude of the intensity variations and the characteristic period.

The results obtained by Sveen are in agreement with some of those reached by Gnezdilov (1972), who showed that the power

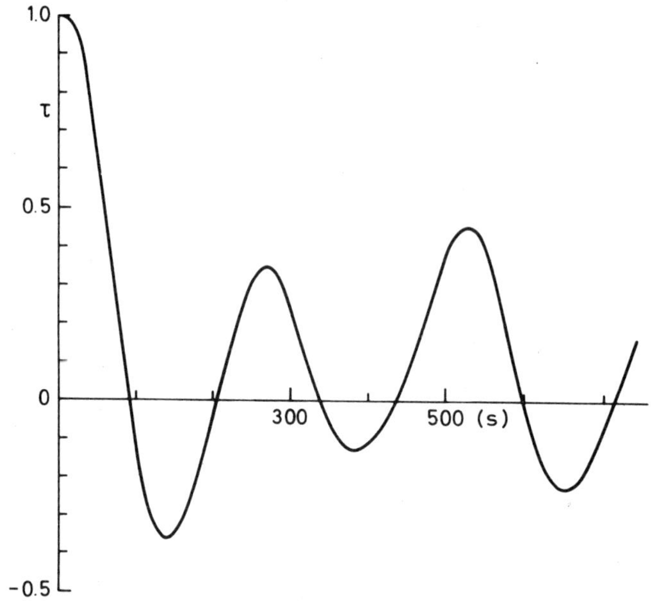

Fig. 7.1. Autocorrelation curve covering the time interval from 12 to 18 h UT on June 15, 1957. The period of the intensity variations is about 265 s or roughly 4.5 min. (Sveen, unpublished.)

spectra of noise storm fluctuations could take different forms. Some times they were smooth and suggested that the fluctuations had a noise character. In other cases maxima appeared against a noisy background. The behaviour of these maxima at different frequencies and in different time intervals suggested that the intensity variations were generated in the regions where the radiation at the different frequencies was emitted. For example, Gnezdilov (1974) investigated a storm occurring on September 19 - 20, 1966. Data were available at 111, 204, 234, 287 and 375 MHz. On September 19 quasi-periodic fluctuations were observed with periods of 1000, 470 and 290 s on a frequency of 111 MHz. They were not seen at higher frequencies. On September 20 the fluctuations had shifted to the frequency range 204 - 234 MHz with slightly different periods (697, 480 and 280 s). A new shift occurred on September 20 in connection with a very strong radio burst (spectral type IV) and the development of a second noise storm source.

In a noise storm observed between June 9 and 12, 1969, at 202 MHz, Gnezdilov (1974) found an oscillation period of 860 s, as can be seen from Fig. 7.2. The peak at 860 s stands out quite clearly above the noise level.

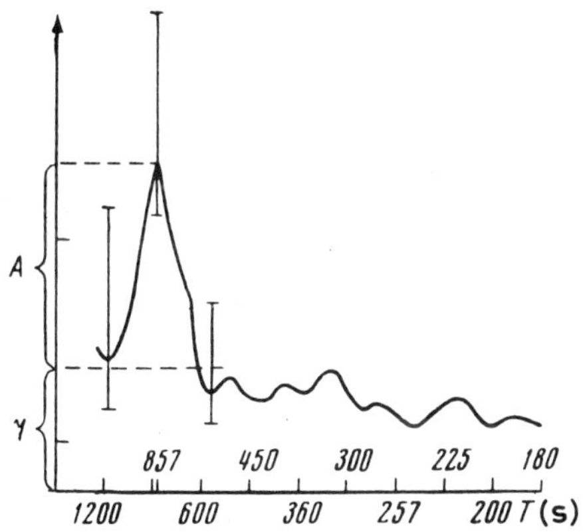

Fig. 7.2. Power spectrum of continuum fluctuations during the interval 4:30 to 7:51.5 UT on June 11, 1969 (Gnezdilov, 1974).

Periodic and quasi-periodic variations in the continuum intensity of noise storms certainly exist, but the phenomenon has, as yet, not been much explored.

7.2. RELATIONS BETWEEN CONTINUUM FLUCTUATIONS AND OTHER VARIABLE PHENOMENA

A possible relation between fluctuations in noise storm continuum and solar activity in the radio range was investigated by Gnezdilov (1974). Two strong and long-lasting noise storms observed at 202 MHz were selected for study. During the storms only one source region appeared on the solar disk. The storms occurred between June 9 - 12, 1969, and November 12 - 18, 1970.

The activity of the continuum fluctuations was described by the index

$$S = \sum_i A_i/2N_i \qquad (7.2\text{-}1)$$

where A_i is the value of the ith individual maximum in the power spectrum and N_i is the value of the "noise" component in the power spectrum. When $A_i/2N_i < 0.2$, Gnezdilov considered the maximum to belong to the "white noise" and it was neglected.

The burst activity in the radio range was characterized by the index

$$R = \sum_j (\Delta T - \Delta t_j)(b_j^1 + b_j^2 + b_j^3)/\Delta T \qquad (7.2\text{-}2)$$

In (7.2-2) ΔT is the time interval under consideration and Δt_j is the time lag of the jth burst relative to the starting point of the time interval ΔT. The symbols b_j^i represent numbers which are proportional to the logarithm of the maximum of the jth burst associated with the active centre above which the noise storm occurred. Three frequencies were chosen and b_j^1, b_j^2, b_j^3 were evaluated for 9500, 3100 and 650 MHz respectively, according to the following scheme:

b_j^i	Flux density 10^{-22} W m^{-2} Hz^{-1}
1	$10^0 - 10^1$
2	$10^1 - 10^2$
3	$10^2 - 10^3$
4	$10^3 - 10^4$

The quantity $(\Delta T - \Delta t_j)/\Delta T$ is a measure of the possible influence of the jth radio burst on the noise storm radiation in the time interval ΔT.

In Fig. 7.3 the results of the analysis of the noise storm of November 1970 are plotted. There is a close correspondence between the changes in the indices R and S from one day to another, but there is no correlation with the variation in radio

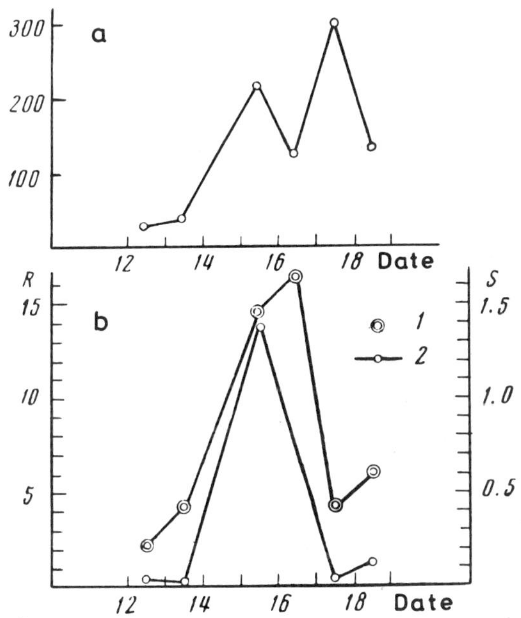

Fig. 7.3. Results from the noise storm of November 12-18, 1970: (a) radio flux at 202 MHz in units of 10^{-22} W m^{-2} Hz^{-1}; (b) activity indices (1) S for fluctuations and (2) R for radio bursts (Gnezdilov, 1974).

flux. The same result was found for both noise storms which were studied. The correlation between R and S is evident from Fig. 7.4.

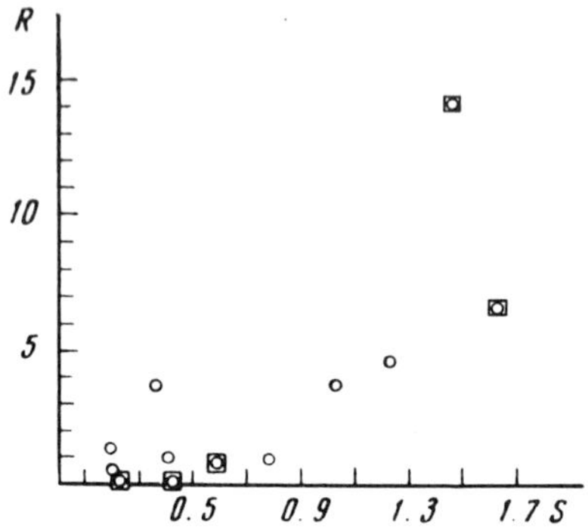

Fig. 7.4. Correlation between indices R and S: circles represent data from June 9-12, 1969; squares represent data from November 12-18, 1970 (Gnezdilov, 1974).

It is reasonable to take the evidence presented by Gnezdilov as an indication of a relationship between burst activity at high frequencies and fluctuations in the noise storm continuum. Different mechanisms can be thought of producing a relation of this type, but it seems necessary to undertake further investigations before definite proposals are made.

Variations in the background continuum of 34 noise storms observed at 208 MHz were compared with simultaneous observations of the magnetic field of sunspots by Eliseyev *et al.* (1974). Magnetic field determinations were made from observations of the FeI line at 6302.5 Å. It was found that periods of large fluctuations in radio flux of the noise storms were accompanied by similar variations in the magnetic field; indeed, there seemed to be good correlation between variations in radio flux and magnetic fields.

Further researches into the subjects treated in this section seem desirable because only few investigations have been made until now and there are reasons to believe that the fluctuations in the noise storm continuum are of great value for the interpretation and theoretical understanding of the noise storm phenomenon.

7.3. EXTREMELY RAPID, PERIODIC VARIATIONS

Rapid oscillations in the radio flux are usually associated with type IV bursts, but in some cases variations of the same nature have been found in noise storm emission.

Abrami (1971) made extensive observations with high time resolution at 237 MHz and detected periodic intensity variations in many cases. In two instances the periodic variations were associated with noise storm emission. The first case occurred on October 6, 1969, when two series of weak, regular oscillations were recorded lasting for 83 s and with periods 0.7 - 1.3 s. The second case was observed on December 26, 1969. The event lasted for 53 s and the period was 3.5 - 5 s.

Oscillations were also observed by Sastry (1969) at 25 MHz. On several occasions he found that enhanced radiation showed distinct periodicities. Periods between 3 and 8 s seemed to be characteristic for the variations.

The emission of bursts may assume a periodic character. Tarnstrom and Philip (1972) give examples in which spike bursts (flash bursts) were emitted quasi-periodically. But periodic emission of bursts does not seem to be a general property of noise storms.

An event of special interest was studied by Elgarøy and Sveen (1973). In the days around October 25, 1972, a noise storm was observed which extended even to the frequency region 500 - 550 MHz. In this band high-resolution spectral observations were undertaken. On October 25 the 225 MHz solar flux increased considerably, starting at 13:52 UT. This increase in flux was accompanied by an interesting change of the 500 - 550 MHz records: extremely short-lasting bursts of wide bandwidth and of narrow bandwidth appeared for about 1 h.

The wide band bursts had a bandwidth of the order of magnitude of 40 MHz. As measured at one frequency the average duration amounted to 0.09 s. The bursts occurred in "showers" lasting for about 1 or 2 min and were separated by quiet intervals of about the same length. Similar fluctuations in the rate of occurrence of bursts are known from observations of normal type I storms (see section 6.1).

Relatively often the burst activity assumed a periodic character, and successive bursts occurred at very regular time intervals. Two examples from the records are reproduced in Fig. 7.5.

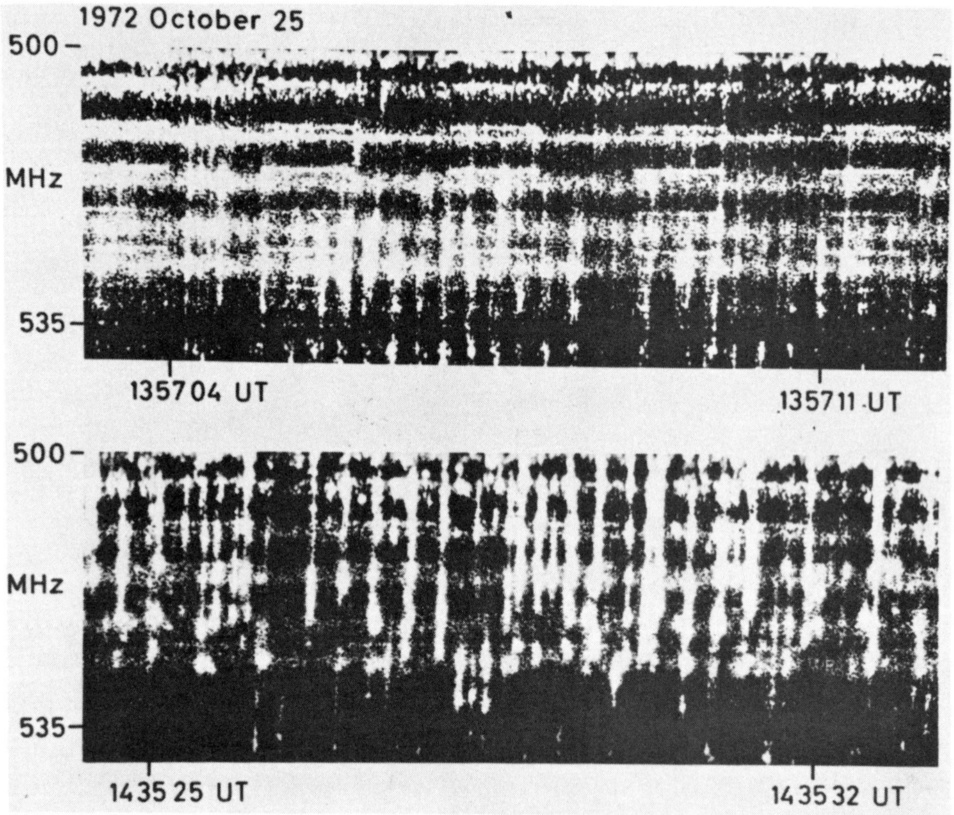

Fig. 7.5. Periodic and quasi-periodic burst emission. Two different periods occur in the upper part of the figure. At 13:57:04 UT the period is about 0.14 s. At 13:57:06 UT the period is around 0.27 s.
(Elgarøy and Sveen, 1973.)

The most surprising feature of the 500 - 550 MHz records was the large number of bursts with the same duration as the wide band bursts, but with a bandwidth which was restricted to a mere 1 - 2 MHz. Occasionally the narrow band bursts formed small chains with a well-defined separation between individual members. The period was determined in 15 cases, and the same periodicity was found as for the wide band bursts. This is shown in Fig. 7.6.

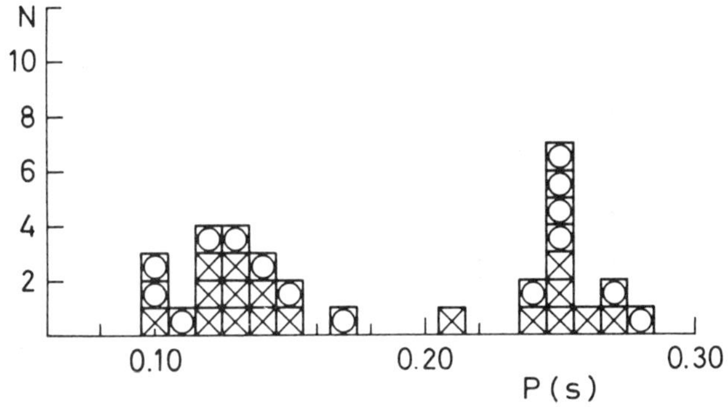

Fig. 7.6. Distribution of periods for wide band bursts x and for bursts of extremely narrow bandwidth o (Elgarøy and Sveen, 1973).

It appears from Fig. 7.6 that the time interval between neighbouring bursts (the period *P*) takes different values, ranging from 0.1 s to about 0.3 s. There is a grouping of periods around two values of *P* which are harmonically related, but one should not attach too much significance to this point before further evidence has accumulated.

The periodic and quasi-periodic phenomena described in the present chapter are probably intimately connected with the generating mechanism of the radiation and should be taken into account when it comes to a theoretical interpretation of the noise storm radiation.

8. METRIC NOISE STORMS AND RELATED PHENOMENA

8.1. METRIC AND DECAMETRIC NOISE STORMS

Traditionally noise storms, which are frequently also denoted type I storms, have been regarded as a metre wave phenomenon. The spectral information presented in section 2.1 together with Fig. 6.5, which shows the distribution of type I chains with frequency, substantiates this view. However, in order to arrive at a better understanding of the noise storm phenomenon one should take into account the characteristics of radiation also emitted at other frequencies than those occupied by the typical noise storm. In this connection it seems appropriate to use the denotation decametric noise storms, even though the latter differ in several respects from metric storms.

At metric wavelengths noise storms generally consist of enhanced background continuum emission with short-lasting bursts of narrow bandwidth superposed. The radiation is usually highly circularly polarized. Some times one may talk about storms of type III bursts (Wild, 1957) and storms of fast drifting type I bursts (Elgarøy, 1961). Different metric storms may show notable variations from one case to another, but in the majority of the cases they possess several quite characteristic features which make them recognizable as noise storms.

When observations are made at frequencies lower than 100 MHz, one finds that there normally appears to be a region around 30 - 40 MHz in which a change of character of the radiation takes place. This change seems to justify the use of the notion decametric noise storms.

From observations in the frequency range 25 - 100 MHz Ellis (1969) found the following types of bursts:

(1) fast drift storm bursts with an average frequency drift velocity of 1.9 MHz s^{-1}, an average bandwidth of 0.03 MHz and an average duration of 0.6 s;

(2) drift pair bursts with $df/dt = 1.2$ MHz s^{-1} and average bandwidth 0.45 MHz (cf. section 6.3);

(3) split pair bursts with $df/dt = 0.08$ MHz s^{-1}, average bandwidth 0.05 MHz and average duration 1.4 s (cf. section 6.4);

(4) chains of split pair bursts.

The above types of bursts represent fine structure in the decametric range. In addition to these types one has the flare associated type III bursts, type V bursts, type II bursts and type IV bursts. But these are of less interest for the comparison with metric storm activity.

It was noted by de la Noë and Boischot (1972) that numerous single, narrow-banded bursts of brief duration identical to one element of a split pair occurred. Such bursts were called "stria" bursts.

Stria, split pair and triplets have a pronounced tendency to be grouped together in chains which drift from high to low frequencies in the same way as type III bursts. Such chains were denoted type IIIb bursts by de la Noë and Boischot (1972). Type IIIb bursts frequently precede normal type III bursts. One example is shown in Fig. 8.1. This might indicate a harmonic relation between type IIIb and type III bursts. But some spectral characteristics suggest that the type IIIb burst is more likely a precursor of the type III burst, and the precursor hypothesis was favoured by de la Noë and Boischot (1972).

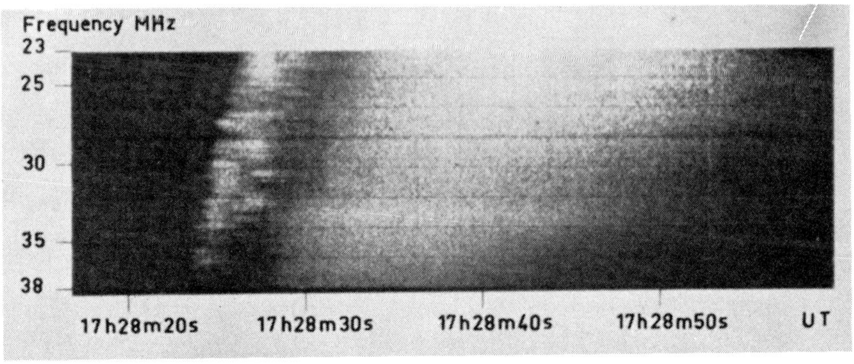

Fig. 8.1. Type IIIb burst followed by a strong type III burst on August 2, 1970 (de la Noë and Boischot, 1972).

According to Stewart (1975) it is more likely that type IIIb/type III pairs are fundamental/second harmonic type III pairs with the striae confined to the fundamental component. One quite clear example of this was presented by Stewart. A variant of the type III bursts, named inverted-U burst, was observed with intense striae in the fundamental and no striae in the second harmonic. Takakura and Yousef (1975) found that in most cases the second harmonic of a type IIIb burst seemed to be a normal type III burst, but they also mention that more rarely type IIIb bursts occurred as a fundamental and second harmonic pair. In such cases there was evidence that the striae were repeated in the harmonic. Mosier and Fainberg (1975) found from observations of high resolution and high sensitivity that narrow-banded bursts as occur in type IIIb, were not confined to a precursor relationship.

Chains of striae bursts were scrutinized by Baselyan *et al.* (1974b) who observed with high sensitivity and resolution at frequencies from 12.5 to 25 MHz. They made a distinction between ordinary stria bursts and diffusive stria bursts. The diffusive stria bursts had essentially longer durations than the ordinary type. Diffusive bursts also formed chains, denoted type IIId by Baselyan *et al.* (1974b). Both type IIIb and type IIId bursts might show echo-type features. The rate of occurrence of type IIId bursts was about five times less than type IIIb bursts.

Baselyan *et al.* studied 39 double, normal type III bursts which showed a harmonic relationship. A number of 75 type IIIb bursts preceding a normal type III burst were then analysed, and it was found that the type IIIb corresponded to emission at the fundamental frequency, whereas the following type III bursts occurred at the second harmonic. As concerns type IIId bursts, these were found to follow normal type III bursts, and in seven cases for which good observations existed, it was found that the normal type III burst occurred at the fundamental frequency and the type IIId burst was the second harmonic.

The average frequency drift velocity in the band from 12.5 to 25 MHz was determined by Baselyan *et al.* by measuring the delay in the maximum of the radio emission at the lower frequency as compared with the higher one. Only bursts on the fundamental frequency were investigated. The results are shown in Fig. 8.2 and demonstrate that type IIIb bursts have a faster frequency drift rate than ordinary type III bursts. The average drift rate of type III bursts was 1 MHz s^{-1}, whereas for type IIIb bursts it was 2 MHz s^{-1} with a somewhat larger spread in the values. Accordingly, type IIIb bursts are excited by faster disturbances than type III bursts. This is in disagreement with the results of de la Noë and Boischot (1972), but appear to be in accord with the data of Ellis (1969). The difference in results between Baselyan *et al.* and de la Noë and Boischot may possibly be attributed to different methods of measurements. Whereas de la Noë and Boischot used the envelope of the leading edge of the type III and type IIIb chains, Baselyan *et al.* referred the measurements to the time of occurrence of maximum

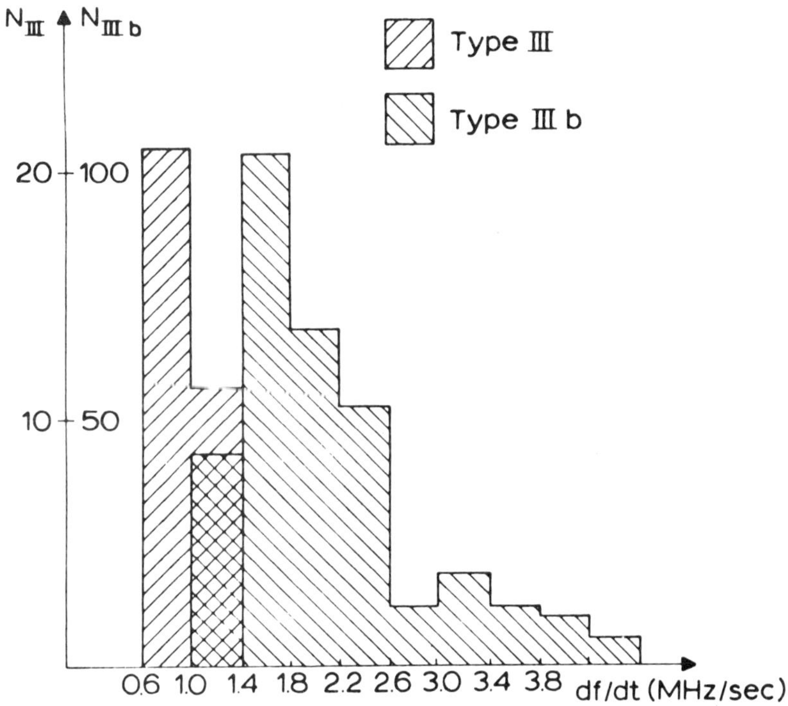

Fig. 8.2. Dependence of the number of bursts of type III and type IIIb on the drift rate for the fundamental component (Baselyan *et al.*, 1974b).

intensity. Baselyan *et al.* suggested that type IIIb and type IIId bursts could be considered as a fine structure of the ordinary type III bursts at the fundamental and at the second harmonic of the plasma frequency respectively. The pecularities in the plasma wave transformation leading to fine structure of type III bursts then should begin at stream velocities about twice the velocities of streams leading to ordinary type III bursts.

New observations remain to be undertaken in order to clear up the somewhat conflicting results as regards decametric chains of bursts. However, the existing knowledge is sufficient for a discussion of the relationship between metric and decametric noise storms.

The relation between metric and decametric noise storms was discussed by Boischot *et al.* (1970). It was found that type I storms at metre-wavelengths in general (i.e. in roughly 90% of the cases) were accompanied by decametric type III storms, and the suggestion was made that the two types of storms were generated by the same energetic electrons streaming outwards from the photosphere or chromosphere.

There is little doubt that the metric type I storm is accompanied by a decametric storm, but the nature of the decametric counterpart is not always the same. The problem was considered in more detail by Möller-Pedersen (1974). It was fond that type I storms are infrequent in the decametric range. Most frequently type IIIb and type III bursts occur at frequencies below those occupied by the type I storm. In extreme cases storms consisting of type IIIb bursts only, or type III bursts only, are associated with the metric noise storm, but in many decametric storms type IIIb and type III bursts occur together. In cases of "mixed" activity, the proportion of type III and type IIIb bursts remains roughly constant throughout one day. Also during some type III storms drift pairs and single drifting bursts may be observed together with fast drift storm bursts and bursts having the appearance of a hook on spectral records.

Möller-Pedersen searched for a possible centre-limb effect in the characteristics of decametric storms. Because of the influence of the ionosphere, position measurements are difficult at long wavelengths. Möller-Pedersen therefore used the position of metric storms determined at 169 MHz and assumed that the position of the simultaneous decametric storm was not too far away from this position. Then the parameter p was determined:

$$p = \frac{\text{number of type III}}{\text{number of type III + type IIIb}}.$$

The results for 39 days of observations are shown in Fig. 8.3.

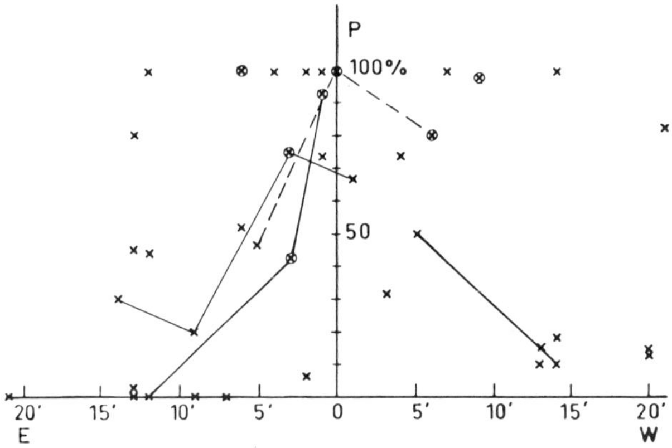

Fig. 8.3. Centre-limb variation of the composition of decametric storms as represented by the parameter p. Interconnected points indicate days of observation with intervals of one or two days. ⊗ indicate storms containing drift pair bursts. (Möller-Pedersen, 1974.)

It is seen from Fig. 8.3 that type IIIb storms are frequently associated with noise storm centres relatively far from the central meridian and that type III storms with or without drift pair bursts correspond to metric storms near the central meridian. The effect comes clearly out in Fig. 8.4.

According to these statistical results the average behaviour of a decametric noise storm should be to start near the eastern limb as a pure type IIIb storm. Then type III bursts appear, and near the central meridian almost only type III bursts occur together with drift pairs and some fast drift storm bursts. The process reverses after central meridian passage.

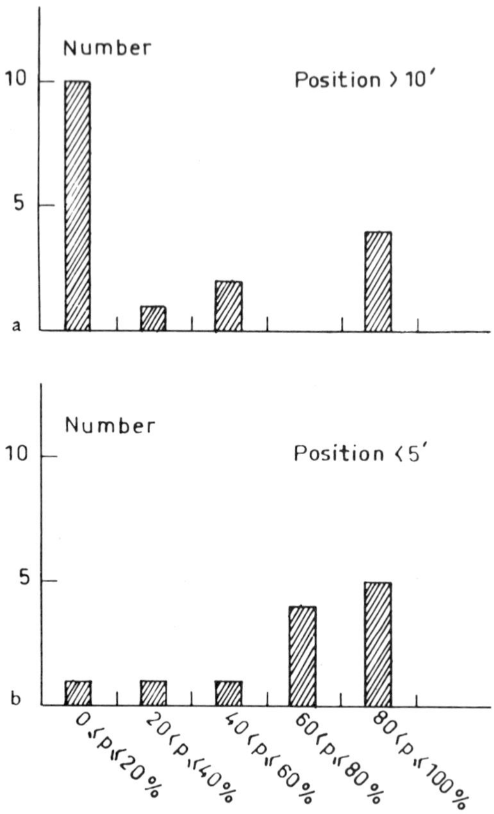

Fig. 8.4. Histograms of the distribution of the composition of storms as given by the parameter p for storms with metric positions far away from the central meridian *(above)* and close to the central meridian *(below)* (Möller-Pedersen, 1974).

A possible interpretation of these results may be, as proposed by Möller-Pedersen, that all types of decametric bursts are excited simultaneously, but are emitted in different directions. Type IIIb bursts are observed at intermediate or large longitudes and are radiated only at large angles to the radial direction. Type III storm bursts are confined to moderate distances from the centre of the solar disk and are accordingly emitted in a cone which is radially oriented.

Drift pairs are more strictly confined to longitudes near the central meridian and are more strongly beamed in the radial direction.

The results described above lead to the conclusion that type I storms at metric wavelengths are very closely connected with noise storms in the decametric wavelength region and that the latter storms may have somewhat varying characteristics depending on the distance of the source from the central meridian. Then, either the same electrons which produce the type I storm also generate the decametric bursts higher up in the corona, which would imply that particles of velocity about 0.3 times the velocity of light are responsible for both phenomena. Or, particles with relatively low energy may generate the type I storm, but the conditions in the type I region are suitable for further acceleration to energies required for the production of type III bursts. In this connection one may mention the phenomenon of weak type III bursts growing out of the type I chains. Evidence in the same direction is provided by observations of type I bursts accompanied by very tiny type III bursts (Weber, 1976, private communication).

There is no reason to think that the storm phenomenon is limited to metric and decametric wavelengths. Observations at hectometric wavelengths by Fainberg and Stone (1970a) by means of equipment aboard the RAE-1 satellite revealed storms of type III bursts persisting for more than half a solar rotation. The frequency range of observation was from 5 to 0.2 MHz corresponding to heights of 10 - 100 solar radii. The type III storms contained bursts which were less intense and occurred over a more limited frequency range than isolated type III bursts. The rate of occurrence of type III storm bursts reached a maximum at central meridian passage of the source. Then a burst occurred on the average every 10 s. The frequency drift velocity corresponded to an exciter speed of 0.38 c, which is in good agreement with the exciter speed of 0.33 c at distances closer to the sun (Fainberg and Stone, 1970b). Comparison of the hectometric storms with simultaneous decametric observations leaves little doubt that storms in the two regions are closely related.

In a general sense one may now consider noise storms to cover phenomena occurring in regions at small heights above the solar surface and extending out to distances of at least 100 $R_\odot$ and probably larger still.

8.2. TYPE I STORM ACTIVITY AND EMISSION AT SHORT WAVELENGTHS

On the high-frequency side noise storms may extend to frequencies of 400 - 500 MHz, i.e. into the decimetric wavelength range. But at centimetric wavelengths noise storms are absent. The dominating variable activity in this region is the slowly varying component which is closely related to sunspot groups and plages. The slowly varying component accordingly shows a 27-day periodic variation.

The question of an evolutionary relation between type I storms and the slowly varying component at high frequencies was investigated by Kai and Sekiguchi (1973). Positions and flux densities of type I storms were determined with an interferometer working at 160 MHz. During 2 years, from 1970 to 1972, about 120 storms were observed. For the slowly varying component (*S*-component) positions and intensities were available at frequencies of 9400 MHz (3 cm), 3750 MHz (8 cm), 3297 MHz (9.1 cm), 1429 MHz (21 cm) and 698 MHz (43 cm).

Observed sources of the *S*-component were considered to be associated with type I sources if their positions were within ±5′ arc from the storm centre in the east - west direction. Since type I storms are directive, only storms which were recorded in a time interval of 4 days before and 3 days after central meridian passage were considered.

One typical example from the material is reproduced in Fig. 8.5. The figure shows that there is a rapid increase in flux density of the *S*-component at 3 and 8 cm which is followed by a strong maximum in the 160 MHz storm activity about 2 days later. This behaviour was present in 19 out of 21 cases in which a type I storm developed on the disk and was associated with the 3 cm *S*-component. The average delay between maximum in the *S*-component at centimetric wavelengths and type I storm maximum activity was 1 - 2 days.

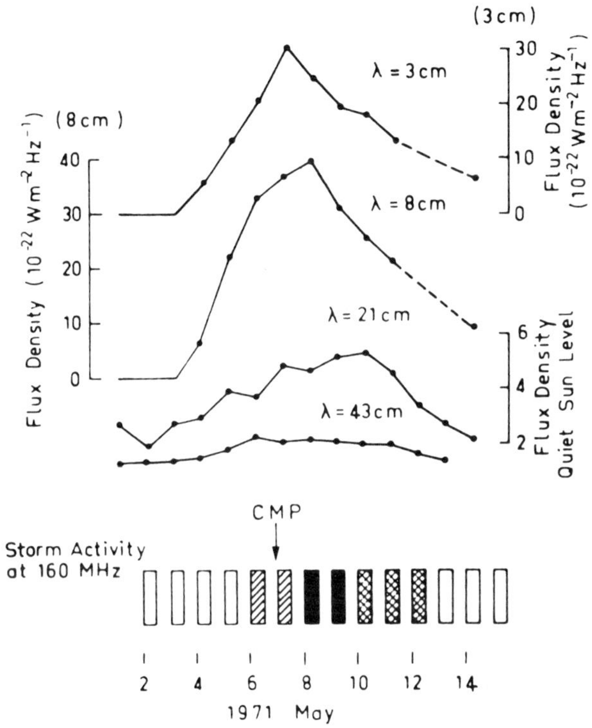

Fig. 8.5. Average daily activity at 160 MHz of a type I storm and flux densities of the *S*-component associated with the storm. Central meridian passage of the associated plage region is indicated by CMP. The 160 MHz storm activity is represented as weak (hatched), moderate (cross-hatched) and strong (black). Open rectangles indicate no storm activity. (Kai and Sekiguchi, 1973.)

Further studies of the material showed that there was a very good correlation between the storm activity and the flux density of the centimetric *S*-component. A threshold effect seemed to be present; *S*-components with 3 cm flux smaller than 7-8 $\times 10^{-22}$ W m^{-2} Hz^{-1} were usually not associated with 160 MHz storm activity. These results clearly emerge from the diagram in Fig. 8.6.

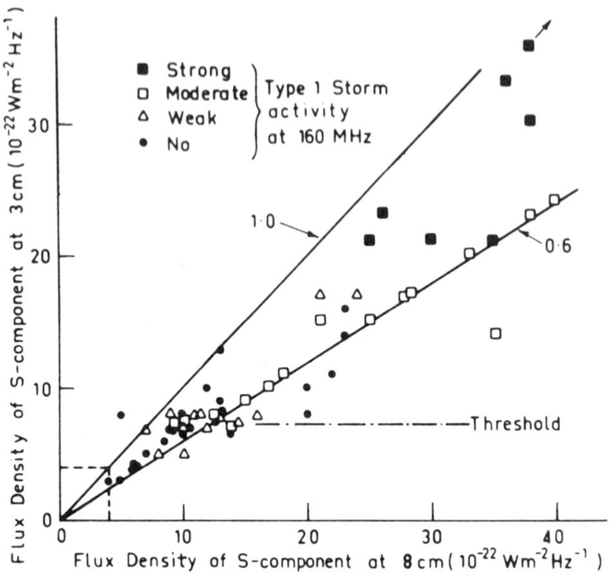

Fig. 8.6. Type I storm activity at 160 MHz plotted in a coordinate system of 3 cm flux density versus 8 cm density of the *S*-component. The two solid lines are plotted for ratios of (3 cm flux)/(8 cm flux) equal to 1 and 0.6. (Kai and Sekiguchi, 1973.)

Whereas proton events are associated only with *S*-components showing a large ratio of flux density at 3 cm to flux density at 8 cm, storm activity is found for ratios between 0.6 and 0.9.

The slowly varying component has been explained as thermal radiation from coronal condensations, and the spectrum shows that above a frequency of about 5000 MHz the sources are optically thin. Accordingly the flux density at 3 cm depends on the optical depth τ. Kakinuma and Swarup (1962) found that τ is dominated by gyro resonance absorption at the second and third harmonics of the gyro frequency. Therefore τ depends mainly on the magnetic field strength, and the variations in the magnetic field in the condensation will be reflected in the 3 cm flux densities. A field of about 1000 G is required to make $\tau = 1$ at 3 cm assuming an electron density of 10^9 cm^{-3}, an electron temperature of 10^6 K and a scale height of 10^4 km. Kai and

Sekiguchi then explain their observational results in the following way: strong magnetic fields in the corona are necessary for the generation of type I storms. The sharp increase in 3 cm flux of the *S*-component followed by strong storm activity after a delay of about 2 days shows that increase in the field strength first occurs in the lower corona where the condensation forms, probably because of newly emerging magnetic flux. Then the intense magnetic fields penetrate progressively into higher layers. When they reach the 160 MHz plasma level they start to activate the type I storm. The upward movement of the magnetic field may be accompanied by magneto-hydrodynamic disturbances which accelerate electrons. This mechanism implies a velocity of the magnetic field movement of 1 - 2 km s^{-1}. The occurrence described here may be a necessary but not sufficient condition for the generation of noise storm radiation.

8.3. TYPE I STORMS AND STATIONARY TYPE IV BURSTS

Bursts connected with large solar flares may have a complex spectrum. Boischot (1958) isolated a special type of flare-induced burst which he called type IV burst. The burst showed an outward movement in the corona at a velocity of the order of 1000 km s^{-1}, it covered a wide band of frequencies and all frequencies appeared to come from the same position. Little fine structure was present, and the radiation was weakly circularly polarized in the extraordinary sense.

Further observations have shown that moving type IV bursts are infrequent, although some very spectacular examples have been recorded. They belong to the first phase of a complete type IV event.

One may roughly make a distinction between two phases in the evolution of type IV events (Pick, 1961). In the first phase radiation occurs at centimetric and decimetric waves and may reach the metre-wave region. The duration of this phase rarely exceeds a quarter of an hour. In the second phase, which lasts about one or several hours, the emission occurs essentially at metric wavelenghts. The continuum does usually not extend to centimetre waves. Perhaps the characteristic that best separates the two phases is the directivity of the emission. The first phase is not directive, whereas radiation in the second phase is rather strongly directive in nature. In the first phase the emission is weakly circularly polarized in the extraordinary mode, but in the second phase the waves are circularly polarized in the ordinary mode. During the first phase the radiation may be very rich in fine structure which is difficult to detect because of the weak contrast with the underlying continuum (a review of this fine structure was given by Slottje, 1972).

Many type IV events show prolonged second phase emission which develops into a continuum storm lasting for days and which may gradually change in character and transfer into an ordinary noise storm. The continuum storm is highly circularly polarized in the ordinary mode, strongly directive, but with few type I bursts and usually poor in fine structure, although notable exceptions regarding the last point have been observed at the Oslo Solar Observatory (Elgarøy, unpublished).

Observations of the spatial and polarization structure (Kai, 1970) of the stationary part of type IV bursts at 80 MHz revealed no features which could indicate a distinction between this type of emission and ordinary noise storms. Bipolar structure was found in stationary type IV bursts as well as noise storms, and the source sizes and the separation between the centroids of the oppositely polarized components were the same in the two cases (source size: $6' \times 6'$ arc, separation between R-L components $3'$ arc).

A very nice example illustrating several characteristics of type IV bursts was described by Wild (1969b). Radioheliograph observations at 80 MHz revealed a first phase of the event, the moving part, as an expanding magnetic arch beyond the solar limb. Three concentrated sources developed along the arch: one unpolarized, and two polarized sources of opposite polarity near the feet, attributed to plasma radiation. The second, stationary, part was seen later as a separate highly polarized source on the disk above the projected position of the flare that had previously triggered the expansion of the arch. The source was typical of the stationary type IV event and was spatially independent from the sources that formed the first part. Evidence of bipolar structure of the stationary source was present, but the accompanying source of opposite polarity was sporadic in nature and generally much weaker.

Type IV events may be quite complicated. Stewart and Sheridan (1971) described a case in which six sources were present, and they concluded that the radiation came from two expanding and one stationary magnetic arch.

It is generally supposed that the stationary sources of type IV events are due to plasma radiation generated by electron streams (Krüger, 1972). Since the radiation occurs at the plasma level, the emission will be directive in nature. Pick (1961) demonstrated quite clearly that type IV emission and type I storms in general were associated with the same centres of solar activity.

Except for the type I bursts, there are only small differences between type I storms and continuum storms. It therefore seems that continuum storms and the background continuum of type I storms are closely related as regards the physics of the two phenomena.

8.4. RELATION BETWEEN TYPE I STORMS AND TYPE III BURSTS

There are large differences between type I storms and bursts of spectral type III as concerns spectra, duration, polarization and several other properties. Type III bursts are excited by disturbances which travel at a speed of approximately 10^5 km s^{-1} out through the corona. The small degree of polarization of type III bursts show that they are generated in regions of weak magnetic fields, whereas type I radiation is highly polarized and arrives from regions of strong magnetic fields.

However, there are indications that the two phenomena may be related in some way. Type III bursts are at times observed to grow out from type I chains, and Malville (1961) found that noise storm activity in general used to occur at the upper frequency limit of type III bursts. Further to this, type I storms are closely connected with type III storms at decametric and hectometric wavelengths, and there seems to be a continuous transition of burst types from pure type I to pure type III.

Kai (1970) was able to define the relationship between the two types of radiation somewhat more precisely. Investigating the spatial position of the sources and the polarization of type I and type III radiation at 80 MHz it was found that the type III sources were unipolar and avoided the location of the type I storm centres. The latter were frequently bipolar. Kai concluded that type I emission originates in strong magnetic fields of opposite polarity, probably fields which close near sunspots. Type III bursts are generated in regions with weak magnetic fields, possibly in neutral sheets as marked by crosses in Fig. 3.5.

If it is assumed that type I emission as well as type III bursts are generated by plasma waves, the above conclusion may be substantiated in a simple way (Kai, 1970). Since type III bursts show a small degree of polarization, both modes probably escape from the source region. The following condition then must be satisfied:

$$\frac{\omega_H}{\omega} < 1 - \frac{\omega_p^2}{\omega^2}, \tag{8.4-1}$$

where ω, ω_H, ω_p represent the angular frequency, gyro-frequency and plasma frequency respectively. The dispersion equation for longitudinal plasma waves may be written ($\omega_H^2 \ll \omega_p^2$)

$$\omega^2 = \omega_p^2 + k^2 v_t^2, \tag{8.4-2}$$

where k is the wave number and $v_t = (3\kappa T/m)^{\frac{1}{2}}$ is the r.m.s. thermal velocity of the coronal electrons. Combining (8.4-1) and (8.4-2) and using $v_0 \simeq \omega/k$, one obtains

$$\frac{\omega_H}{\omega} < \frac{v_t^2}{v_0^2} \tag{8.4-3}$$

Assuming a stream velocity v_0 of 10^5 km s^{-1} and a temperature of

10^6 K, equation (8.4-3) leads to $B < 0.15$ G when the frequency is 80 MHz.

If it is assumed that the strong circular polarization of type I emission in the ordinary mode is caused by a propagation effect, the source must be below the stop region and one must have

$$\frac{\omega_H}{\omega} > 1 - \frac{\omega_p^2}{\omega^2}. \qquad (8.4\text{-}4)$$

In combination with (8.4-2) and using a value of v_0 of 2×10^4 km s^{-1} (Takakura, 1963), one obtains $B > 3.5$ G. It is seen that the field strengths in the source regions of type I and type III emissions may differ by at least one order of magnitude.

A very clear demonstration of the relationship between type I storms at metric wavelengths and type III storms at decametric and hectometric wavelengths was provided by observations undertaken in the time period August 17 - 22, 1968 (Stewart and Labrum, 1972). A strong type I storm was observed at frequencies mainly above 75 MHz, but extending as low as 50 MHz at times. On August 20 the spectrum showed a continuous succession of type III bursts starting in the frequency range 25 - 80 MHz, usually below 60 MHz. There were indications of type III storms also on August 18 and 19. The bursts then appeared to start at frequencies below 25 MHz. Combining radioheliograph observations at 80 MHz with radio spectral data and optical data, Stewart and Labrum derived the model of the source region depicted in Fig. 8.7.

The association between type I and type III activity could best be explained by assuming that the energetic electrons which excited the type III bursts were accelerated near the observed position of the 80 MHz source (marked by a star in Fig. 8.7). Stewart and Labrum suggested that magneto-hydrodynamic waves travelled along strong fields of a bipolar sunspot group. If they were sufficiently energetic they might reach the top of a closed-loop structure and trigger a pinch instability at the cusp of a helmet magnetic structure. The model in Fig. 8.7 provides a simple geometrical explanation of the observed frequency and

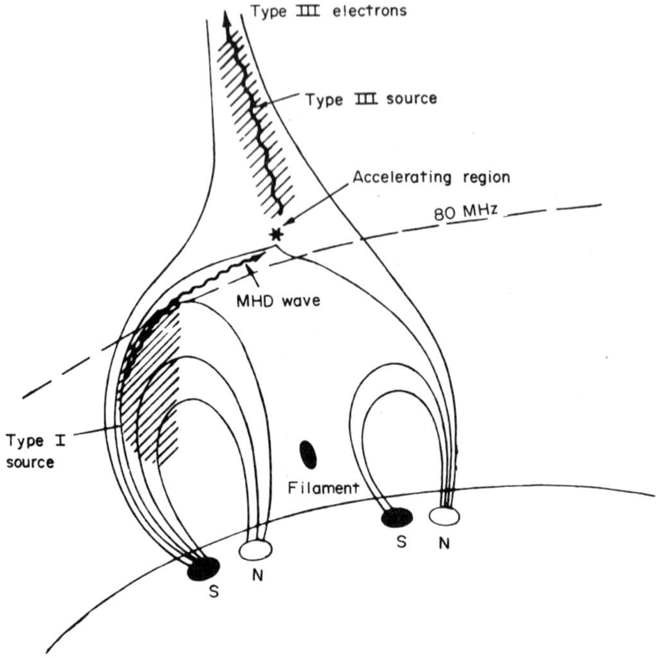

Fig. 8.7. Proposed model for type I and type III sources (Stewart and Labrum, 1972).

spatial relationship between associated type I and type III storm activity. The type III emission originates at the plasma level in a region of weak magnetic field lines. The type I source is at the plasma level in a region of strong, closed magnetic fields. It is seen that the type III source is above a stable dark filament. Such filaments are associated with inversion lines of the longitudinal photospheric magnetid field (Howard, 1959) and indicate the presence of a neutral magnetic plane higher in the corona. Coronal structures in which type III bursts are excited show a good correlation with plage filaments (Mercier, 1973). From the geometry it follows that the disturbances generating the type III bursts are induced at a level which is near or slightly above the height from which the type I emission originates. The type III bursts should accordingly have starting frequencies below, or at the lower frequency limit of the concurrent type I storm.

One remark is necessary in connection with the model in Fig. 8.7. It is generally assumed that electron clouds generate type III bursts. But such particle clouds seem not to be able to travel in the central part of a streamer. Analysing how electrons travel in an active region streamer which contains a current sheet according to the model of Pneuman (1972), it was found by Smith and Pneuman (1972) that the electrons could not move radially across the transverse magnetic field which arises in the current sheet, nor drift out of the sheet with a sufficient velocity to explain type III bursts. The most likely place for the electrons to travel is therefore in a region somewhat distant from the current sheet. Even though the Pneuman model is known to be wrong (Weber and Rosenberg, 1974), it still appears that the electrons must travel near the edges rather than in the centre of streamers. It is fully possible to modify the model in Fig. 8.7 in accordance with this.

More complicated relationships may exist between type I storms and type III bursts, in particular when the latter are of the impulsive type occurring in close connection with flares (Sheridan *et al.*, 1973; Kai and Sheridan, 1974; Kai and Nakajima, 1974). But there is no doubt that the connections which do exist (or do not exist) are of value for an understanding of the type I phenomenon. For instance, Gordon (1970) proposed a mechanism in which the acceleration of the electrons responsible for type III bursts took place in the type I source region. Since radioheliograph observations showed that the type I sources and the type III sources may be at distances of 0.5 $R_\odot$ or larger, the mechanism proposed by Gordon appears less likely.

8.5. METRIC TYPE I STORMS AND ENERGETIC ELECTRONS IN THE INTERPLANETARY SPACE

It is known that type III bursts are correlated with electron streams with particle energies in the range of 10 - 100 keV (Lin, 1970; Palmer and Lin, 1972; Lin *et al.*, 1973).

Lin and Anderson (1967) and Lin (1970) divided the energetic electron events observed near the earth in two classes: simple and complex. The complex events originated in solar active regions with intense type I radio noise activity.

Sakurai (1971a) reconsidered the classification of energetic electron events proposed by Lin and Anderson and found that the distinction between the two types was most likely due to the strong directivity of type I emission; complex events occurred in connection with flares near the central part of the solar disk, simple events were associated with flares at large distances from the central meridian where the chances for observing type I emission are less good. One might therefore conclude that all flares which produced energetic electron events were generated within active regions characterized by intense type I activity.

An explanation of this finding was outlined by Sakurai. In the type I source region beams of electrons with a velocity at least twice the velocity of the ambient electrons are generated continuously (Takakura, 1963), and the region may be considered as a reservoir containing an enormous number of energetic electrons. Suprathermal electrons may be a prerequisite for rapid acceleration of electrons to energies of 10 - 100 keV in small flares (importance between 1F and 1B). Some of the electrons accelerated in the flares are ejected into outer space, exciting type III emission during the flight through the corona, and are observed as distinct electron events near the earth. Some of the flare electrons may be injected into regions of higher density and strong magnetic fields and generate microwave impulsive bursts and X-rays. Also the type I emission may be intensified due to the supply of accelerated energetic electrons.

Weak solar flares of importance $\leq SB$ do not generate electron beams which are observable as distinct electron events at the earth's orbit (Sakurai, 1971b), but when they frequently occur in a type I noise active region some of the suprathermal electrons may be further accelerated to speeds higher than 0.1 c. These may be responsible for decametric and hectometric type III storms and they are also observed as an increase in the background flux of energetic electrons.

The essential facts in the relation between type I radiation and energetic electrons in the earth's neighbourhood appear to be: type I storms are strongly correlated with type III storms at decametric wavelengths. Decametric type III storms reveal the presence of a channel of escape of energetic electrons into space. These electrons have been observed. Whether the same electrons produce the type I storm and the type III storm (Boischot *et al.*, 1970), or magnetohydrodynamic waves from the type I source trigger a pinch instability at the cusp of a helmet structure (Stewart and Labrum, 1972), or energetic noise storm electrons are further accelerated in a flare (Sakurai, 1970a, b) or some other mechanism may explain the observations, is a matter for discussion.

8.6. ACTIVE REGIONS AND INTERPLANETARY SECTOR STRUCTURE

Observations with satellites and space probes have revealed the existence of an interplanetary magnetic field which is organized in broad and relatively stable sectors. Within a sector the polarity of the magnetic field is about constant. It is generally accepted that the interplanetary magnetic field is of solar origin, but the opinions have differed as to the mechanism behind the sector structure. The question is whether the interplanetary field arises from a mean photospheric field which is organized in photospheric sectors of the same polarity or if it comes from a more limited region of the solar atmosphere.

Martres *et al.* (1970) correlated radio observations of the solar corona made at metric and centimetric wavelengths with sector structures observed for two periods in 1964 and 1966 (Wilcox and Ness, 1965; Ness and Wilcox, 1967). The first period corresponds to the solar minimum period while the second corresponds to the rise of a new solar activity cycle. Radio observations were available at 169, 408 and 9300 MHz. It was then found that the meridian passage of sector boundaries always coincided with the crossing of a coronal condensation observable at metre wavelengths. When the transit time from the sun to the earth, amounting to 4 days, was taken into account, it was concluded that the boundary of the sector was always located to the west of the coronal condensation. Furthermore, it appeared that the coronal condensations corresponded to recurrences of faculae or "memories" (memory denotes coronal condensations for which no visible phenomena were observed at the disk during the same solar rotation). Coronal memories may be interpreted as streamers (Axisa *et al.*, 1971). Strong eruptive centres were never observed near the sector boundaries, but their presence influenced the longitudinal extent of the interplanetary sectors. Active centres enlarged the sectors and displaced the eastern boundary of the sector.

Sakurai and Stone (1971) studied a period in 1968 using radio data at 169 and 408 MHz in combination with data on the interplanetary magnetic field supplied by Ness and Fairfield. It was found that sector boundaries observed during the period which was studied were connected with type I noise active regions. The number of such regions were four or five per solar rotation and there seemed to be a one-to-one correspondence between these active regions and the sector boundaries. A model that may account for the observations is reproduced in Fig. 8.8. The photospheric magnetic fields connected with the origin of the sector boundaries seem to play an essential role in the formation of type I noise active regions. In view of this Sakurai and Stone suggested that superthermal electrons released from such regions might stream outward along the associated sector boundary and give rise to hectometric solar radio emission in the interplanetary medium. Thus it appears that an open magnetic structure above the active centre is a condition that has to be satisfied for the generation of type I activity to occur.

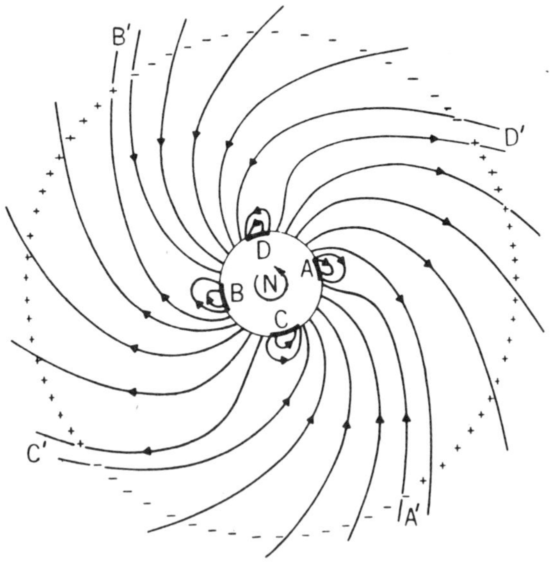

Fig. 8.8. A model of the relationship between type I noise active regions and the sector structure of the interplanetary magnetic field. *A*, *B*, *C* and *D* indicate type I noise active regions. *A'*, *B'*, *C'* and *D'* give the sector boundaries. (Sakurai and Stone, 1971.)

Both the results of Martres *et al.* (1970) and those of Sakurai and Stone (1971) are compatible with the hypothesis that the interplanetary sector structure is associated with a limited region in the solar atmosphere (nozzle hypothesis). But sector structure is not only associated with type I noise active regions. The investigations performed during periods of low solar activity demonstrate that one also has a quiet period sector structure which is correlated with a coronal memory (streamer), which indicates the presence of open magnetic field lines.

Martres *et al.* (1970) proposed that radial magnetic fields emanate from the stable coronal condensation observed after sector boundaries. The radial expansion of the field originating from the condensation is able to occupy a sector of the interplanetary space and maintain a flow of gas – the solar wind.

An increasing number of eruptive centres enlarges the size of the interplanetary sectors. The explanation may be as proposed by Martres *et al.* (Fig. 8.9). Magnetic field lines form loops above the active centre. Assume that one loop extends higher than the Alfven surface and is annihilated by the mechanism described by Petschek (1966). The pressure ahead of loop 2 then decreases and it is convected outwards. A series of field line reconnections between lines from the active centre and those from the condensation successively facilitate the convection of a large number of field lines into interplanetary space. In this way combination of field line reconnection and

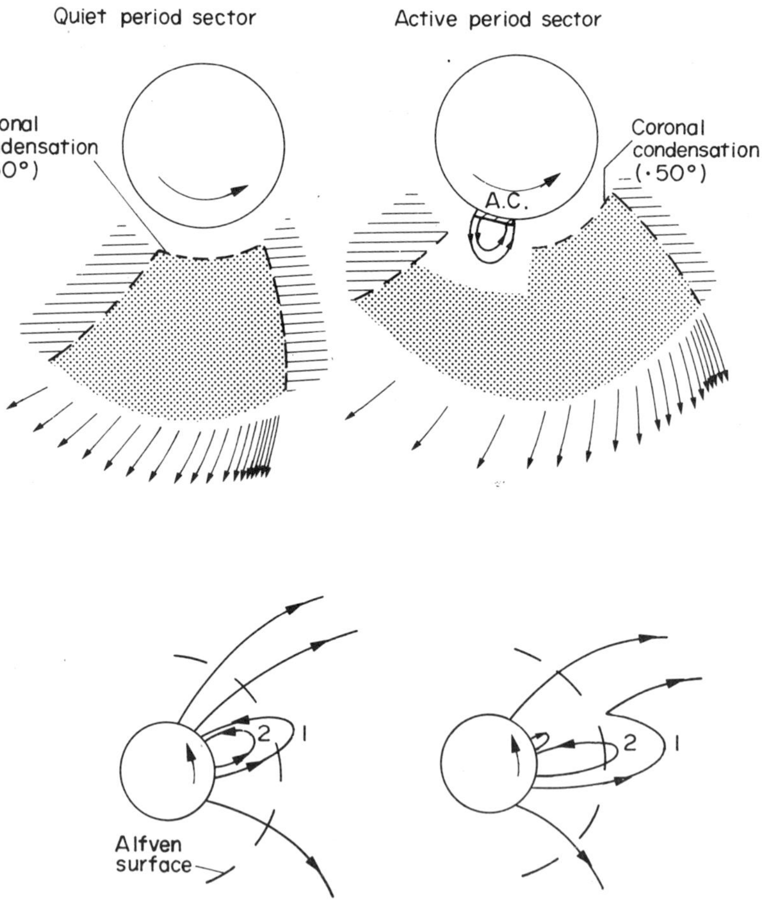

Fig. 8.9. Quiet and active period sectors and the evolution of an interplanetary sector (Martres *et al.*, 1970).

solar rotation may effectively enlarge the original smaller sector. According to this picture the magnetic field near the eastern boundary will be less uniform than the field in the neighbourhood of the western boundary, and the field near the western boundary will be the stronger, as is confirmed by observations (Ness and Wilcox, 1967).

According to the observations and interpretations described here, the effects of noise storm regions may be traced straight out to 1 AU. The primary physical processes are confined to relatively small regions in the corona, but the secondary effects are surprisingly widespread.

9. DISCUSSION OF SOME OBSERVED FEATURES OF NOISE STORMS

9.1. INTRODUCTION

It is commonly said that although noise storms represent the most frequently occurring type of non-thermal radio emission from the sun, it is the one that is theoretically least understood. One may ask one cardinal question: Where does the energy which generates the noise storm emission come from? The answer is that one does not know. But there have been several proposals, and one of them, or some of them, may one day prove to be correct. Even though the theory of noise storms is not at all well established, progress has been made in the past years, and one may reach a far more satisfactory stage within a few years. Noise storms probably give rise to the most difficult and complicated questions from a theoretical point of view; other burst emissions may be more readily explained. But progress in the understanding of other burst types is not without consequences for the theory of noise storms, and the time may now have come to proceed to the more difficult problems in solar radio astronomy. During the next solar maximum, developments in observations and theory may lead to considerable progress in our understanding of the most important manifestation of solar activity in the radio range – the noise storm phenomenon.

It seems appropriate to point out that even though there are important problems that are poorly understood in connection with noise storms, this does not mean that nothing is explained. Indeed, several questions have been solved and in other cases the ideas to a solution already exist. It is also tempting to say that one of the fascinating and challenging aspects of noise storms is the fact that there is ample room for new observations and new ideas which may eventually, together with the large volume of knowledge already existing, lead to a consistent and satisfactory explanation of the whole phenomenon.

In the previous chapters observational results have been presented without too much view to the possible interpretation of the data obtained. In the following this part of the problem is to be considered.

9.2. BURSTS AND BACKGROUND CONTINUUM

Hitherto it has more or less *a priori* been assumed that type I storms consist of a wide band, relatively steady continuum emission with numerous narrow band, short duration bursts superposed. But earlier theories suggested that the background continuum may be due to a superposition of a large number of small storm bursts (McCready *et al.*, 1947; Ryle and Vonberg, 1948; Wild, 1951; Vitkevich, 1957; Takakura, 1959; Fokker, 1960; Kai, 1962; Titulaer, 1967). Indeed, on high-resolution recordings one may frequently observe that many bursts overlap, and bursts which appear single may be the superposition of several elementary bursts. One may then regard the background continuum as the sum of an immense number of elementary bursts. The question was treated quantitatively by Takakura (1959), and it appeared that one could always synthesize real noise storms by choosing appropriate values for the parameters determining the density and the amplitude distribution of the elementary bursts. It was possible to account for both the background level and the occurrence of separate storm bursts. Similar results were obtained by Fokker (1960), but he was not convinced that the whole of the continuum was due to overlapping of elementary bursts (Fokker, 1965b). A model noise storm was constructed by Titulaer (1967) by distributing a large number of elementary bursts at random over the time - frequency domain. The time derivatives of the intensity $d \log I/dt$ at two different frequencies were correlated for different frequency separations and the decrease of the correlation for increasing frequency separation was determined. Titulaer suggested that a similar correlation analysis of a real noise storm should make it possible to determine the mean bandwidth of the elementary bursts from the observed correlation curve. This figure would be characteristic for a given storm and might be compared with the bandwidth of isolated bursts that occur during more quiet intervals in the storm period. However, it is not possible in this way to prove the hypothesis that noise storms are entirely built up of large numbers of elementary bursts.

There are now several observations which provide evidence against the hypothesis that the whole storm phenomenon may be broken down to the occurrence of numerous elementary bursts. One very important result is that the extension of the region from which the background continuum originates is not the same as the scattering range of the individual bursts. Individual bursts occur in a few privileged regions of the storm centre (Daigne, 1968; Bougeret, 1972). Occasionally these are without any continuum region.

Another argument against the superposition hypothesis is that storm bursts have a tendency to occur in clusters and in chains revealing a high degree of ordering, which is in disagreement with a random distribution of their amplitudes. Also the systematic change in spectral properties and polarization of storm bursts observed by Chernov *et al.* (1972) argues in the same direction. Furthermore, during periods of relative calmness, when bursts are almost absent, the continuum intensity level coincides approximately with the lower limit to which the intensity occasionally drops during periods when bursts are frequent. One may also mention that the frequency spectrum of the burst intensity averaged over short periods does not appear to be related to that of the background continuum (Kundu, 1965). The creation of continuum sources in stationary type IV events and the similarity of these sources to the noise storm sources (Kai, 1970) also supports the view that storm bursts and background continuum are distinct phenomena.

The division of the storm emission in background continuum and a burst component now appears to be generally accepted. This does not, of course, exclude the possibility that strong concentrations of storm bursts may produce sporadic enhancements of the general intensity level.

9.3. ENERGIES INVOLVED IN NOISE STORM RADIATION

Observations of the flux density of single storm bursts have been made at different frequencies by Wild (1951), Hauge (1956), Fokker (1960), de Groot (1966) and Steinberg *et al.* (1974).

The observations show that type I bursts may typically reach a flux density at maximum of 10^{-21} W m^{-2} Hz^{-1} in the 150 - 200 MHz region, and strong bursts have maximum flux in the range $>10^{-20}$ W m^{-2} Hz^{-1}. Bursts at frequencies around 100 MHz are somewhat more intense, and the larger ones may exceed a flux of 10^{-19} W m^{-2} Hz^{-1} at their maximum. It appears that a flux density of 5×10^{-21} W m^{-2} Hz^{-1} may be regarded as characteristic for the maximum intensity of a reasonably strong storm burst.

One may now determine the electromagnetic energy emitted by the burst source

$$W_{em} = F\Delta t\Delta f R^2 \Omega \exp(\tau_0), \qquad (9.3\text{-}1)$$

where F is the observed flux, Δt the duration and Δf the bandwidth of the burst, R is the distance from the earth to the sun, Ω is the solid angle into which the waves are emitted and τ_0 is the optical depth of the source. The product of the bandwidth and the duration of type I bursts depends only to a small degree on the frequency at which the bursts occur. It has usually been assumed that the waves are emitted into a solid angle of π steradians, but according to the results of Steinberg *et al.* (1974), who found a polar emission diagram with a half-power beamwidth around 25°, a solid angle of 0.15 steradians is more realistic. The optical depth may be about 2 for waves generated at a frequency close to the plasma frequency at the place of origin and propagating through the corona to the earth according to Wild *et al.* (1963). With $F = 5 \times 10^{-21}$ W m^{-2} Hz^{-1}, $\Delta t = 0.3$ s and $\Delta f = 4$ MHz one then obtains $W_{em} = 1.5 \times 10^{15}$ erg.

As regards the dimensions of the type I burst source, little information is available, but an estimate may be made. Suppose the radiation originates at a frequency close to the plasma frequency: the group velocity of the electromagnetic waves is given by (magnetic field neglected)

$$v_g = c(1 - X)^{\frac{1}{2}}. \qquad (9.3\text{-}2)$$

Since $X \simeq 1$, v_g is considerably smaller than the velocity of light, and may roughly be equal to 0.25 c. Assuming a rise time of 0.1 s for the burst, this would suggest a source dimension not larger than 7500 km. The energy density of the electromagnetic energy then becomes $E_{\text{em}} \simeq 10^{-12}$ erg cm^{-3}. If a plasma wave mechanism is assumed, the conversion efficiency from longitudinal waves to transverse electromagnetic waves may be of the order of 10^{-6} (Wild *et al.*, 1963). The energy density in the plasma waves would then be correspondingly larger, $E_l \simeq 10^{-6}$ erg cm^{-3}.

The magnetic field in the noise storm region is poorly known, but according to Kai (1970) $B > 3.5$ G, giving a magnetic energy density amounting to $E_m > 0.5$ erg cm^{-3}.

Assuming an electron density in the storm region of 10^8 cm^{-3} and a temperature of 10^6 K, the thermal energy density of the plasma becomes $E_{\text{th}} = n K T \simeq 0.1$ erg cm^{-3}.

For a reasonably strong type I burst the following energy relations then appear to be valid:

$$E_m > E_{\text{th}} \gg E_l \gg E_{\text{em}} \,. \qquad (9.3\text{-}3)$$

The energy of the plasma waves responsible for a type I burst would according to the above estimates correspond to the thermal energy contained in a coronal volume of linear dimensions around 100 km.

Data concerning the spectrum of the continuum background were presented in section 2.1. Using the average spectrum determined by Smerd (Fig. 2.2) it is found that the integrated flux is about 3×10^{-12} W m^{-2}. The cone into which the continuum emission is radiated may be somewhat larger than for the single type I bursts, say the half-power beamwidth is 50° corresponding to a solid angle of about 0.6 steradians. If the optical depth equals 2, it is found that during a typical storm an energy of 4×10^{17} erg is emitted from the source region per second, and the corresponding total energy in the form of plasma waves may be

4×10^{23} ergs s^{-1}. The waves may be generated within a volume of about 10^{30} cm^3; thus as regards energy the background continuum is unimportant compared with the magnetic and thermal energy in the source region.

If 10 type I bursts are generated per second, about one-tenth of the energy leaking out from the corona in the form of noise storm radiation is confined to the burst component. Noise storms therefore represent a slow release of energy from large coronal volumes and only a fraction of the energy stored in this volume in the form of magnetic energy or thermal energy is released even when the storm has a relatively long duration. This situation is different from the one encountered in the case of flare-related type III bursts. In such events a large fraction of the stored energy is released during a very short time.

The change in character of the noise storms from metric to decametric wavelengths is likely to be an effect of the pressure relations in the solar corona. At a height of about 1.8 $R_\odot$ above the photosphere the solar wind pressure becomes larger than the magnetic pressure due to the tangential component of the magnetic field and the field lines open.

Since noise storms represent negligible energies as compared with the different energy reservoirs in the solar atmosphere, energy considerations do not lead to a specific theory for noise storms. On the other hand, theories for the noise storm emission must be able to survive a confrontation with the energy relations deduced from the observations.

9.4. DURATION OF TYPE I BURSTS

The observational results concerning the duration of type I bursts were described in section 2.3. From Fig. 2.14 it is seen that the average burst duration increases somewhat towards lower frequencies. Since the determinations of the height of type I emission provide reasons to assume that the radiation is produced at the plasma frequency, it is plausible to suggest that the bursts are generated by plasma waves. If this is the case, it is tempting to relate the burst duration to collisional damping in the corona.

A wave propagating in the z-direction is described by its electric vector **E** in the following form (Wild *et al.*, 1963):

$$E = E_0 \exp[(\omega' t + k' z)] \exp[-i(\omega t + kz)], \qquad (9.4\text{-}1)$$

where k is the wave number and ω the angular frequency. The first factor represents damping or growth of the wave. For electron plasma waves the effect of damping is expressed in terms of temporal damping by the imaginary part of the frequency. When the collision frequency $\nu << \omega$, one has (Westfold, 1949)

$$(\omega + i\omega')^2 = \omega^2 - i\nu\omega \qquad (9.4\text{-}2)$$

or

$$\omega' = -\tfrac{1}{2}\nu .$$

The mean intensity of the wave is then given by

$$\overline{EE^*} = \tfrac{1}{2}E_0{}^2 \exp(-\nu t). \qquad (9.4\text{-}3)$$

According to this the bursts should show an exponential decay, and during a time $t = 1/\nu$ the intensity should fall by a factor $1/e$. For practical purposes one may assume that $\Delta t \simeq 1/\nu$, where Δt is the half-power width of the time profile.

For the collision frequency one may adopt the value given by Scheuer (1960) increased by a factor of 20% as pointed out by Chambe and Lantos (1971). Most of the correction is due to the proportion of helium ions in the corona. The corrected formula then becomes

$$\nu = 4.36 N_e T^{-3/2}\left[17.72 + \ln(T^{3/2}/f)\right]. \qquad (9.4\text{-}4)$$

Using (9.4-4) the values given in Table 9.1 were calculated.

Table 9.1

Collisional damping times for T = 10^6 K

$f = f_p$ (MHz)	N_e (cm^{-3})	$1/\nu$ (s)
400	2×10^9	0.006
300	1.1×10^9	0.011
200	4.9×10^8	0.024
100	1.2×10^8	0.092
50	3.1×10^7	0.357
25	0.8×10^7	1.428

It is clearly seen that this gives too short durations for all frequencies except perhaps for the two lowest ones.

When there is a magnetic field present, the expression for the damping coefficient is modified (Zheleznyakov, 1965), and may be written

$$\gamma = \nu \frac{(1-Y^2)^2 - 2XY^2\sin^2\theta}{(1-Y^2) + XY^2\sin^2\theta} \qquad (9.4\text{-}5)$$

where X and Y are the usual magneto-ionic parameters and θ is the angle between the wave vector and the magnetic field. When $\theta = 0$, $\gamma = \nu$ and no correction is needed. In the case $\theta = 90^\circ$ one also has $\gamma \simeq \nu$ since $Y \simeq 0.1$ and $X \simeq 1$ in the source region.

It is obviously a too rough approximation to assume $\Delta t = 1/\nu$. A somewhat better approach would be to investigate the decaying portion of the time profile, and if an exponential decay is found, the damping constant may be determined.

Elgarøy and Hauge (1958) determined the average time profile of storm bursts at 200 MHz. The profile appeared to follow a function of the form $I = I_m(1 + At^2)^{-\frac{1}{2}}$, where I is the intensity, A a constant and t the time as counted from the instant of maximum intensity. The actual damping constant for the decaying part of the average of 10 bursts appears to correspond to a collision frequency of 14 s^{-1}. Using (9.4-4) one obtains a temperature of $T \simeq 2 \times 10^6$ K. This may be a quite reasonable value for the temperature in the source region. However, there are large uncertainties connected with this method. The effect of scattering has been neglected and also the effect of a finite source size. The situation may be somewhat better when bursts at low frequencies are considered, but here also difficulties are encountered. Aubier and Boischot (1972) discussed the method in connection with type III bursts and found arguments which bring doubt on the significance of coronal temperatures derived from the decay constant based on the hypothesis of collisional damping of the plasma waves.

Takakura (1963) discussed the duration of type I bursts and found it likely that the duration of the bursts was too long to be explained by collisional damping of plasma waves. The duration might rather be determined by the lifetime of the beam of electrons which was supposed to excite the plasma waves responsible for the bursts. If the bursts are generated by a beam of electrons which excite plasma waves which are transformed to radio waves by some process, four different kinds of lifetime must be discussed:

(1) deflection time t_D due to collisions with thermal electrons;

(2) energy exchange time t_E due to collisions with thermal electrons;

(3) energy loss time t_1 to excite coherent plasma waves;

(4) energy loss time t_2 to excite incoherent plasma waves.

The shortest of these is the one of importance for the duration of the bursts. When the velocity of the electrons in the beam exceeds the most probable speed of the thermal electrons, the deflection time t_D is less than the energy exchange time t_E (Spitzer, 1956); $t_E > t_D$. The deflection time t_D is approximately the time taken for a particle to be deflected by 90°, or, stated in another way, for the beam of electrons to be broken up. Takakura further showed that t_D is less than both t_1 and t_2. According to this the conclusion is that t_D is the relaxation time of importance when bursts are generated by a stream of electrons of low (a few times the thermal velocity of electrons) velocity.

The deflection time has been given by Spitzer (1956) and is equal to

$$t_D = \frac{m^2 v_0{}^3}{8\pi e^4 N_e \ \ln \Lambda[\Phi(v_0/v_t) - G(v_0/v_t)]} \tag{9.4-6}$$

where m is the electron mass, e is the charge of proton, v_0 is the beam velocity, N_e is the electron density and v_t is the most probable speed of thermal electrons $(2\kappa T/m)^{\frac{1}{2}}$. For temperatures in the range of 1 - 2 million degrees and $N_e \simeq 10^8 - 10^9\ \mathrm{cm}^{-3}$, $\ln \Lambda \simeq 20$. It is a slowly varying function of density and temperature and has been tabulated by Spitzer (1956). The function $[\Phi(v_0/v_t) - G(v\ /v_t)]$ has a value between 0.9 and 1 when $v_0/v_t > 2$. Using a value for this function of 0.95 together with $\ln \Lambda = 20$ and expressing N_e by the corresponding plasma frequency, one obtains from (9.4-6),

$$t_D \simeq 2.5 \times 10^{-12}(v_0{}^3/f_p{}^2) \tag{9.4-7}$$

When the velocity of the exciting electron beam is small, the generated longitudinal waves have a frequency slightly larger than the plasma frequency. This follows from the dispersion relation for plasma waves travelling in the direction of the magnetic field:

$$\omega^2 = \omega_p^2 + (3\kappa T/m)k^2, \tag{9.4-8}$$

where ω_p is the angular plasma frequency and k is the wave number $(2\pi/\lambda)$. Using the resonance condition $v_0 = \omega/k$ the dispersion equation may be written

$$f_p^2 = f^2[1 - (3/2)(v_t/v_0)^2]. \tag{9.4-9}$$

From the observed burst durations one may now determine the velocity v_0 of the exciting electron beam. This must be done by iterating the expression for t_D and the dispersion equation. It is convenient to express v_0 in units of v_t, hence $x = v_0/v_t$ is introduced.

Using values of Δt_{max}, $\Delta t_{average}$ and Δt_{min} taken from Fig. 2.14, Table 9.2 has been calculated.

Table 9.2

Values of $x = v_0/v_t$ derived from observed burst durations.

Frequency (MHz)	$T = 10^6$ K			$T = 0.5 \times 10^6$ K		
	x_{min}	$\overline{x}$	x_{max}	x_{min}	$\overline{x}$	x_{max}
25	-	-	-	-	1.7	2
100	1.5	2.1	3.5	2.1	3.0	5.0
200	1.5	3.0	4.2	2.1	4.2	6.0
300	2.1	3.3	4.7	3.0	4.7	6.7
400	2.6	4.0	5.7	3.6	5.7	8.0

For plasma waves to be excited, $v_0 > 3^{\frac{1}{2}}v_t$, and it is seen that at the lowest frequencies (below about 50 MHz) it is not possible to excite plasma waves on the assumptions used here and with $T = 10^6$ K or $T = 0.5 \times 10^6$ K. It seems necessary to undertake a more detailed calculation for low frequencies. However, the difficulty encountered here may also be regarded as being in agreement with the observational evidence; type I bursts are rare at frequencies below about 50 MHz and the decametre storm departs in several important respects from the metre-wave storm, as has been pointed out in section 8.1.

It is seen from Table 9.2 that the velocity v_0 increases with increasing frequency. This shows that the acceleration is larger at smaller heights above the photosphere, and may indicate a relation between v_0 and the strength of the magnetic field in the source region.

Table 9.2 also shows that when a temperature of 10^6 K is assumed, the condition $v_0 > 3^{\frac{1}{2}}v_t$ does not hold for the most short-lasting bursts at frequencies below about 200 MHz. It is therefore tempting to suggest that the temperature in the 100 and 200 MHz sources is less than 1 million degrees.

If the temperature amounts to 0.5×10^6 K in the 100 - 200 MHz source regions and increases to 10^6 K at the lower height from which the 400 MHz radiation originates, the average vertical temperature gradient in the corona would be of the order of -0.2 K km^{-1}. A gradient of this magnitude may well exist; in fact, steeper gradients may be expected in active regions.

A temperature of half a million degrees is lower than most observations give for active coronal regions. One may then suggest that bursts are generated in filamentary structures with lower temperatures than the surroundings. This is not unreasonable, since a magnetic field along a loop may very effectively decrease the energy transport across the field lines. In this connection it is noteworthy that Letfus (1965) found that the presence of filaments with a relatively low temperature at larger distances from the core of coronal condensations might explain some observed differences in temperature derived from optical and from radio observations.

Since the most probable thermal velocity of the electrons is given by $v_t = 5.5\, T^{\frac{1}{2}}$ km s^{-1}, the beam velocities derived from Table 9.2 range from roughly 10^4 km s^{-1} to about 3×10^4 km s^{-1}. The implication of this is that type I bursts should be generated by electron streams of low velocity in the solar corona. But if the duration is determined by the collisional damping of plasma waves, somewhat higher stream velocities would be needed. However, there is an upper limit of 6×10^4 km to the stream velocity since a type III burst would be generated if $v_0 \geq 0.2$ c (Wild *et al.*, 1959).

Malville *et al.* (1967) observed spike bursts (flash bursts) at 230 and 345 MHz and found decay times of 0.09 and 0.06 s respectively. The bursts occurred during a type IV event, but spike bursts have the same properties as the most short-lived type I bursts. Spike bursts are also observed in type I storms. The generating mechanism may therefore be the same.

The discussion of the results illustrates the situation as regards short-lasting bursts. Malville *et al.* found that the observations could be explained either by the deflection-time if the beam velocity was about 1.6×10^4 km s^{-1}, or by collisional damping provided the temperature in the source region was in the range from 2.4 to 2.9 million degrees. In the latter case beam velocities of 4×10^4 km s^{-1} to an upper limit of 6×10^4 km s^{-1} appeared to be necessary. But on the basis of their data alone, Malville *et al.* were not able to decide which mechanism might be the most efficacious, and this is a good illustration of the general situation. As yet it has not been possible to arrive at a definite answer to the question of what determines the duration of a type I burst.

Fung and Yip (1966b) considered the duration of type I bursts if they were generated by a gyro-synchrotron mechanism. It was concluded that the radiating system could not be a bunch of electrons because a storm burst then would have a duration which was too short, less than 0.05 s. The effect of scattering was neglected. Fung and Yip suggested that the bursts were generated by helical electron streams. The burst duration would then amount to

$$\Delta t = \frac{\ell}{v_{\parallel}} \tag{9.4-10}$$

where ℓ is the length of the stream and $v_{\parallel}$ the velocity of the stream parallel to the magnetic field. Take $v_{\parallel} = 0.1$ and 0.7 c and durations of 0.5 and 0.8 s. One then obtains stream lengths between 1500 and 24000 km in the case of the lowest velocity and lengths of about 10000 to 170000 km for the highest velocity.

A formal theory for the generation of type III bursts was suggested by de Jager and van't Veer (1958). The excitation of electromagnetic radiation was supposed to occur at a point $P(f_1, t_1)$ that followed a straight line in the f,t-plane. The instantaneous shape of the band of excited frequencies was gaussian, and the radiation intensity decayed exponentially with time. The resulting intensity distribution could then be written in the form

$$I(f,t) = \int_{-\infty}^{t} dt_1 \exp\left[-(f-f_1)^2/b^2-(t-t_1)/\tau_2\right]. \tag{9.4-11}$$

The excitation, the bandwidth b, the decay time τ_2 and the drift rate a in $f_1 = at_1$ were assumed to be constant. The relative shapes of the time profiles generated in this way become

$$I(y,p) = \left[1+(\mathrm{sign}\, y)\ \mathrm{erf}\ y/p\right]\exp(-y) \tag{9.4-12}$$

where y is a linear function of t and $p = b/a\tau_2 = \tau_1/\tau_2$ and τ_1 is the time necessary for the excitation source to travel a frequency interval b (de Groot, 1966). Some curves derived in this way are given in Fig 9.1. One may neglect the mechanism behind the excitation and regard the method as a convenient means to express the time profile of single bursts by a number — the value of the parameter p. Both in the case of type I bursts as well as in the case of type III bursts the majority of the time profiles correspond to values of p between 1.0 and 1.5. According to de Groot there is a preference for p-values of 1.2 to 1.3. Thus the half-width of the excitation function is about equal to the relaxation time τ_2.

Regardless of the mechanism suggested to generate type I bursts, the study of time profiles is complicated by the influence of scattering of the transverse electromagnetic waves during propagation through the corona. The scattering introduces

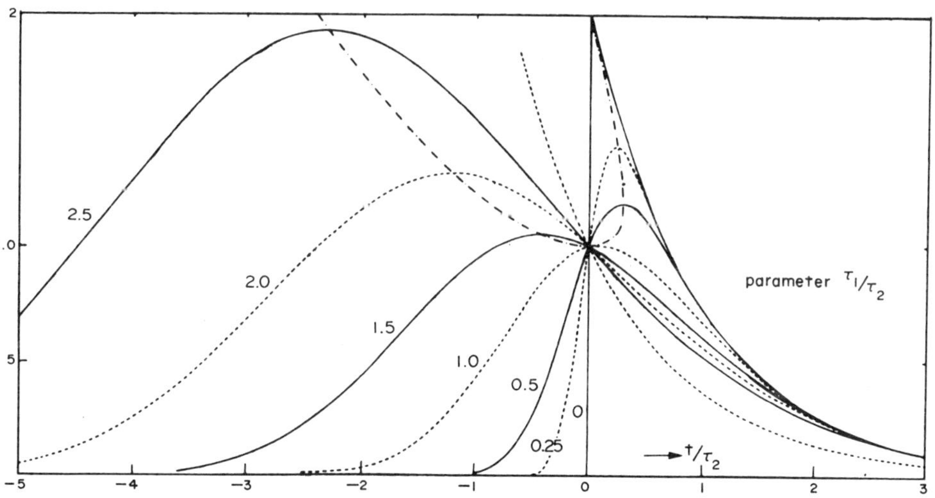

Fig. 9.1. Theoretical time profiles of type III bursts, with the ratio of the excitation time and the decay time as a parameter (de Groot, 1966).

a widening of the burst in time. As was shown in Fig. 5.6, the transient response of the corona has a profile that is similar to many profiles found from observations. The decaying part is nearly exponential in form, and the half-power widths of the transient response given in Table 5.3 are comparable with the observed burst durations, at least for the higher frequencies. However, Fig. 5.6 and Table 5.3 were based on the assumption of an isotropic emitter, which may not be realistic for type I bursts. The first results from the STEREO experiment (Steinberg *et al.*, 1974) showed that type I bursts were strongly directive. The beamwidth appears to be smaller than 25°. Further information on this property may come when more of the material obtained during the experiment has been analysed.

Calculations of the transient response of the corona were performed by Caroubalos *et al.* (1972) for an isotropic emitter and for two directive sources. Denoting the time taken by the transient response to fall from 90 to 10% of its maximum value, the results shown in Fig. 9.2 were found.

The transient response τ_t as a function of frequency and observing angle Θ is small for a source emitting isotropically when the frequency is well above the plasma frequency, but when $f \to f_p$, τ_t increases considerably. This is clearly also an outcome of the computations undertaken by Riddle (1974). For f = 80 MHz and f_p = 78 MHz, or $X = f_p{}^2/f^2 = 0.95$, the effect of scattering was, indeed, considerable.

In the case of directive sources, Fig. 9.2 shows that the values of τ_t decreases with increasing directivity.

If type I bursts are generated by electron streams with velocities in the range from 1.5×10^4 to 6×10^4 km s^{-1}, i.e. from approximately $2v_t$ to $8v_t$ in a corona of a temperature around 2 million degrees, the dispersion relation (9.4-9) shows that values of $X = f_p{}^2/f^2$ between 0.625 and 0.977 may occur. This corresponds to $f \simeq 145$ MHz and $f \simeq 116$ MHz respectively in Fig. 9.2. In this frequency range the average half-power duration of type I bursts is about 0.5 s, and it is seen from Fig. 9.2 that the effect of scattering should not be neglected even if the emission is strongly beamed. The influence is

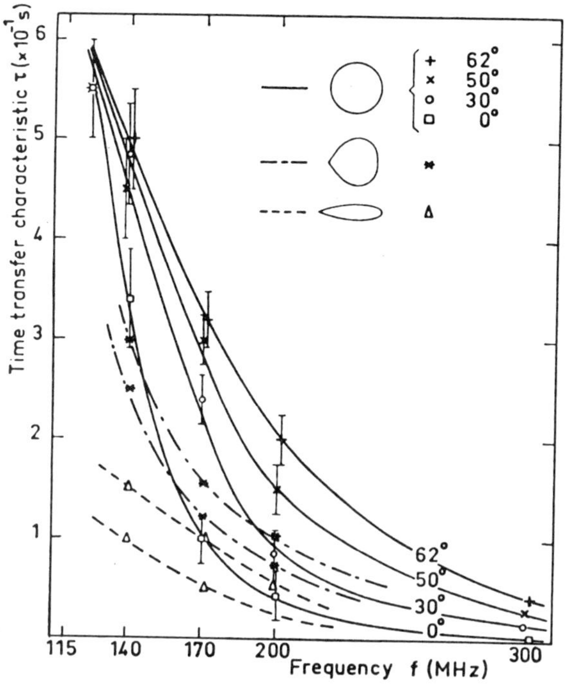

Fig. 9.2. Variation of the transient response $\tau_t(f,\Theta)$ for an isotropic and two directive sources. The critical frequency f_p is 115 MHz. (Caroubalos *et al.*, 1972.)

moderate for an exciter of small velocity and strongly beamed emission, but for larger beam velocities and wider emission polar diagrams the scattering effects are considerable.

In addition to scattering one should also consider possible effects of multi-path propagation in the source region (Bougeret, 1973b; Heyvaerts, 1975), but as yet no quantitative work on this subject has been performed in connection with type I bursts.

The interpretation of type I burst profiles appears to be far more complicated than supposed a decade or two ago. Some progress has been made, and one is now in a better position to ask relevant questions. Further studies may lead to improved knowledge of the microstructure of the solar corona.

9.5. BANDWIDTH OF TYPE I BURSTS

The narrow bandwidth of type I bursts is a very prominent feature on spectral recordings. Observational data were presented in section 2.4. It appears that the average bandwidth Δf and the observing frequency f follow the relation $\Delta f \simeq 0.02f$. Type I bursts are almost monochromatic.

If the bursts are generated by electron streams of low velocity exciting plasma waves, their bandwidth may be explained by some simple considerations (Elgarøy and Eckhoff, 1966). The dispersion equation (9.4-9) can be written

$$f_{p_1}^{\ 2} = f_{max}[1-(3/2)(v_{t1}/v_{01})^2], \quad f_{p_2}^{\ 2} = f_{min}[1-(3/2)(v_{t2}/v_{02})^2], \qquad (9.5\text{-}1)$$

where f_{max} is the highest frequency and f_{min} is the lowest frequency which is emitted in the burst at a chosen moment. As is evident from equations (9.5-1) three variables determine the burst duration: f_p, v_t and v_0.

The dimension of the exciting disturbance may be small in the direction of the magnetic field lines (cf. Takakura, 1963), perhaps of the order of a few hundred kilometers or less. It is reasonable to assume that f_p and T can be regarded as constant over such small lengths in the z-direction. But in the plane perpendicular to z there may be a variation both in electron density and temperature if the bursts are generated in stream tubes (filamentary structures, fibres). In the first instance the variation of T across the tube may be neglected. The bandwidth of the burst may therefore be due to a variation of the electron density across the source and a dispersion Δv_0 in the velocity of the beam particles.

In the discussion of burst durations it was found that possible values of $x = v_0/v_t$ were as given in Table 9.2, and a source temperature of 0.5×10^6 K may be favoured. Assuming that $\Delta v_0/v_0$ is constant and small whatever the frequency of the burst, and, moreover, that $\Delta N_e/N_e$ is constant, the bandwidths to be expected when different values of the variables are assumed can be calculated. The results are shown in Table 9.3.

TABLE 9.3

Calculated bandwidths of type I bursts

Freq. (MHz)	Temp. (10^6 K)	$\bar{x}$	Widening in MHz due to velocity dispersion $\Delta v_0/v_0$			Widening in MHz caused by ΔN_e $\Delta N_e/N_e$		
			5%	10%	15%	2%	3%	4%
100	0.5	3.0	1.3	1.9	3.0	0.9	1.4	1.8
200	0.5	4.2	0.7	1.6	2.5	1.9	2.8	3.8
300	0.5	4.7	1.0	2.5	3.1	2.9	4.4	5.8
300	1.0	3.3	2.1	4.5	7.5	2.8	4.2	5.6
400	0.5	5.7	1.3	2.4	3.7	3.9	5.8	7.8
400	1.0	4.0	1.7	3.8	5.5	3.8	5.7	7.6

For the frequencies 300 and 400 MHz all values have been determined using two values of $\bar{x}$, namely those corresponding to $T = 0.5 \times 10^6$ and 10^6 K.

It appears that $\Delta N_e/N_e$ must be less than 4%, or the electron distribution in the source introduces too large bandwidths at the higher frequencies. Moreover, a velocity dispersion of 15% is somewhat on the higher side when it comes to explain the bursts in the 100 MHz range. Therefore one must have $\Delta v_0/v_0 < 15\%$ and $\Delta N_e/N_e < 4\%$.

In Fig. 2.18 the mean values of the observed burst bandwidths were plotted together with the intervals within which the majority of bursts were found. A good fit with observations is obtained with values of $\Delta v_0/v_0 = 10\%$, $\Delta N_e/N_e = 2\%$ and $T = 0.5 \times 10^6$ K.

Considering the lower values of the observed bandwidths at the different frequencies (which is $\simeq 2$ MHz between 100 and 200 MHz and 4 MHz between 300 and 400 MHz), it is seen from Table 9.3 that values of $\Delta v_0/v_0$ up to 10% may be permitted. According to this the smallest bandwidths do not impose too strict limitations on the velocity dispersion.

The most important outcome of these calculations is that they demonstrate that very uniform conditions must prevail in the burst-emitting source. The active region corona is very inhomogeneous; therefore the source must be relatively small, and it is an attractive thought that it may be confined to magnetic tubes in which the field imposes a high degree of uniformity.

If the bursts are generated in coronal magnetic tubes with radial symmetry over a cross-section of the tube, and the electron density increases towards the axis of the tube, whereas the temperature decreases and reaches a minimum along the axis of the tube, the effects of the density and temperature variations on the dispersion relation will counteract.

The condition for the effect of a density variation to be cancelled by a temperature gradient may be calculated:

$$f_{p_1}^{\,2}/[1-(3/2)(v_{t_1}/v_0)^2] = f_{p_2}^{\,2}/[1-(3/2)(v_{t_2}/v_0)^2]. \qquad (9.5\text{-}2)$$

From this one finds

$$\Delta f_p/f_p \simeq -\, 3v_t\Delta v_t/2v_0^{\,2} - 3v_t^2), \qquad (9.5\text{-}3)$$

where $\Delta f_p = f_{p_1} - f_{p_2}$ and $\Delta v_t = v_{t_1} - v_{t_2}$.

Since v_0/v_t may be in the range from 3 to 6 and $\Delta v_t/v_t = \frac{1}{2}\Delta T/T$, one gets

$$\begin{aligned} \Delta f_p/f_p &\simeq -\,(1/10)(\Delta T/T), \quad v_0 = 3v_t, \\ \Delta f_p/f_p &\simeq -\,(1/46)(\Delta T/T), \quad v_0 = 6v_t. \end{aligned} \qquad (9.5\text{-}4)$$

The relations (9.5-4) show that a temperature variation of, say, 50% across the source may balance out a density variation $\Delta N_e/N_e$ between 10% and 2%, depending on the beam velocity v_0: The effectiveness of this mechanism is not large, but it may nevertheless be of some importance. As the effectiveness increases strongly with lower values of v_0, one should expect that short-lasting bursts should show an average bandwidth which tended to be smaller than for the long-lasting ones. Although no correlation between burst duration and bandwidth has been detected, some results found by Eckhoff (1966) support this suggestion.

The frequency profile of storm bursts may take various shapes (Elgarøy, 1961). In a sample of 22 bursts, 11 bursts had a symmetric profile, 2 showed a low-frequency cut-off and 9 had a high-frequency cut-off. The symmetric bursts showed good agreement with a gaussian profile. Owing to the small observational sample, conslucions could not be drawn as to the frequency of occurrence of the different frequency profiles. The results are in agreement with those obtained by de Groot (1966).

From a theoretical point of view one may suggest that skew frequency profiles are caused by Landau damping and by the propagation conditions in the inhomogeneous corona.

In a short-lasting burst, i.e. a burst generated by a low-velocity disturbance, Landau damping may cause a depression of the high-frequency wing. According to Berz (1956), Landau damping becomes appreciable at $x = v_0/v_t = 3.4$, when the waves are damped after 300 periods, and already at $x = 3.2$ the damping occurs in 50 periods in the case of no magnetic field. Therefore one should expect that short-lasting bursts should be asymmetric, whereas those of longer lifetimes have a gaussian profile. However, in both cases an additional "shaping" of the bursts may take place during the propagation through the corona.

Tidman and Dupree (1965) discussed emission in the case when a flux of superthermal electrons exists in a thermal plasma. The velocity distribution of the superthermal electrons was such that $df(v)/dv < 0$, where $f(v)$ denotes the distribution function. The field of longitudinal plasma waves is then peaked around f_p because Landau damping is small in this frequency region. In this theory a resonance peak occurs as the result of a velocity distribution with a long tail, and de Groot (1966) suggested that the mechanism may be applied to type I bursts. Bandwidths of the correct magnitude then occurs for electron velocities $v_e \simeq 2(3\kappa T/m)^{\frac{1}{2}}$. In order to obtain sufficient energy to produce a type I burst a volume of dimensions $10^4 - 10^5$ km is necessary, which is somewhat larger than one would like to have.

Vereshkov (1974) considered type I bursts to be the result of conversion of plasma waves, which arise as the result of an instability at the front of a hydromagnetic pulse, and acoustic waves into transverse waves. For the bandwidth of the bursts he found that the heterogeneity of the conditions in the region of generation was the dominating factor. Since in his theory $f \simeq f_p = (N_e e^2/\pi\ m)^{\frac{1}{2}}$, Vereshkov used the simplified Baumbach-Allen equation $N_e(\rho) = \text{const}\ (\rho)^{-6}$ and found

$$\Delta f/f = \tfrac{1}{2}(\Delta N_e/N_e) = \tfrac{1}{2}|dN_e/d\rho|\ \lambda/N_e(\rho) \simeq 0.1 \qquad (9.5\text{-}5)$$

Here λ is the wavelength of an acoustic wave, taken to be equal to 3×10^9 cm. The resulting bandwidth is certainly larger than observed, but by adjusting the parameters it seems possible to arrive at reasonable results without being unrealistic.

Thermal Doppler broadening of plasma waves was suggested as responsible for the bandwidth by de Jager and van't Veer (1958). If relevant, this mechanism leads to temperatures of about 0.4×10^6 K in the region where type I bursts are generated (Elgarøy, 1961).

Fung and Yip (1966a) proposed a mechanism involving cyclotron radiation from streams of electrons for the origin of noise storm emission. Energies in the range of 10 - 100 keV were assumed. For type I bursts bandwidths of $\Delta f/f \simeq 0.09$ were obtained, and for great pitch angles $\Delta f/f < 0.03$ appeared possible.

As has been demonstrated, several different mechanisms have been proposed to account for the bandwidth of type I bursts, but it is not possible to select a particular theory from the study of the bandwidth alone. Many pieces of evidence must fit together in order to arrive at a theory for the type I radiation.

9.6. FREQUENCY SPLITTING

The observational evidence concerning frequency splitting in type I chains and in bursts was described in sections 6.2 and 6.4. Frequency splitting is also a well-known phenomenon in type II bursts. Different mechanisms have been proposed to account for the splitting. Here those which may be relevant to type I radiation will be discussed.

The approximately simultaneous appearance and the similarity of the elements in split bursts indicate that a single source may radiate at two and sometimes at three frequencies. Fung (1966) proposed a mechanism which may explain the occurrence of double bursts. A helical electron stream moving through the magnetized coronal plasma may radiate coherent cyclotron waves. Radiation in the ordinary mode at frequencies near the second harmonic of the cyclotron frequency may be observable. Then one may assume that when a stream moves radially out from the sun at a large angle with the direction to the observer, radiation may escape along a direct path and one that is reflected at a lower point in the corona. Because of the Doppler effect the radiation along the direct path occurs at a higher frequency than that which has been reflected. A double burst would result with the element at the lower frequency somewhat delayed as compared with the higher frequency element. When the separation between the elements is a small fraction of the observing frequency, as is found from the observations, the separation should increase linearly as a function of the longitudinal particle speed. The particle speed should be reflected in the frequency drift df/dt of the elements, but, as shown in Fig. 6.21, this is contrary to observational evidence. Thus this mechanism leads to a prediction which is at variance with the observations.

It is possible that the splitting of the bursts is associated with resonances in the coronal plasma or in the electron streams travelling in the corona. Tidman *et al.* (1966) discussed plasma resonances to be expected in the presence of a weak magnetic field.

From the Appleton - Hartree equation the following critical frequencies ($n = 0$) are found for electromagnetic waves in the case of longitudinal propagation:

$$\omega \simeq \omega_p \mp \tfrac{1}{2}\omega_H$$

when it is assumed that $\omega_H \ll \omega_p$. The minus sign applies to the ordinary mode and the plus sign to the extraordinary mode. For transverse propagation ($\theta = 90^\circ$) the refractive index is zero for

$$\omega = \omega_p, \quad \omega \simeq \omega_p + \tfrac{1}{2}\omega_H.$$

Observations of triple splitting in some bursts may suggest radiation at the critical frequencies $\omega = \omega_p$ and $\omega \simeq \omega_p \mp \frac{1}{2}\omega_H$ for electromagnetic waves. However, only the two higher frequencies would be able to escape. One could omit this difficulty by assuming that some radiation was emitted at the harmonic frequencies $\omega = 2\omega_p$ and $\omega = 2\omega_p \mp \omega_H$. But then a triple burst at the harmonic frequency should be associated with a double burst at the fundamental frequency, and there is no evidence for this (Ellis and McCulloch, 1967). In addition the two elements near the plasma frequency should show opposite polarization, which has not been observed.

Electrostatic resonances may also be considered. They occur where the refractive index approaches infinite or where it changes very rapidly with frequency. The refractive index for longitudinal waves in a cold magnetized plasma is infinite when (Akhiezer *et al.*, 1967)

$$1 - \frac{\omega_p^2 \cos^2\theta}{\omega^2} - \frac{\omega_p^2 \sin^2\theta}{\omega^2 - \omega_H^2} - \frac{\Omega_p^2 \cos^2\theta}{\omega^2} - \frac{\Omega_p^2 \sin^2\theta}{\omega^2 - \Omega_H^2} = 0, \quad (9.6\text{-}1)$$

where Ω_p and Ω_H are the angular ion plasma and gyro frequencies respectively. Waves with a large refractive index have a small phase velocity and may interact effectively with the plasma particles. When charged particles pass through the plasma, this may lead to instabilities. Neglecting the contribution of ions in (9.6-1), one obtains the two resonance frequencies

$$\omega_1^2 = \tfrac{1}{2}(\omega_p^2 + \omega_H^2) + \tfrac{1}{2}[(\omega_p^2 + \omega_H^2)^2 - 4\omega_p^2\omega_H^2\cos^2\theta]^{\frac{1}{2}}, \quad (9.6\text{-}2)$$

$$\omega_2^2 = \tfrac{1}{2}(\omega_p^2 + \omega_H^2) - \tfrac{1}{2}[(\omega_p^2 + \omega_H^2)^2 - 4\omega_p^2\omega_H^2\cos^2\theta]^{\frac{1}{2}}. \quad (9.6\text{-}3)$$

In the case of longitudinal propagation $\theta = 0^\circ$, one gets the following resonance frequencies:

$$\omega_1 = \omega_p, \quad \omega_2 = \omega_H.$$

When $\theta = 90^\circ$ equation (9.6-2) gives (assuming $\omega_H^2 \ll \omega_p^2$)

$$\omega_1 = (\omega_p^2 + \omega_H^2)^{\frac{1}{2}} \simeq \omega_p + \tfrac{1}{2}\frac{\omega_H^2}{\omega_p}.$$

Equation (9.6-3) should not be used close to $\theta = 90^\circ$. When $\theta \to 90^\circ$ one may use (Körper, 1957)

$$\omega_2 = \frac{(\Omega_p^2 + \Omega_H^2)^{\frac{1}{2}}}{\left(1+\frac{\omega_p^2}{\omega_H^2}\right)^{\frac{1}{2}}} \qquad (9.6\text{-}4)$$

The latter expression may be modified and written

$$\omega_2 \simeq \frac{\omega_p\omega_H}{(\omega_p^2 + \omega_H^2)^{\frac{1}{2}}(M/m)^{\frac{1}{2}}} + \delta \qquad (9.6\text{-}5)$$

where M is the ion mass and δ is a small number. It is seen that the electrostatic resonances which may be of interest are as follows

$$\theta = 0^\circ: \quad \omega_1 = \omega_p, \; \omega_2 = \omega_H,$$

$$\theta = 90^\circ: \quad \omega_1 = \omega_p + \tfrac{1}{2}(\omega_H^2/\omega_p), \quad \omega_2 = \frac{\omega_p\omega_H}{(\omega_p^2 + \omega_H^2)^{\frac{1}{2}}(M/m)^{\frac{1}{2}}}.$$

Tidman *et al.* (1966) showed that the emission around the electron plasma frequency ω_p may be split in two components for some velocity distributions. They considered the case when a Maxwellian electron velocity distribution coexisted with a tenuous flux of superthermal electrons in a plasma which was weakly magnetized. If the superthermal electrons had an anisotropic distribution with $T_\perp > T_\parallel$ and were drifting along the magnetic field, an enhanced longitudinal wave spectrum with wave vectors parallel and perpendicular to the magnetic field would be excited. Emission may then occur at $\omega = \omega_p$ and $\omega = \omega_p + \frac{1}{2}(\omega_H^2/\omega_p)$.

The above results were applied by Ellis and McCulloch (1967) to interprete split bursts. Since there was observational evidence for emission of split bursts near the plasma frequency as well as at the harmonic of the plasma frequency, a plasma radiation process appears likely. The two elements of split bursts may then be identified with radiation near ω_p and near the hybrid frequency $(\omega_p^2 + \omega_H^2)^{\frac{1}{2}} \simeq \omega_p + \frac{1}{2}(\omega_H^2/\omega_p)$. Absorption may account for a smaller intensity and bandwidth of the element with frequency near ω_p.

Triple bursts can be accounted for by the resonance and combination frequencies: $\omega_1 = \omega_p$, $\omega_2 = \omega_p + \frac{1}{2}(\omega_H^2/\omega_p)$ and

$$\omega_3 = \frac{\omega_p \omega_H}{(\omega_p^2 + \omega_H^2)^{\frac{1}{2}}(M/m)^{\frac{1}{2}}} + \omega_p + \frac{1}{2}(\omega_H^2/\omega_p) \simeq \frac{\omega_H}{(M/m)^{\frac{1}{2}}} + \omega_p + \frac{1}{2}(\omega_H^2/\omega_p).$$

The frequency separation between the elements are related by

$$\omega_2 - \omega_1 = (\omega_3 - \omega_2)\,\frac{M/m}{2\omega_p}. \qquad (9.6\text{-}5)$$

For a frequency of 25 MHz and an upper frequency interval of 0.1 MHz, the lower frequency interval becomes, according to (9.6-5), $f_2 - f_1 = 0.37$ MHz. The lower frequency interval is larger than indicated by observations which show only a minor difference between the upper and the lower intervals. A possible explanation is that part of the lowest element is supressed by self-absorption at and below the plasma frequency.

A coronal magnetic field varying from 10 G at the 60 MHz plasma level to 2 G at the 25 MHz level explains the observations. For an observed splitting of 8 MHz in the 200 MHz range (Elgarøy, 1961), a field of 20 G is necessary.

According to Baselyan *et al.* (1974a) the splitting of diffusive pairs, which occurs at the harmonic of the plasma frequency, is about twice as large as for the fundamental pairs. This is in accordance with the theory outlined here. Moreover, only the ordinary mode should escape since the radiation is emitted at frequencies lower than the cut-off frequency for the extraordinary magneto-ionic mode, $\omega_e = \frac{1}{2}\omega_H + \frac{1}{2}(4\omega_p^2 + \omega_H^2)^{\frac{1}{2}}$. The bursts should then be polarized in accordance with the observations of de la Nöe and Boischot (1972) who found that the elements were always polarized in the same circular sense, and the degree of polarization was usually high.

The element at the lowest frequency had a tendency to arrive somewhat before the others. This may be the result of a propagation effect.

Different line-splitting mechanisms were considered by Melrose and Sy (1971) who suggested a non-magnetic splitting mechanism for drift pair solar bursts. Introducing the notions t, l and s for transverse, longitudinal and sound waves respectively, they considered the generation of t-waves. Generation of t-waves can result from the coalescence process $l + s \rightarrow t$ and from the induced decay process $l \rightarrow t + s$. Conservation of energy and momentum requires

$$\omega_t = \omega_l + \omega_s, \quad \mathbf{k}_t = \mathbf{k}_l + \mathbf{k}_s, \tag{9.6-6}$$

$$\omega_t = \omega_l - \omega_s, \quad \mathbf{k}_t = \mathbf{k}_l - \mathbf{k}_s. \tag{9.6-7}$$

When the l-waves are generated by a stream, their distribution is directed in the streaming (forward) direction and $|\mathbf{k}_t| \ll |\mathbf{k}_l|$ when the stream velocity $v_0 \ll c$. In the case $|\mathbf{k}_s| \gtrsim |\mathbf{k}_l|$ (9.6-6) and (9.6-7) show that the emission occurs in the backward direction, whereas the condition $|\mathbf{k}_s| \lesssim |\mathbf{k}_l|$ leads to forward emission. Provided s-waves with $|\mathbf{k}_s| \gtrsim \omega_p/v_0$ are concentrated in the backward direction and s-waves with $|\mathbf{k}_s| \lesssim \omega_p/v_0$ are concentrated in the forward direction, (9.6-6) gives backward emission at the higher frequency and (9.6-7) favours forward emission at the lower frequency. An intermediate trace may be generated when either of the two other possibilities are realized.

The mechanisms described by (9.6-6) and (9.6-7) implies a frequency split given by

$$2\omega_s \simeq \omega_p 2v_s/v_0\left[1+(v_e/\omega)^2\right]^{-\frac{1}{2}} \tag{9.6-8}$$

where $v_s = (m/M)^{\frac{1}{2}}v_e$ is the sound velocity and mv_e^2 is the electron temperature in ergs. The splitting obtained from

(9.6-8) is in accordance with observations when the stream velocity v_0 is several times v_e.

According to Melrose and Sy s-waves can be generated in the backward direction by the stream, and the distribution of s-waves needed to account for the asymmetry in the emission could be produced as a result of non-linear scattering of the s-waves. A condition for the generation of the s-waves is that the density N_1 in the stream satisfies the condition $N_1 > N_e v_s / v_0$.

The relation (9.6-8) predicts larger splitting for larger values of v_0, which is in disagreement with the results indicated by the observations of Ellis and McCulloch (1967). It permits unpolarized bursts to be split, but this is not necessary since Ellis (1969) and de la Nöe and Boischot (1972) find that split bursts are polarized.

It is tempting to conclude that the mechanism proposed by Ellis and McCulloch (1967) involving electrostatic resonances in magnetized plasma at present best may account for the properties of split bursts.

10. INTRODUCTION TO NOISE STORM THEORIES

10.1. EARLY WORK

In Chapters 1-8 a large number of results taken from the impressive volume of observational data concerning noise storms were presented. Together with advances in observational techniques and new data gained from the observations, a development has occurred in the theoretical understanding of the noise storm phenomenon. However, owing to the complexity of the problems much work remains to be carried out on the theoretical side. It is also true that the experimentalist should furnish the theorist with more comprehensive data of higher accuracy and quality than has yet been possible.

It is immediately obvious that thermal radiation cannot account for noise storm emission. This is seen from simple reasoning. The flux density S due to a source of uniform brightness temperature T_b subtending a solid angle Ω is given by

$$S = 2\kappa T_b \Omega / \lambda^2, \qquad (10.1\text{-}1)$$

where κ is Boltzmann's constant and λ the wavelength. Assuming a background continuum flux of 5×10^{-21} W m^{-2} Hz^{-1} at a frequency of 200 MHz and a source size of 2′ arc, the brightness temperature amounts to about 1.3×10^9 K. For type I bursts smaller source sizes may be realistic, and it may be necessary to explain brightness temperatures of the order of 10^{12} and perhaps 10^{15} K.

The observed brightness temperature along a given ray trajectory is given by

$$T_b = \int_0^\tau T \exp(-\tau)\, d\tau, \qquad (10.1\text{-}2)$$

where T is the radiation temperature at the optical depth τ.

Consider a volume of isothermal gas of radiation temperature T and optical thickness Δ. One gets

$$T_b = T[1 - \exp(-\Delta)]. \qquad (10.1\text{-}3)$$

For large optical thickness the emission from the gas is equivalent to that from a black body and $T_b = T$. For small optical thickness $T_b = T\Delta$. Since the optical thickness $\Delta \propto T^{-3/2}$, T_b will drop according to $T^{-\frac{1}{2}}$ when the optical thickness is small. This occurs in the corona at the typical heights where noise storm radiation is generated. Consequently a sudden localized heating causes a decrease in the brightness temperature of the plasma, and an explanation of storm emission in terms of thermal radiation is ruled out. The situation is different in the layers closer to the solar surface, and thermal bursts at centimetric wavelengths occur. The non-thermal character of noise storm emission has been recognized for a long time.

Earlier theories suggested that the non-thermal radiation from the sun might be explained as gyro radiation from fast electrons (Kiepenheuer, 1946). But the gyrating electrons would radiate in the extraordinary mode and the emission would not escape from the corona since there is a stop region at the level $Y = 1 - X$ above the layer where the generation of gyro radiation occurs. Furthermore, radiation in the extraordinary mode is incompatible with the observations which have shown that the ordinary mode is received.

Kruse *et al.* (1956) suggested that the above difficulty could be overcome because a gyrating electron of high energy would radiate both in the ordinary and in the extraordinary mode except in directions parallel to and perpendicular to the magnetic field. It was supposed that the gyrating electrons in the plasma would radiate as much energy as in free space. This assumption does not hold because reabsorption of energy is very important.

The problems of gyro radiation were further considered by Twiss and Roberts (1958). They showed that the radiation is predominantly radiated in the extraordinary mode, but that the ordinary mode is also weakly excited. At harmonics of the gyro frequency the fraction of energy radiated in the ordinary mode was a few per cent of that radiated in the extraordinary mode. Twiss and Roberts found that many of the data concerning type I bursts suggested second harmonic gyro radiation. If so, one would still expect the radiation to occur in the ectraordinary mode, contrary to observational evidence. To overcome this difficulty one must either assume that the absorption in the extraordinary mode is far in excess of that in the ordinary mode, or that the plasma frequency at the place of origin of the radiation is so high that the second harmonic in the extraordinary mode is not able to escape. The latter condition implies that $f_p < 2f_H < 2^{\frac{1}{2}}f_p$. But in order to explain the energy flux in large type I disturbances, electrons with an average energy of 3 MeV were necessary. Then the narrow bandwidth of type I bursts would be difficult to explain. In order to circumvent the insufficient flux from non-relativistic electrons, one may suggest the possibility of negative absorption. This was done by Twiss (1958). He showed that this phenomenon may occur in several situations, also when the predominant radiation process is gyro radiation from non-relativistic electrons. The hypothesis that type I bursts are generated by gyro radiation was, however, left unconfirmed. Further discussions of amplified gyro radiation were undertaken by Wild *et al.* (1963), Smerd (1965, 1968) and by Fung and Yip (1966a). Takakura (1956) also suggested gyro radiation as a possible mechanism responsible for type I noise storms. To increase the effectiveness of the gyro mechanism Takakura suggested that the electrons gyrate in groups which have small dimensions (about 15 cm) compared with the wavelength.

A dominating trend in attempts to explain the high-level radiation involves coherent plasma waves. Shklovsky (1946) proposed that enhanced solar radio emission was caused by macroscopic oscillations in the coronal plasma at the plasma frequency $f_p = (N_e e^2/\pi m)^{\frac{1}{2}}$. The oscillations were excited by streams of charged particles. Martyn (1947) also suggested plasma oscillations as a possible cause. The oscillations were thought to be excited by the breakdown in electric polarization because of turbulent motion in a magnetic field. Later on Bohm and Gross

(1949a,b, 1950) discussed a number of problems in connection with plasma oscillations. They found that the conditions in the solar corona were favourable for the existence of plasma oscillations and that there were good chances for the radiation to escape, primarily in the presence of a magnetic field.

Haeff (1949), Bailey (1950) and Buneman (1950) proposed different plasma mechanisms. Bailey showed that an ionized gas in which streams of electrons are drifting can, under certain conditions, act as an amplifier of electric fields. Haeff more specifically studied the amplification of longitudinal waves in two streams of electrons with different velocities. Buneman considered the generation and amplification of waves in dense charged beams under crossed electric and magnetic fields in laboratory experiments, but did not investigate to what extent this was representative of conditions in the solar atmosphere.

Denisse and Rocard (1951) introduced shock waves in the theory. They studied the characteristics of a shock wave propagating in an ionized medium without magnetic field, and found that plasma oscillations might be excited. The bandwidth appeared to be narrow, and the short duration of bursts might result from intrinsic instability of the excitation process.

Ginzburg and Zheleznyakov (1959b) analysed critically different possibilities for the generation of type I bursts. They concluded that the bursts probably were produced by coherent radiation generated in a plasma pervaded by a magnetic field. The bursts might be emitted during the damping of free oscillations in the coronal plasma. They considered the excitation mechanism to be obscure, but it was possibly due to magneto-hydrodynamic shock waves.

Denisse (1960) argued that the observed directivity and refraction effects in connection with noise storms showed that the noise source was situated at a height in the corona which is close to the critical frequency. The radio noise was therefore likely to be caused by fast particles exciting longitudinal plasma waves. The fast particles might be ejected from chromospheric flares. One may suppose that some of the emitted corpuscles are trapped in suitable magnetic field configurations, probably associated with coronal streamers. The trapped particles diffuse slowly along streamers and excite plasma waves at frequencies near the plasma frequency. Favourable conditions occur when the waves propagate along magnetic lines of force in the direction of increasing density, i.e. towards the solar surface. They are then coupled to the ordinary magneto-ionic wave mode which propagates in the same direction, but which may be reflected in dense regions. It is important that the waves are not absorbed or too much attenuated during the reflection process. This may explain that noise storms are confined to the metre-wave region. On frequencies below about 500 MHz the rays may reach the layer of zero refractive index and be reflected with smaller loss than on higher frequencies. Narrow-band short-duration type I bursts were explained by Denisse as sporadic specular reflections (focussing) in a small frequency band.

The majority of the studies referred to in this section were not elaborated in too much detail. But ideas were presented which might be followed up in more thorough studies of noise storms. In a sense one may say that the first theories provide the background material which was necessary for further theoretical developments, which also depended on progress in general plasma physics and experimental results.

10.2. MODELS OF STORM REGIONS

Before discussing detailed theories for noise storms, a few remarks on observationally determined models of the storm region appear appropriate. Some models were presented in section 3.8, and Fig. 8.7 shows a model proposed by Stewart and Labrum (1972) in order to explain type I storm activity and associated dekametric type III activity.

Another model, depicted in Fig. 10.1, illustrates the first important point to be stressed: type I sources occur in magnetic loops formed by closed fields directly associated with a particular sunspot group, or larger-scale fields extending from the leading spot of the local group to other active regions.

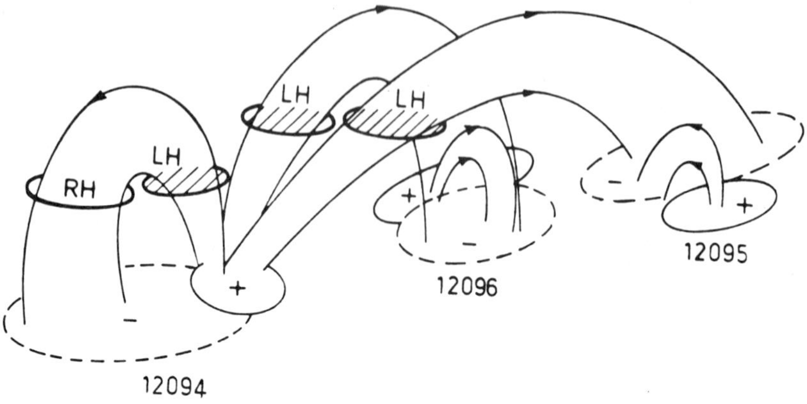

Fig. 10.1. Model proposed to account for observed positions of type I storm centres relative to associated optical features (the numbers refer to McMath plage regions) (Kai and Sheridan, 1974).

The magnetic loops are probably inhomogeneous as indicated by scattering effects. One possibility is that the loops are made up of filaments (fibres, stream tubes) of small cross-sections forming part of the coronal microstructure. The situation may, of course, not always be as complicated as shown in Fig. 10.1, but this figure explains the multiple sources observed in some noise storms, and it also offers the possibility to explain erratic shifts in source position because one source may decay and another become active.

Secondly, it should be emphasized that the connection between metric storms and decametric storms may be explained by models combining a region of closed magnetic loops with a region of open field lines. One model was presented in Fig. 8.7, and remarks on the basis of this model were made. Here it should only be added that closed-loop structures which are seen at eclipse photographs often extend out to about 1 $R_{\odot}$. Using a density model such as 2 x Newkirk (1961), this height corresponds to an electron density giving a plasma frequency of 45 MHz, which is approximately the frequency region in which transition from metric to decametric storm characteristics takes place.

Another model offering the possibility to explain type I storms and dekametric phenomena such as impulsive (flare associated) type III bursts and storm type dekametric continuum and type III bursts, is displayed in Fig. 10.2. Gergely and Kundu (1975) found that some type III bursts were displaced from the continuum source and started at higher frequencies and were more intense than the type III bursts which showed source positions coinciding with the continuum source. They were termed "off fringe" type III bursts and probably correspond to the class which usually is called impulsive type III bursts. "On fringe" type III bursts correspond to storm type dekametric fast drift bursts (type III bursts), and it is possible that a large part of the dekametric continuum may be resolved in such bursts when equipment of high resolution is used. The empirical model in Fig. 10.2 gives an idea of the association of intense centimetre and decimetre wave sources with type I storms. At the highest frequencies there is only continuum emission without type I bursts. At lower frequencies (around 400 MHz) type I bursts appear superposed on the continuum, while below about 100 MHz the frequency of occurrence of type I bursts diminishes, and a large number of dekametric type III bursts is observed, possibly together with some continuum emission. The location of the "off fringe" type III sources is such that particles accelerated in chromospheric flares may have direct access to open field lines. Particles exciting "on fringe" type III bursts may be connected with acceleration processes which are somehow associated with the noise storm phenomenon.

It is useful to bear in mind the empirically derived models of noise storm regions when theories of storms are discussed.

10.3. SOME THEORETICAL CONCEPTS

In theoretical studies of mechanisms for generation of radio noise one may encounter notions and concepts which need some words of explanation. In most cases one finds the necessary information in books by Zheleznyakov (1970), Tsytovich (1970, 1972) and Kaplan and Tsytovich (1973).

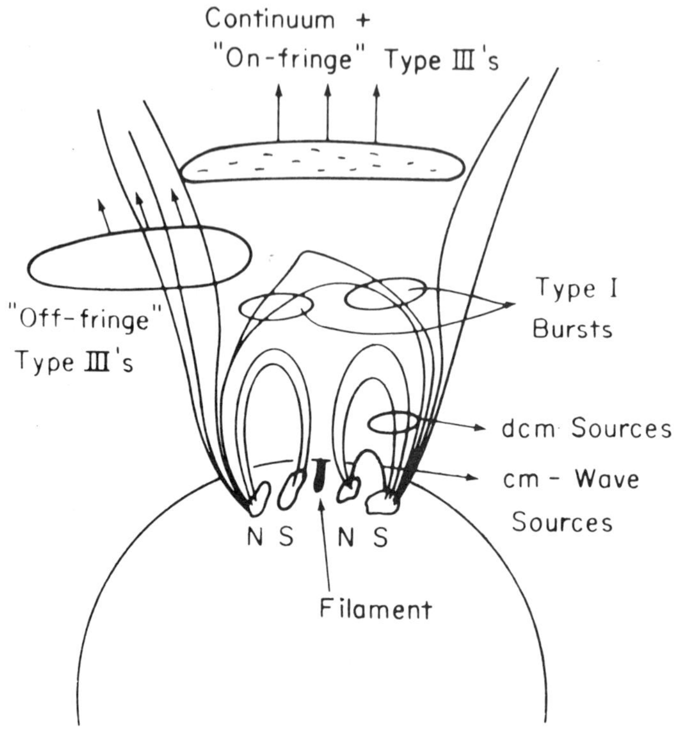

Fig. 10.2. Model of storm region indicating positions of storm type (on fringe) and impulsive (off fringe) type III bursts together with metric type I regions and sources of intense continuum emission at short wavelengths (Gergely and Kundu, 1975).

In order to facilitate reading of the following sections, some selected concepts from the literature mentioned above are presented here.

Decay and Fusion of Waves

Non-linear interactions between waves in a plasma lead to a transfer of energy between waves of the same mode and between different modes.

An electron moving in the electric field of two waves with different wave vectors $\mathbf{k}_1$ and $\mathbf{k}_2$ and frequencies ω_1 and ω_2 can be described by the equation

$$m \frac{d^2 r}{dt^2} = e\mathbf{E}_1 \cos[\omega_1 t - (\mathbf{k}_1 \mathbf{r})] + e\mathbf{E}_2 \cos[\omega_2 t - (\mathbf{k}_2 \mathbf{r})]. \qquad (10.3\text{-}1)$$

$\mathbf{E}_1$ and $\mathbf{E}_2$ are the amplitudes of the fields of the waves. The solution of (10.3-1) for $\mathbf{r}$ contains terms involving the product of the field strengths $\mathbf{E}_1$ and $\mathbf{E}_2$. The corresponding motion has frequencies $\omega_1 + \omega_2$ and $\omega_1 - \omega_2$. The motion of the electron proceeds with successive accelerations and decelerations and leads to emission of waves with frequencies of $\omega_1 + \omega_2$ and $\omega_1 - \omega_2$. This is the decay interaction or fusion (coalescence) of waves.

Energy and momentum must be conserved in the process

$$\omega = \omega_1 + \omega_2, \qquad \mathbf{k} = \mathbf{k}_1 + \mathbf{k}_2. \qquad (10.3\text{-}2)$$

Several different processes involving fusion or decay of different waves may occur in the solar plasma.

Differential and Integral Scattering

A plasma ion creates a polarization cloud of electrons surrounding the ion and moving around with it. If a wave is incident upon the "coat", the wave is scattered and forms other waves. The incident wave may have frequency ω' and wave vector $\mathbf{k}'$. For the scattered wave the respective quantities are ω and $\mathbf{k}$. During the scattering the ion looses part of its energy $\varepsilon - \varepsilon'$ and momentum $\mathbf{p} - \mathbf{p}'$. In the process the total energy and total momentum are conserved:

$$\varepsilon - \varepsilon' = \varepsilon(\mathbf{k}) - \varepsilon'(\mathbf{k}'), \qquad (10.3\text{-}3)$$

$$\mathbf{k} - \mathbf{k}' = \mathbf{p} - \mathbf{p}'. \qquad (10.3\text{-}4)$$

Since the momentum of the wave $\mathbf{k}$ is much smaller than the particle momentum,

$$\begin{aligned} \omega - \omega' &= c\left[(p^2 + m^2c^2)^{\frac{1}{2}} - (p'^2 + m^2c^2)^{\frac{1}{2}}\right] \\ &= c\left\{(p^2 + m^2c^2)^{\frac{1}{2}} - [(\mathbf{p} - \mathbf{k} + \mathbf{k}')^2 + m^2c^2]^{\frac{1}{2}}\right\} \qquad (10.3\text{-}5) \\ &\simeq (\mathbf{k} - \mathbf{k}')\boldsymbol{v}. \end{aligned}$$

In the case of scattering by ions, v is equal to the thermal ion velocity v_T. Then $\omega - \omega' \simeq |\mathbf{k} - \mathbf{k}'|v_T$. If the direction of the wave vector changes appreciably, then

$$\omega - \omega' \simeq kv_T \simeq \omega \frac{v_T}{v_\phi}, \qquad (10.3\text{-}6)$$

where v_ϕ is the phase velocity of the incident wave. In the case of undamped waves, $v_\phi >> v_T$ and the change in frequency during scattering is small. This result is valid for all cases of non-linear scattering by thermal ions, also when the type of wave is changed in the scattering process.

When the change in frequency is known, the difference in the absolute magnitudes of **k** and **k'** can be determined. If the wave type remains unchanged during the scattering, one has

$$\omega(k) - \omega'(k') \simeq (k - k') \frac{\partial\omega}{\partial k} \simeq (k - k')v_g. \qquad (10.3\text{-}7)$$

From (10.3-6) and (10.3-7) it is found that

$$\frac{k - k'}{k} \simeq \frac{v_T}{v_g}. \qquad (10.3\text{-}8)$$

If $(k - k')/k << 1$, the process is called differential scattering. When $(k - k')/k \simeq 1$, one has a case of integral scattering.

In differential scattering both the frequency and the absolute value of the wave number change by a small amount, but the direction of the wave vector may change considerably.

When a longitudinal wave is scattered by a thermal ion and the scattered wave remains a longitudinal wave, the scattering is always integral for $v_\phi < 130v_t$ in an isothermal plasma.

If the longitudinal wave changes into a transverse wave during scattering by the "coat" of the ion, the change in frequency is still small as can be seen from (10.3-6). But the wave number of the transverse wave is much smaller than that of the longitudinal wave. The concept of differential scattering can be introduced also in this case provided all wave numbers are expressed in the same scale.

Induced Processes in Plasma

Equations governing non-linear processes in a plasma may be derived from physical considerations based on the concept of induced process. The idea of induced processes was first introduced by Einstein in connection with a two-level system. But the energy levels in any system are discrete. Then it suffices to limit the discussion to two chosen levels. Statistical equilibrium requires that any two states chosen are in equilibrium with one another.

In a two-level system with energy levels $\varepsilon_1 > \varepsilon_2$, one may have spontaneous transitions from level 1 to level 2 when the system is without external influence. Then a spontaneous emission probability per second, u_{12}, may be defined. If external radiation with frequency near to that of spontaneous emission acts on the system, transitions from state 1 to state 2 as well as from 2 → 1 are induced. The probability of the induced processes depends on the radiation intensity. Denoting the probability of induced emission by w_{12} and the probability of induced absorption by w_{21}, one has

$$w_{12} \propto |\mathbf{E}(\mathbf{k})|^2, \; w_{21} \propto |\mathbf{E}(\mathbf{k})|^2 \qquad (10.3\text{-}9)$$

where $\mathbf{k}$ is the wave number corresponding to the frequency ω_{12} and $\mathbf{E}(\mathbf{k})$ is the electric field of the wave which induced the transitions.

The radiation intensity is proportional to the number of quanta $N(\mathbf{k})$. Thus

$$w_{12} = \tilde{w}_{12}N(\mathbf{k}), \; w_{21} = \tilde{w}_{21}N(\mathbf{k}). \qquad (10.3\text{-}10)$$

In the presence of induced processes one may now write

$$\frac{\partial n_1}{\partial t} = [u_{12} + \tilde{w}_{21}N(\mathbf{k})]n_1 + \tilde{w}_{21}N(\mathbf{k})n_2 = -\frac{\partial n_2}{\partial t}, \qquad (10.3\text{-}11)$$

where n_1 and n_2 are the populations of the two energy levels. Application of (10.3-11) to the case of statistical equilibrium leads to a relationship between u_{12}, $\tilde{\omega}_{12}$ and $\tilde{w}_{21}$, namely

$$\tilde{w}_{12} = u_{12} = \tilde{w}_{21}. \quad (10.3\text{-}12)$$

Therefore the total probability of spontaneous and induced emission is $u_{12}[1+N(\mathbf{k})]$, and the probability of absorption is $u_{12}N(\mathbf{k})$.

The interaction of electromagnetic waves with a two-level system may now be considered. The differential probability of spontaneous emission $u_{12}(\mathbf{k})$ can be introduced

$$u_{12} = \int u_{12}(\mathbf{k})\frac{d\mathbf{k}}{(2\pi)^3} \quad (10.3\text{-}13)$$

Consider a wave $\mathbf{k}$ interacting with the two-level system. One must then write (10.3-11) for the differential probabilities and integrate over all $\mathbf{k}$. Then one has

$$\frac{\partial n_1}{\partial t} = - u_{12}n_1 + (n_2 - n_1) \int u_{12}(\mathbf{k})N(\mathbf{k}) \frac{d\mathbf{k}}{(2\pi)^3}. \quad (10.3\text{-}14)$$

The radiation field varies during the interaction

$$\frac{\partial N(\mathbf{k})}{\partial t} = u_{12}(\mathbf{k}) \; (1 + N(\mathbf{k}))n_1 - u_{12}(\mathbf{k})N(\mathbf{k})n_2$$

$$= u_{12}(\mathbf{k})(n_1 - n_2)N(\mathbf{k}) + u_{12}(\mathbf{k})n_1. \quad (10.3\text{-}15)$$

The number of particles per cubic centimetre is n.

Take a wave packet $N(\mathbf{k})$ with frequency near ω_{12} and wave vector near $\mathbf{k}_0$. The total number of quanta is given by

$$N = \int N(\mathbf{k}) \frac{d\mathbf{k}}{(2\pi)^3}.$$

Integrating over $\mathbf{k}$, (10.3-14) and (10.3-15) may be replaced by

$$\frac{\partial n_1}{\partial t} = - u_{12}n_1 + (n_2 - n_1)u_{12}(\mathbf{k}_0)N = - \frac{\partial n_2}{t} \quad (10.3\text{-}16)$$

$$\frac{\partial N}{\partial t} = u_{12}(\mathbf{k}_0)(n_1 - n_2)N + u_{12}n_1. \quad (10.3\text{-}17)$$

If the intensity of the wave packet is sufficiently large, induced processes are dominant. Then

$$\frac{\partial N}{\partial t} = \gamma N \text{ with } \gamma = u^0_{12}(n_1 - n_2) \text{ and } u^0_{12} = u_{12}(k_0). \qquad (10.3\text{-}18)$$

If $n_1 < n_2$ the wave is absorbed, and if $n_1 > n_2$ it is amplified. When $n_1 > n_2$ it is usually said that the temperature is negative.

In the case $n_1 > n_2$, the number of quanta increases exponentially.

The growth rate of the instability is independent of time only if the radiation intensity is sufficiently low. Saturation may occur. If one neglects spontaneous processes, the change in population is determined from (10.3-16):

$$\frac{\partial n_1}{\partial t} = -\gamma N,$$

$$\frac{\partial (n_1 - n_2)}{\partial t} = -\ 2(n_1 - n_2)u^0_{12}N = -2\gamma N. \qquad (10.3\text{-}19)$$

This shows that the population difference decreases with time leading to the saturation effect.

The two-level system is analogous to the non-linear interaction of three random phase wave packets. Consider two beams of transverse waves and their interaction with plasma waves. The beam of high-frequency waves corresponds to atoms in the upper level. The beam of low-frequency waves is the analogue to atoms in the lower level. One may say that quanta with energy $\hbar\omega_1$ are converted into quanta with energy $\hbar\omega_2 (\hbar\omega_1 > \hbar\omega_2)$ with the emission of longitudinal waves. One then has a physical correspondence with the case of a two-level system. There is a wide range of applications of the formalism outlined here.

Induced Scattering

Consider transformation of electron plasma waves (l-waves) into transverse (t-waves) by scattering on thermal ions (subscript i). The scattering by thermal ions is described by the probability per unit time $w_i^{tl}(\mathbf{k},\mathbf{k}')$ that a plasmon in the range $d\mathbf{k}/(2\pi)^3$ at $\mathbf{k}$ is scattered into a photon in the range $d\mathbf{k}'/(2\pi)^3$ at $\mathbf{k}'$. The evolution of $N^t(\mathbf{k}')$ due to scattering by ions is described by

$$\frac{\partial N^t(\mathbf{k}')}{\partial t} + v_g^t(\mathbf{k}')\frac{\partial N^t(\mathbf{k}')}{\partial s} =$$

$$\int w_i^{tl}(\mathbf{k}',\mathbf{k})\left\{[N^l(\mathbf{k}) - N^t(\mathbf{k}')] - \frac{\hbar[\omega^t(\mathbf{k}')-\omega^l(\mathbf{k})]}{\kappa T_i} N^l(\mathbf{k})N^t(\mathbf{k}')\right\}, \tag{10.3-20}$$

where s is the distance along the ray path and v_g^t is the group velocity. In (10.3-20) the linear term in N^l describes spontaneous scattering $l \rightarrow t$, and the term which is linear in N^t describes spontaneous scattering $t \rightarrow l$. The remaining term describes induced scattering. Induced scattering is important when the intensity is high. Kaplan and Tsytovich (1969) showed that induced scattering dominates when

$$T^t \gtrsim \frac{v_\phi}{2^{\frac{1}{2}}v_T} T_i. \tag{10.3-21}$$

Take $v_\phi \simeq 50\,000$ km s^{-1}, $T_i \simeq 10^6$ K and consequently $v_T \simeq 100$ km s^{-1}. Then $T^t \gtrsim 3 \times 10^8$ K. It is quite likely that induced scattering is important in plasma theories of noise storm radiation.

Number of Waves, $N^\sigma(\mathbf{k})$

It is frequently useful to introduce the number $N^\sigma(\mathbf{k})$ of waves of type σ per unit phase volume. Denoting the energy density of these waves by W^σ one may write

$$W^\sigma = \int \hbar\omega^\sigma(\mathbf{k})N^\sigma(\mathbf{k})\frac{d\mathbf{k}}{(2\pi)^3}, \tag{10.3-22}$$

where $d\mathbf{k} = k^2 dk\, d\Omega$ is the volume element in wave-vector space and $\omega(\mathbf{k})$ represents the dependence of the frequency of the waves on the wave number. The constant $\hbar$ is the Planck constant divided by 2π. In several papers units of measurements are used which make this constant equal to unity.

Plasma Turbulence

Under cosmic conditions there are frequently sources which excite different forms of waves in a plasma. If the power of the emission sources is large (excitation due to instabilities), the level of the plasma wave energy increases rapidly and non-linear interactions occur. The energy extracted from the source when one kind of waves in a narrow range of wave numbers is excited, becomes redistributed over the whole phase space of the original wave mode and is also transferred to other wave modes: plasma turbulence is created. Plasma turbulence is the state of a plasma in which one or several wave modes are excited in a wide range of frequencies and wave numbers. In plasma turbulence the "eigenfrequencies" are of great importance.

Quasi-linear Relaxation

Various forms of plasma instabilities (beam, loss cone, anisotropy, etc.) represent maser effects. They are due to an inversion in energy level populations. In the case of beam instability the inversion occurs in $\mathbf{p}_{\parallel}$, for the loss cone in $\mathbf{p}_{\perp}$ and for the anisotropy instability in θ.

The excitation of plasma waves because of these instabilities leads to a rapid flow of energy from the exciting particles and into plasma waves. Therefore a fast "equalization" of the inversion in the energy levels occurs, and the instability disappears (levelling of the "hump" of the beam, filling of the "pouring out cone", etc.). This process is called quasi-linear relaxation. It involves the interaction between the exciting disturbance and the excited waves.

Rayleigh Scattering, Combination Scattering

Longitudinal electron plasma waves may be transformed to electromagnetic waves by scattering on random inhomogeneities of the plasma. The permeability of the plasma fluctuates with variations in the electron concentration δN. In an isotropic plasma ($\mathbf{B}_0 = 0$) the variation in electron density may be written $\delta N = \delta N' + \delta N''$. The term $\delta N' = \delta N_i$ is connected with the quasi-neutral fluctuations of the plasma density $\delta\rho_i \simeq M\delta N_i$. The second term is connected with variations in the electron concentration accompanying fluctuations in electric charge $\delta\rho = -e\delta N''$. Since variations in plasma density are slow, scattering of longitudinal waves on $\delta N'$ is not accompanied by any significant change in frequency (Rayleigh scattering). Scattering on charge fluctuations $\delta N''$ is a combination of plasma waves and is accompanied by a considerable ($\Delta\omega \simeq \omega_p$) change in frequency (combination scattering). The first mechanism has been invoked to explain transformation of longitudinal waves to radio waves with frequency near to the electron plasma frequency. The second mechanism produces transverse waves near the second harmonic of the electron plasma frequency.

11. PLASMA WAVE THEORIES OF TYPE I BURSTS

In order to obtain sufficiently high brightness temperatures coherent radiation from streams of suprathermal particles have been considered. One class of theories invokes an electrostatic instability excited by a stream propagating along the magnetic field.

11.1. TAKAKURA'S THEORY

The ideas about the origin of type I bursts became more definite when Takakura (1963) presented a theory for such bursts. Two important aspects of the theory were that a plasma wave mechanism was suggested and that the waves were excited by electron streams of low velocity. Furthermore, the streams were generated in the source region of the bursts. Since then the concept of short-lived electron streams generating type I bursts has been applied in several other theories.

Takakura's theory is frequently referred to in the literature, and it has been valuable for subsequent work on the theoretical interpretation of noise storms. In short it may be explained in the following way: coherent plasma waves are excited by a stream of electrons with a group velocity amounting to a few times the thermal speed of coronal electrons. The beam is created during the collision of two wave packets of Alfvén waves propagating along a magnetic field in opposite directions. The duration of the bursts is given by the lifetime of the electron beam, and the transformation of longitudinal plasma waves to transverse waves occurs by Rayleigh scattering. Some points of the theory need to be considered in more detail.

Acceleration

Takakura suggested that thermal electrons might be accelerated when two Alfvén waves collide. The situation is as depicted in Fig. 11.1. Two Alfvén waves propagate in opposite directions along the magnetic field. Thermal electrons may be accelerated by being reflected back and forth once at the colliding wave fronts. The minimum pitch angle θ of electrons which may be reflected is given by

$$\sin^2\theta = B_0/B, \qquad (11.1\text{-}1)$$

where B_0 is the static magnetic field and B is the maximum field at the wave front. The higher the field in the wave front, the larger is the number of electrons which may be reflected. Electrons which do not satisfy (11.1-1) penetrate the wave: they belong to the loss cone. The mechanism proposed here is in principle the mechanism proposed by Fermi to accelerate particles to high energies. The speed of the Alfvén waves is given by

$$v_A = B/(4\pi M N_i)^{\frac{1}{2}} \qquad (11.1\text{-}2)$$

where M is the mass of the proton and N_i the density of protons. If now v_A is about equal to the most probable speed of thermal electrons v_t, a beam of electrons occurs in front of wave number 2 with a velocity in the $-z$ direction which is equal to about $3v_t$. The distribution depends on the time after the collision and on the distance from the wave front. For times of 0.2 - 0.4 s after the collision a beam of electrons with an effective thickness of about 100 km precedes wave number 2. The velocity is about 3.1 - 3.2 v_t in the $-z$ direction. In this way a positive gradient in the distribution function $f'(v_z)$ is derived which is

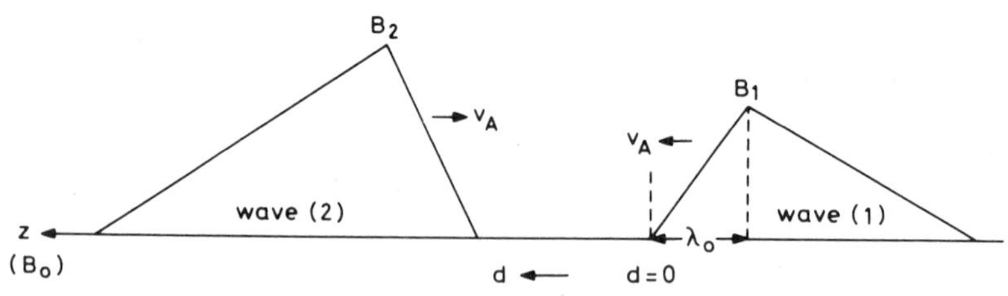

Fig. 11.1. Alfvén waves propagating in opposite directions along magnetic lines of force (Takakura, 1963).

of the order of $10^{-2}/v_t^2$ in a layer of thickness around 100 km.

As was shown in section 9.4, electron beams generated in this way have lifetimes which are in agreement with the measured duration of type I bursts. In order to obtain beam velocities v_0 of correct magnitude a magnetic field giving an Alfvén velocity $v_A = v_0/3$ is necessary. In the 200 MHz radio source one then finds $B \simeq 50$ G, and at 80 MHz a field of 15 G must be present. These are rather high field strengths.

Excitation of Coherent Plasma Waves

When the velocity distribution of the electrons $f(v)$ has a hump at $v \geq 3^{\frac{1}{2}}v_t$ where $f'(v)$ is positive, coherent plasma waves may be excited. Plasma waves may grow up when (Bohm and Gross, 1949b)

$$\frac{\pi}{2}\omega_p\left(\frac{\omega_p}{k}\right)^2 f'\left(\frac{\omega}{k}\right) > \frac{1}{2t_c}, \qquad (11.1\text{-}3)$$

where t_c is the mean collision time of thermal electrons. If one takes $\omega \simeq \omega_p$, $\frac{\omega}{k} = v_0$ and writes $f'(v_0) = \xi v_t^{-2}$, where ξ is a numerical factor (11.1-3) reduces to

$$\xi > \left(\frac{v_t}{v_0}\right)^2 \frac{1}{\pi t_c \omega_p}. \qquad (11.1\text{-}4)$$

With $v_t/v_0 \simeq 1/3$, $t_c = 0.03$ s and $f_p = 200$ MHz $(\omega_p = 4\pi 10^8\ \text{s}^{-1})$ one obtains $\xi > 10^{-9}$. This condition is well satisfied with the acceleration mechanism proposed by Takakura. A value of ξ of the order of 10^{-2} may be obtained at the collision of two Alfvén waves. The magnetic field has been neglected in the treatment of the plasma waves which may be permitted provided the waves are excited parallel to the magnetic field.

Transformation to Radio Waves

Takakura assumed that the excited plasma waves were transformed to transverse waves by the mechanism of Rayleigh scattering proposed by Ginzburg and Zheleznyakov (1958). Then the energy flux P from waves scattered in a volume V is given by

$$P = \frac{ne^4 N_e V}{6m^2 c^3} E_0^{\ 2}, \qquad (11.1\text{-}5)$$

where P is in erg s^{-1}, n is the refractive index for the radio waves, N_e is the electron density and E_0 is the amplitude of the electric field strength in the plasma wave.

Using the quasi-transversal approximation for ordinary waves together with the dispersion equation for plasma waves in the field free case, one finds

$$n = (3/2)^{\frac{1}{2}}(v_t/v_0). \qquad (11.1\text{-}6)$$

The field strength E_0 is not easy to determine. It was estimated by Takakura by assuming that the energy gain of the plasma waves due to the electron beam was balanced by energy loss due to thermal collisions. The expression for P could then be transformed to

$$P \simeq 2 \times 10^3 \frac{e^2 v_0^{\ 2} \omega_p^{\ 2} N_e V}{c^3} \left(\frac{v_t}{v_0}\right)^5 \xi^4, \qquad (11.1\text{-}7)$$

where P is in erg s^{-1}. The flux escapes in a cone Ω which may be determined by refraction and for which one has $\Omega \simeq \pi n^2$ provided $n^2 \ll 1$. Equation (11.1-7) may then be written

$$F \simeq 1.3 \times 10^3 \frac{e^2 v_0^{\ 2} \omega_p^{\ 2} N_e V}{\pi c^3 R^2 \Delta f} \left(\frac{v_t}{v_0}\right)^3 \xi^4. \qquad (11.1\text{-}8)$$

Here R is the distance between the sun and the earth. Assume a frequency of 200 MHz, $v_0 = 1.6 \times 10^9$ cm s^{-1}, $v_t = 5.5 \times 10^8$ cm s^{-1}, $N_e = 5 \times 10^8$ cm^{-3}, $\Delta f = 5$ MHz, $\xi = 10^{-2}$ and $V = 2 \times 10^{27}$ cm^3. Then the flux at the earth becomes

$$F = 5 \times 10^{-21} \text{ W m}^{-2} \text{ Hz}^{-1}.$$

This corresponds to the intensity of a relatively strong type I burst. In (11.1-8) absorption of the radio waves in the corona has been neglected. Since the thickness of the source is about 10^7 cm, the cross-section must cover an area of 2×10^{20} cm^2, which is definitely on the large side. One would then expect a source size slightly above 3' arc in the absence of scattering. Scattering would make the source still wider, so the theory predicts a source size which is at the limit or rather somewhat above the limit which appears acceptable.

Directivity and polarization

Since the waves originate at a level where the refractive index is small, refraction introduces a directivity. In a plane-layered plasma the waves will propagate in a cone directed along the gradient of the refractive index. For small values of the refractive index n, the solid angle of the cone is about

$$\Omega \simeq \pi n^2 \tag{11.1-9}$$

and the aperture angle of the cone, 2ψ, is found from

$$\sin\psi \simeq n. \tag{11.1-10}$$

With a value of v_t/v_0 of about 1/3, one determines n from (11.1-6) and Ω from (11.1-9). It is then seen that the emission occurs within a solid angle of $\pi/6$ or the aperture angle $2\psi \simeq 50^\circ$. For higher values of v_0 the emission becomes still more beamed; as an example, $\Omega \simeq \pi/24$ and $2\psi \simeq 25^\circ$ when $v_0 = 6v_t$. This treatment is rather simplified since magnetic fields have been neglected, but it demonstrates that beaming of the emission is to be expected, as is also actually found from observations.

The polarization properties of storm bursts may be explained as a propagation effect, as was done by Takakura. The condition for the ordinary mode only to escape is

$$1 < X < 1 - Y, \tag{11.1-11}$$

where X and Y are the well-known magneto-ionic parameters. Using the dispersion equation (9.4-9) together with the condition $v_0 = 3v_A$, where v_A is given by (11.1-2), it can be shown that X may be expressed in the following way

$$X \simeq Y^2/(Y^2 + 0.1). \tag{11.1-12}$$

From (11.1-11) and (11.1-12) one obtains

$$Y^2/(Y^2 + 0.1) > 1 - Y \text{ or } Y > 0.4. \qquad (11.1\text{-}13)$$

This is a necessary condition for the waves to be completely polarized in the ordinary magneto-ionic mode. The situation is illustrated in Fig. 11.2.

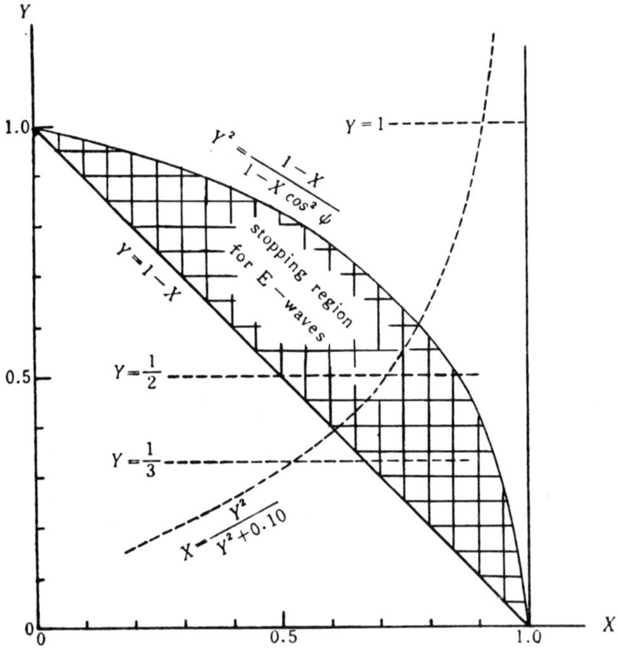

Fig. 11.2. Type I burst sources are situated on the dashed curve. For the ordinary wave only to escape the source position should be above the line $Y = 1 - X$. (Takakura, 1963.)

The magnetic field which was necessary to give a sufficient Alfvén velocity for the electrons to be accelerated to about $3v_t$, was found to be about 50 and 15 G for the 200 and 80 MHz sources respectively. Then $Y \simeq 0.7$ at 200 MHz and $Y \simeq 0.5$ at 80 MHz, and the condition for the ordinary wave only to escape is satisfied. Resonance absorption in layers where $Y = 1/2$ and $Y = 1/3$ increases with increasing angle between the wave and the magnetic field, and the escaping ordinary waves are limited to a cone of opening angle around 30° (Ginzburg and Zheleznyakov, 1959a; Kakinuma and Swarup, 1962).

When the condition (11.1-13) is not satisfied both wave modes may propagate, but resonance absorption at the level $Y = 1/3$ must be taken into account (Fig. 11.3). For waves propagating at small angles with the magnetic field (less than 20°) the absorption of both wave modes is small, and unpolarized bursts may result. For larger angles the ordinary mode only escapes. Partially polarized bursts occur only under restricted conditions.

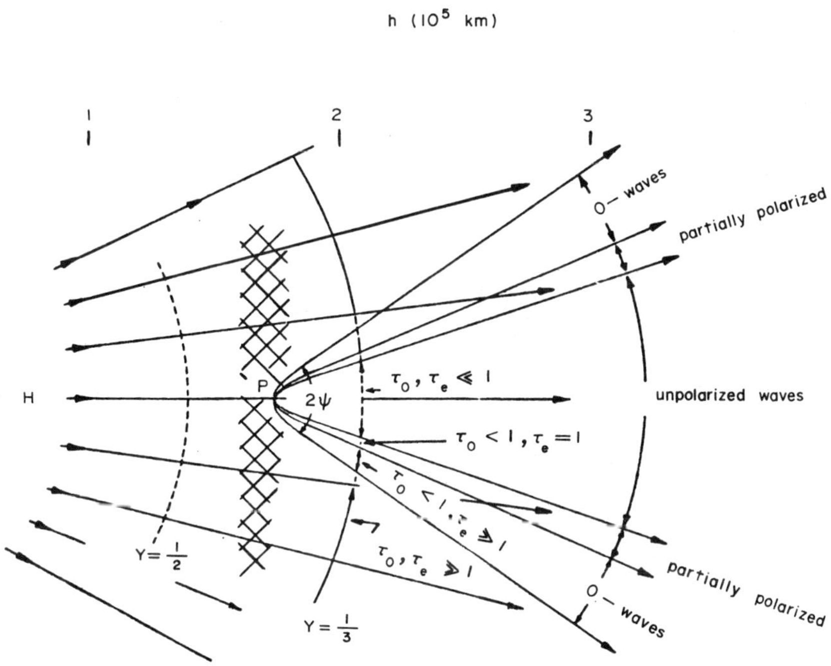

Fig. 11.3. Polarization of escaping waves (Takakura, 1963).

Type I storms are concentrated near central parts of the solar disk. When storms are observed near the limb, the source height is larger than for sources near the central meridian (cf. section 1.5). Takakura suggested that limb sources were generated in dense clouds (which might have filamentary structure) far above the plasma level and the level $Y = 1/3$. Both modes escape, and the bursts are unpolarized, as is observed. Unpolarized bursts may be generated in such regions also when the source is near the centre of the solar disk, but their intensity is likely to be small and they may be difficult to observe together with strongly polarized bursts. When the magnetic field is weak, the velocity of the beam generated in the way proposed by Takakura becomes smaller, and plasma waves may not be set up. The answer given by Takakura was that since the medium surrounding the source has a smaller density, the Alfvén velocity here is relatively high, and the Alfvén velocity in the source may be determined by an average density.

The question of the background continuum was not treated by Takakura. In this connection it is remembered that the posibility of explaining the background continuum as due to the superposition of a large number of bursts has been discussed by several authors (section 9.2).

Conclusions

The theory explains the increasing source height with decreasing frequency because $\omega \simeq \omega_p$. The short duration, narrow bandwidth and possible frequency drift of type I bursts are explained together with the directivity and polarization.

However, several points in the theory may be a matter of discussion. Electric charge separation in the pulse front may prevent the formation of the unstable electron beam (Trakhtengerts, 1966). No really satisfactory treatment of the beam plasma instability has yet been given and the conclusions are largely qualitative. Moreover, non-linear (induced) effects may be important (Sy, 1973a,b). The necessary source size is too large.

It is a merit of the theory that it has provoked further studies of the generation of type I bursts and that it contained many new ideas at the time of publication.

11.2. TRAKHTENGERT'S THEORY

One difficulty mentioned by Takakura was that more electrons than ions were reflected at the pulse front, and a charge separation occurs. This point was taken up by Trakhtengerts (1966), who considered the instability of hydromagnetic waves with charge separation taken into account.

The problem is first to determine the velocity distribution of the electrons and ions when the magnetic field of the hydromagnetic pulse and the electric field produced by charge separation at the wave front are taken into account. Whether this distribution is stable with respect to excitation of plasma waves and electromagnetic waves will now be investigated.

Trakhtengerts found that the instability suggested by Takakura (1963) would be quenched by the electric field, but that a different instability could occur. At the pulse maximum, electron countercurrents appear. A hydrodynamic instability arises and plasma waves are excited.

Distribution Function

Calculation of the velocity distribution of the particles in front of the hydromagnetic wave is somewhat complicated. The reason is that the velocity of the hydromagnetic wave is roughly equal to the thermal electron velocity. Therefore the problem cannot be treated in the hydrodynamic approximation, which requires that the thermal ion and electron velocities are much smaller than the velocity v_A of the hydromagnetic pulse.

Trakhtengerts performed an approximate calculation of the distribution function f for a hydromagnetic wave of prescribed form. It was assumed that the pulse was sufficiently weak and wide so that the magnetic moment was conserved for the electrons

$$\mu = \frac{\frac{1}{2} m v_{\perp e}^{2}}{B} = \text{const.}, \qquad (11.2\text{-}1)$$

where $v_{\perp e}$ is the component of the electron velocity perpendicular to B. The effect of the hydromagnetic pulse is to change the magnetic field $|B| = B(z - v_A t)$. The width d of the pulse was assumed to be within the limits

$$r_{He} << d << \ell, \tag{11.2-2}$$

where $\ell = v_e/\nu$ is the mean free path of electrons and $r_{He} = v_e/\omega_{He}$ is the gyro-radius of an electron. Coulomb collisions "thermalizes" the distribution function with increasing distance from the wave front. Neglecting collisions it might be assumed that a non-Maxwellian distribution function extends to a distance of $\pm\ell$ on both sides of the pulse front.

In a coordinate system tied to the hydromagnetic (MHD) pulse the magnetic moment μ is conserved

$$\mu_j = \frac{\frac{1}{2}m_j v_\perp^2}{B} = \frac{\frac{1}{2}m_j v_{\perp\infty}^2}{B_0} \tag{11.2-3}$$

The total energy when the electric field is taken into account is

$$\tfrac{1}{2}m_j v_\xi^2 + \mu b + q_j\phi = \tfrac{1}{2}m_j v_{\xi\infty}^2 \tag{11.2-4}$$

where $v_\xi = v_\parallel - v_A$, $m_j = M$ and $q_j = e$ for ions, $m_j = m$ and $q_j = -e$ for electrons, $b(\xi) = B(\xi) - B_0$, v_ξ is the longitudinal component of the electron velocity ($\parallel$ to B_0) and ϕ is the electric potential caused by charge separation at the MHD wave front.

A group of particles is reflected with a change in velocity of $\pm 2v_A$. Other particles traverse the pulse without reflection. The condition separating the two groups of particles is that the longitudinal particle velocity vanishes at the maximum of the MHD pulse in a coordinate system moving along with the wave.

$$v_\xi^2 = v_{\xi\infty}^2 - \frac{2\mu_j b}{m_j} - \frac{2q_j\phi_m}{m_j} = 0. \tag{11.2-5}$$

When

$$v_{\xi\infty}^2 = (v_\parallel - v_A)^2 < \frac{2\mu_j b}{m_j} + \frac{2q_j\phi_m}{m_j}$$

the particles are reflected. When

$$v_{\xi\infty}^2 > \frac{2\mu_j b}{m_j} + \frac{2q_j\phi_m}{m_j}$$

the particles penetrate. In these two regions of velocity there will be different forms of the distribution function. The difference consists in the choice of different signs in the expression for the velocity derived from (11.2-3) and (11.2-4) for which one has

$$v_{\parallel\infty} = v_A \pm [(v_\parallel - v_A)^2 + v_\perp{}^2 b/B + 2q_j\phi/m_j]^{\frac{1}{2}}. \quad (11.2\text{-}6)$$

The minus sign is for reflected particles, the plus sign for penetrating particles.

Using (11.2-3) to (11.2-6) and noting that the undisturbed distribution f_0 is Maxwellian corresponding to a temperature T (T is henceforth in energy units), one has in the laboratory system in which the MHD pulse is propagating in the $+\xi$ direction with $\xi \geq 0$,

$$f = \begin{cases} f^- & \text{for } v_A \leq v_\parallel < \infty \text{ and } (v_\parallel - v_A)^2 \leq v_\perp{}^2 b_m/B_0 + 2q_j\phi_m/m_j, \\ f^+ & \text{for } v_A < v_\parallel < \infty \text{ and } (v_\parallel - v_A)^2 \geq v_\perp{}^2 b_m/B_0 + 2q_j\phi_m/m_j. \end{cases} \quad (11.2\text{-}7)$$

The amplitude of the MHD wave is b_m. The form of the distribution function $f^\pm$ is the following:

$$f^\pm = (m_j/2\pi T)^{3/2} \exp\left\{-m_j v_\perp{}^2 B_0/2TB - m_j/2T\left[((v_\parallel - v_A)^2 + v_\perp{}^2 b/B + 2q_j\phi/m_j)^{\frac{1}{2}} \pm v_A\right]\right\}^2 \times$$
$$1\left\{m_j(v_\parallel - v_A)^2/2 + q_j\phi + m_j v_\perp{}^2 b/2B\right\}, \quad (11.2\text{-}8)$$

where

$$1(x) = \begin{cases} 1 \text{ when } x > 0 \\ 0 \text{ when } x \leq 0. \end{cases}$$

The electric field occurs in the wave front because $v_T \ll v_t \simeq v_A$, where v_T is the thermal ion velocity. Therefore more electrons will be reflected than ions (for ions $v_\perp/v_\parallel$ as referred to the wave front will be very small). Because of the charge separation an electric field appears, which tends to pull the electrons through the magnetic barrier whereas it hardly affects the ion motion.

Electric Field at Pulse Front

In order to determine the electric potential $\phi(\xi)$ at the pulse front, the kinetic equation for the distribution function f and the Poisson equation for ϕ must be solved simultaneously. Under stationary conditions the charge flux at infinity on both sides of the MHD wave front should vanish:

$$I_{e_{\text{tot}}} = I_{i_{\text{tot}}}, \qquad (11.2\text{-}9)$$

where $I_e = \int v\, f_e dv$ and $I_i = \int v_{\parallel} f_i dv$ denote the electron and ion flux respectively. Introducing the distribution function (11.2-8) in (11.2-9) and using (11.2-7) a transcendental equation is obtained for ϕ_m. The discussion of this equation shows that the result emerging from the more rigorous treatment can be obtained in a simple way. In the case of interest here, $mv_A{}^2/2T \simeq 1 << Mv_A{}^2/2T$, and since $b_m/B_m << 1$ the following expression is obtained for ϕ_m:

$$e\phi_m/T = (Mv_A{}^2/2T)\, b_m/B_m, \qquad (11.2\text{-}10)$$

where $B_m = B_0 + b_m$.

This equation together with (11.2-7) and (11.2-8), determines the stationary distribution function of particles in a magnetized plasma in the presence of a MHD pulse of finite amplitude. The question is now whether this distribution is stable or not. This may be determined by finding the complex frequency from the dispersion equation for plasma waves: the imaginary contribution to the frequency represents the increment of instability.

No Instability Outside the Pulse Front

When ion motion is neglected, the dispersion equation for plasma waves propagating along the magnetic field has the form

(Zheleznyakov, 1970)

$$1 + 2\pi \frac{\omega_p{}^2}{k} \iint \frac{\partial f/\partial v_{\parallel}}{\omega - kv_{\parallel} - i\nu} v_{\perp}\, dv_{\perp}\, dv_{\parallel} = 0 \qquad (11.2\text{-}11)$$

where k is the wave number and ω is the complex plasma-wave frequency.

Trakhtengerts treated two regions separately: outside the front $|\xi| > d$, where the instability involves reflected particles only and inside the front where the electric field makes the major contribution.

In the first case one finds from the dispersion equation (11.2-11) using (11.2-7) and (11.2-8), the following result for the instability increment γ:

$$\gamma = (\pi/8)^{\frac{1}{2}} \omega_p{}^3/k^2 v_t{}^3 \left\{ (2v_A - \frac{\tilde{\omega}}{k}) \exp[-\frac{m}{2T}(\frac{\tilde{\omega}}{k} - 2v_A)^2 - (\frac{\tilde{\omega}}{k} - v_A)^2 \frac{mB_0}{2Tb_m} - \frac{e\phi_m}{T}\frac{B_0}{b_m}] - \frac{\tilde{\omega}}{k}\left(1 - \exp[-\frac{m}{2T}\frac{B_0}{b_m}(\frac{\tilde{\omega}}{k} - v_A)^2 - \frac{e\phi_m}{T}\frac{B_0}{b_m}]\right) \exp(-\frac{m\tilde{\omega}^2}{2k^2T}) \right\} \qquad (11.2\text{-}12)$$

where $\tilde{\omega}$ is the real part of ω related to k by the dispersion equation $k = \tilde{\omega}[(\tilde{\omega}^2 - \omega_p{}^2)/3\tilde{\omega}^2 v_t{}^2]^{\frac{1}{2}}$, $v_t = (2T/m)^{\frac{1}{2}}$ and ϕ_m is given by (11.2-10). From the equation for γ it is deduced that the electric field at the pulse ϕ_m introduces a reduction in the number of reflected electrons and effectively quenches the instability.

When two hydromagnetic waves travel towards each other, multiple reflections may occur. The longitudinal velocity will then increase, but Trakhtengerts showed that it is even more difficult for instability to occur for the fast particle beams produced through multiple reflections than in the case of single reflections.

The conclusion is that the electric field at the front of the MHD pulse removes the beam instability suggested by Takakura (1963).

Hydrodynamic Instability

Due to the electric field electron countercurrents appear at the pulse maximum. A rough estimate of the velocity of the opposing streams can be made from the relation

$$(1/2)mv_n^2 \simeq e\phi,$$

$$v_n \simeq (2e\phi/m)^{\frac{1}{2}} \simeq v_A(Mb_m/mB_m)^{\frac{1}{2}}, \qquad (11.2\text{-}13)$$

when (11.2-10) is used. If the condition

$$v_A(Mb/mB)^{\frac{1}{2}} >> v_t \qquad (11.2\text{-}14)$$

is satisfied, a hydrodynamic instability arises, resulting in the excitation of plasma waves.

Using the distribution function (11.2-7) and (11.2-8) in the dispersion equation (11.2-11) and integrating with respect to $v_\perp$, one obtains

$$1 + \alpha(P_1 - P_2) = 0, \qquad (11.2\text{-}15)$$

where $\alpha = \omega_p^{\ 2}/\pi^{\frac{1}{2}}v_t$ and P_1 and P_2 are given by

$$P_1 = (B_0/b)(2T/m) \int_0^{e\phi/T} \frac{\exp(-B_0u/b)}{\omega^2 - 2e\phi k^2/m + k^2v_t^{\ 2}u}\left(\frac{2e\phi}{m} - uv_t^{\ 2}\right)^{\frac{1}{2}} du$$

and

$$P_2 = (2T/m) \int_0^{\infty} \frac{\exp(-u)}{\omega^2 - 2e\phi k^2/m - k^2v_t^{\ 2}u}\left(\frac{2e\phi}{m} - uv_t^{\ 2}\right)^{\frac{1}{2}} du$$

Provided

$$|\omega^2 - 2e\phi k^2/m| >> k^2v_t$$

it can be shown that a hydrodynamic instability arises involving two countercurrents. The dispersion equation becomes

$$1 - \frac{\omega_p^{\ 2}k^2v_t(2e\phi/m)^{\frac{1}{2}}B}{\pi^{\frac{1}{2}}(\omega^2 - 2e\phi k^2/m)^2B_0} = 0. \qquad (11.2\text{-}16)$$

The instability will have an aperiodic character with increment

$$\gamma = v_t\left[\frac{e\phi}{T}\,k^2 + \frac{\omega_p k}{v_t}\left(\frac{B}{B_0\pi^{\frac{1}{2}}}\right)^{\frac{1}{2}}\left(\frac{e\phi}{T}\right)^{1/4}\right]^{\frac{1}{2}}$$

in a coordinate system moving with the MHD pulse. For the maximum increment one gets

$$\gamma_{max} \simeq 2^{-\frac{1}{2}}\omega_p\left[\left(\frac{e\phi}{T}\right)^{-1/4}\left(\frac{B}{B_0\pi^{\frac{1}{2}}}\right)^{\frac{1}{2}}\right] \qquad (11.2\text{-}18)$$

and the wavelength becomes

$$\lambda(\gamma_{max}) \simeq (\pi v_t/\omega_p)(e\phi/T)^{3/4}. \qquad (11.2\text{-}19)$$

Plasma waves will be excited in the region where $e\phi/T >> 1$. These, which have $\omega \simeq 0$, must be scattered on high-frequency plasma fluctuations with $\omega \simeq \omega_p$ in order to produce radio waves which may be observed at the earth.

Possible Mechanism for Generation of Type I Bursts

Suppose that hydromagnetic waves are generated at the photospheric level and travel into the corona along magnetic lines of force in the bipolar field above a sunspot group. One condition for the instability to arise is that the width d of the MHD wave front should be much less than the mean free path ℓ of the electrons. This condition begins to hold in the transition region between chromosphere and corona where there is a sharp increase in the temperature. Moreover, the condition (11.2-14) must be satisfied. For $v_A(bM/Bm)^{\frac{1}{2}} >> v_t$ to occur, it is necessary that

$$b_m/B_0 >> 8\pi NT/B_0{}^2 = W_T/W_M, \qquad (11.2\text{-}20)$$

where W_T and W_M are the thermal and the magnetic energy densities. In the case $v_A \simeq v_t$ this condition may be written $b_m/B_0 >> m/M$.

According to the theory of Trakhtengerts one would expect to observe bursts with narrow bandwidth drifting in frequency at a rate corresponding to v_A, the Alfvén velocity. The duration may be determined by the time taken for the pulse to transfer its energy to the instability in question, or by the time taken for the pulse to travel into a region where the conditions for the excitation of the instability no longer holds.

The instability increment may increase when two MHD waves collide because the potentials ϕ of the waves are added to each other. The duration of the bursts generated in this way will be given by the sum of the interaction time for the two pulses and the damping time for the plasma waves.

It is possible that the above results may be applied to explain type I bursts and type I chains. Hanasz (1966) found that most chains had an outward component of velocity. The distribution of radial velocities showed a maximum at around $+200$ km s^{-1} and a spread between $+800$ and -700 km s^{-1}. Since the horizontal component in many cases probably is considerable, it is not unreasonable to assume that the velocity of the disturbance may be in the range of $500-1000$ km s^{-1}. The propagation velocity is equal to the Alfvén velocity

$$v_A = 2 \times 10^4 \frac{B}{f_p} \text{ (km s}^{-1}\text{)},$$

where B is in gauss and f_p in megahertz. Inserting numerical values it is found that the magnetic field strength in the source region may be of the order of 5 - 10 G.

A condition for charge separation to occur and electron countercurrents to appear was that (11.2-2) holds.

$$r_{He} << d << \ell.$$

For a temperature of 1 million degrees, a density of 4×10^8 cm^{-3} and a magnetic field of 5 G, one gets 10 cm $< d <$ 350×10^5 cm. The characteristic dimension of the hydromagnetic wave is about the same as the ion gyro radius:

$$r_{Hi} = \frac{v_A}{\Omega_H} \simeq 10^3 \text{ (cm)},$$

which is in accordance with the above requirements.

The electrons must be accelerated to a velocity which is considerably larger than the thermal electron velocity in the ambient plasma. For the hydrodynamic instability to arise one must have (11.2-14)

$$v_A\left(\frac{bM}{Bm}\right)^{\frac{1}{2}} >> v_t .$$

If $v_A \simeq 1000$ km s^{-1} and v_t the electron thermal velocity corresponding to a temperature of 1 million degrees, one gets

$$(b/B) >> (1/50),$$

and this condition is likely to be met. Lower values of v_A may also occur without drastic consequences for the mechanism. Thus it seems that the mechanism may work under conditions appropriate to those to be expected in a noise storm region in the corona. In an encounter between oppositely propagating pulses the instability increments may increase considerably through summation of the potentials of the two waves.

In this theory the duration of a single burst amounts to

$$\Delta t = (d_1 + d_2)/v_a + \tau,$$

where $d_{1,2}$ are the widths of the colliding pulses and τ is the damping time of the plasma waves. The first part of the expression may be neglected, so the duration depends essentially on the last term. Correct values for the burst duration can be obtained using possible values of the physical parameters in the source.

The plasma waves may be converted to electromagnetic waves by combination scattering. The bandwidth of the radiation will be narrow as observed for type I bursts.

Type I bursts and chains of bursts may now be produced in the following way: normally type I bursts are generated when two magneto-dynamic pulses collide in a filament (stream tube). The pulses loose a considerable part of their energy in such a collision and are unable to excite more bursts. Thus most frequently one burst only is generated. If one of the pulses is stronger, two or more bursts may be excited. A pulse which is particularly strong may generate a chain of type I bursts when it collides with a train of pulses propagating in the opposite direction. The stronger disturbances may be those which travel in the outward direction, and the majority of the chains may get a negative frequency drift velocity. When a strong disturbance occurs in the region near a sunspot, pulses may travel upwards along two or perhaps more tubes. In such cases one may expect chains on neighbouring frequencies. But the frequency separation between simultaneous chains need not have a characteristic value; one would rather expect a continuous spectrum of frequency differences, as was observed by Hanasz (1966). Another consequence is that chains need not start simultaneously, and neighbouring chains may in general be uncorrelated. This is in good agreement with the observations.

From a strong magneto-dynamic pulse emission may be observed also in the time interval between collisions. One may suggest that this accounts for the "bridges" of emission between individual bursts frequently observed in chains.

Conclusions

The model explains that the emission frequency decreases with increasing height, since $\omega \simeq \omega_p$. Refraction then causes a beaming of the emission. The polarization of type I bursts may be explained. Chains come out as a natural consequence of the theory, which also explains the narrow bandwidth and short duration of the bursts.

There is a serious objection to the theory: it fails to explain the high brightness temperature of type I burst sources, the instability produces plasma waves of very low frequency, $\omega \simeq 0$. Trakhtengerts suggested that these waves were transformed to high-frequency waves by scattering on high-frequency thermal fluctuations. But then, as shown by Smith and Sturrock (1971), the radiation temperature of the elctromagnetic radiation will be that of the corona (about a million degrees). This is far below the values derived from observations.

11.3. SY'S THEORY

Further work on the plasma wave theory of type I bursts was undertaken by Sy (1973a). Following Takakura (1963) he assumed that an electron stream with a speed several times the thermal speed of electrons generated coherent plasma waves. Three factors were considered to be of particular significance for a theory:

(a) when the energy densities of radiation and electron plasma waves are high, induced effects are important;
(b) the magneto-ionic modes are weakly coupled and maintain their separate identities;
(c) scattering of the radiation by coronal inhomogeneities increases the observed source size.

On account of the results presented in the chapters on the polarization of noise storm emission and the coronal scattering of radiation from noise storm sources, it appears likely that the second assumption may be wrong and, moreover, that scattering is less effective than previously assumed.

Sy attempted in particular to explain the high brightness temperature of type I bursts and the strong circular polarization in the ordinary mode.

Amplification

In the plasma wave theory of type I bursts, longitudinal waves must be converted to transverse electro-magnetic waves. This occurs by scattering, and the dominant process in the problem under consideration is the non-linear scattering by plasma ions. Amplification occurs when an initially small amount of radiation produced by spontaneous non-linear scattering of electron plasma waves, induces more electron plasma waves to be scattered to produce radiation.

In the case of small energy density of electron plasma waves and an optically thin source, induced effects can be neglected. However, considering possible energy densities as derived from observations, Sy assumed that the sources of type I bursts were optically thick. Then the energy balance between electromagnetic and plasma waves is predominantly determined by induced scattering.

The physics of induced scattering is simple. Consider one thermal, charged particle. Around this particle there will be a polarization cloud. If it is an ion, neighbouring electrons will be attracted and ions repelled. Accordingly there will be a "coat" of electrons with an excessive negative charge around an ion, the size of which being of the same order of magnitude as the Debye radius. When a plasma wave is incident upon the "coat", it is scattered and forms a wave of the same type or another type. This process is called non-linear scattering. A thermal electron forms a similar "coat" but with an opposite charge. Scattering in this case is more complicated. An incident plasma wave will move both the "coat" and the "central" electron. The "coat" and the electron both emit secondary waves, but because the charges are of opposite sign, the two emissions are in antiphase and may frequently cancel one another (Kaplan and Tsytovich, 1973).

Induced scattering occurs with a decrease in the frequency of the scattered waves. Therefore radiation with frequencies less than the average frequency of the electron plasma waves is amplified.

Polarization

In the source region the electron gyro frequency is considerably smaller than the electron plasma frequency in the model assumed by Sy. Then the absorption coefficients (which may be negative in the case of amplification) are related to the efficiency of non-linear scattering. The efficiency of the scattering of electron plasma waves to *o*-mode and *x*-mode radiation depends on $|\mathbf{E}_l\mathbf{E}|^2$. $\mathbf{E}_l$ is the electric field of the plasma waves and **E** the field of the radio waves.

When the electron plasma waves propagate parallel to the magnetic field, $\mathbf{E}_l$ is parallel to **B**. The scattering is most efficient for that magneto-ionic wave mode with the greater component of **E** parallel to **B**. This is the *o*-mode. For the case of propagation perpendicular to **B**, the **E** vector of the *o*-mode is parallel to **B**, while the **E** vector of the *x*-mode is perpendicular to **B**. Even a small excess in the *o*-mode absorption coefficient (negative) can lead to a high net degree of polarization in the *o*-mode. This occurs because the intensity of the amplified radiation emerging from the source depends exponentially on the optical depths.

Simplified Model

An electron stream is assumed to travel along magnetic field lines at a speed v_0, which is several times the thermal speed of electrons, v_t. The stream generates plasma waves propagating along **B**. In the source region $\omega_H \ll \omega_p$ applies. There is an amplifying and an absorbing layer (Fig. 11.4). In the amplifying layer the frequency of the radiation is less than the frequency of the electron plasma waves. Since the coronal electron density decreases in the outward direction, there may be a region where the opposite condition occurs and net absorption takes place. The layers may be a few thousand kilometres thick. The thickness of the absorbing layer ℓ'' must be smaller than that of the amplifying layer ℓ'. Assuming a coronal temperature of 2×10^6 K, the brightness temperature of the radiation at the source T_{bs} was derived by Sy (1973b) and is given by

$$T_{bs} \approx 3.4 \times 10^9 \exp\left(0.4 W \omega_p \ell / N_e\right), \tag{11.3-1}$$

where $\ell = \ell' - \ell''$ and W is the energy density of the plasma waves. With $\ell = 10^8$ cm the energy density W needed to produce a burst with $T_{bs} = 10^{16}$ K is of the order of 10^{-7} erg cm^{-3}. Assuming $v_0 \approx 6 v_t$ a stream density of 3×10^2 cm^{-3} is needed to produce this energy density.

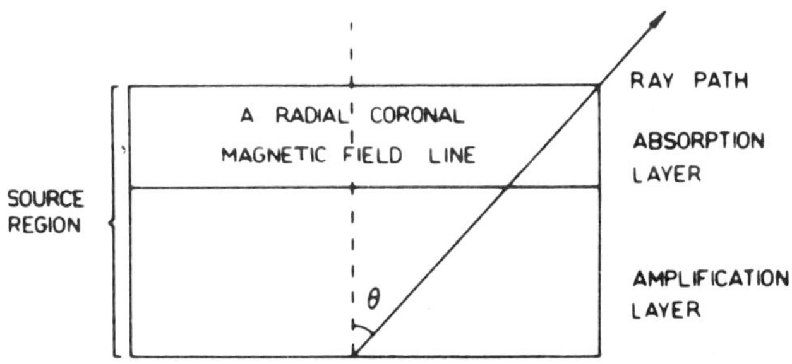

Fig. 11.4. Schematic drawing of the source region (Sy, 1973a).

The net degree of circular polarization in the ordinary mode emerging from the source was estimated by Sy and is found from

$$r_c = [\exp(-\tau_o + \tau_x) - 1][\exp(-\tau_o + \tau_x) + 1]^{-1}, \tag{11.3-2}$$

where τ_o and τ_x are the optical depths of the ordinary and the extraordinary modes including contributions from both layers. The predominance of emission in the ordinary mode is most difficult to explain when the magnetic field is weak. For $\omega_H/\omega_p = 10^{-2}$ Sy found the result given in Fig. 11.5, and it is seen that the emission in the ordinary mode reaches a high degree of circular polarization for angles $\theta \gtrsim 30^\circ$.

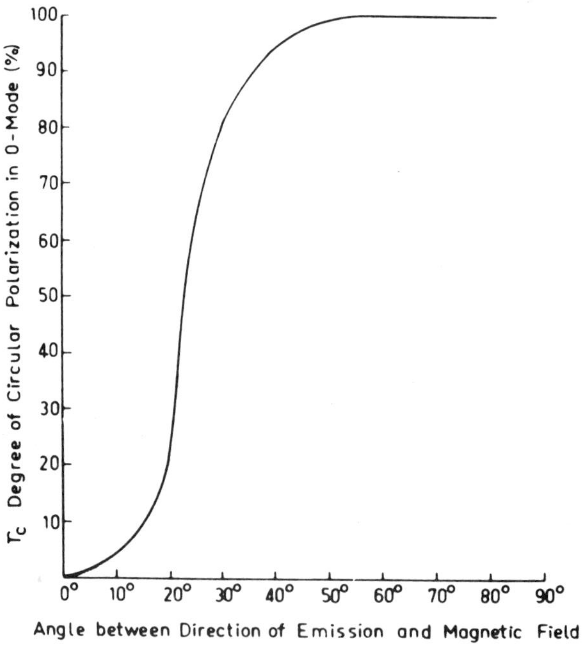

Fig. 11.5. The degree of circular polarization in the ordinary mode as a function of the angle between the direction of emission and the magnetic field (Sy, 1973a).

Conclusions

The theory is developed to account for the properties of type I bursts in terms of plasma radiation. Two important observational results are in focus; the high brightness temperature of type I sources and the almost 100% circular polarization of the radiation in the ordinary mode. Both properties are accounted for in the theory.

The polarization depends on the optical depth of the source which is not well known. An unsolved problem is how electron streams with a velocity $v_0 \simeq 6v_t$ are generated.

11.4. VERESHKOV'S THEORY

Vereshkov (1974) suggested that type I bursts might be the result of the conversion of plasma waves excited at the front of a hydromagnetic pulse and an acoustic wave into radio waves. This may occur when the MHD wave packets overtake acoustic (A) waves propagating in active regions in the solar corona.

The MHD pulse propagates with the Alfvén velocity v_A which may be in the range of 1000 km s^{-1} in the region of radio sources. The velocity of propagation of the sound waves is given by $v_s \approx (T_e/M)^{\frac{1}{2}}$ (where T_e is in energy units) and is of the order of 100 km s^{-1}. Thus one has $v_A >> v_s$.

The A-waves propagate with the wave vector $\mathbf{k}_s || \mathbf{B}_0$. Since the wave packets have small dimensions compared with the scales of the corona and, furthermore, $v_A >> v_s$, the emission may be regarded as occuring from a small source in the corona. The generation of A-waves and MHD waves constitutes a special problem, but it is quite possible that such waves are generated in active regions, for instance in connection with flares.

Conversion Mechanism

When plasma waves are transformed into radio waves through interaction with acoustic waves, the laws of conservation of energy and momentum must hold,

$$\omega_t = \omega_l + \omega_s, \quad \mathbf{k}_t = \mathbf{k}_l + \mathbf{k}_s. \tag{11.4-1}$$

In the present case $\omega_s << \omega_t$, ω_l and $|\mathbf{k}_s| << |\mathbf{k}_t|$, $|\mathbf{k}_l|$. Consider quasi-longitudinal (along B_0) wave propagation. Then the following dispersion relations apply:

$$\left.\begin{aligned} \omega_t^2 &= \frac{\omega_p^2}{1 \pm \omega_H/\omega_p} + c^2 k_t^2, \\ \omega_l^2 &= \omega_p^2 + 3v_t^2 k_l^2, \\ \omega_s^2 &= k_s^2 v_s^2. \end{aligned}\right\} \qquad (11.4\text{-}2)$$

The plus sign is for the ordinary and the minus sign is for the extraordinary mode. From (11.4-1) and (11.4-2) it is seen that a magnetic field is necessary for the conversion to occur, and that the ordinary mode only can be generated. This is an interesting point since observations definitely favour emission in the ordinary mode. Using the conservation laws one may now determine the following relations:

$$\left.\begin{aligned} \omega_t^2 &\approx \omega_p^2 + \frac{3v_t^2}{c^2}\,\frac{\omega_p \omega_H}{1+\omega_H/\omega_p}, \\ k_t &\approx \frac{\omega_p}{c}\left(\frac{\omega_p/\omega_H}{1+\omega_H/\omega_p}\right)^{\frac{1}{2}}, \\ \Delta k_t &\approx \Delta k_l \approx \left(\frac{\omega_p}{\omega_H}\right)^{\frac{1}{2}}\left(1+\frac{\omega_H}{\omega_p}\right)^{\frac{1}{2}} \frac{v_s}{c}\,\Delta k_s, \end{aligned}\right\} \qquad (11.4\text{-}3)$$

where Δk_t, Δk_l and Δk_s are the spectral widths of the waves which interact. The kinetic equation for plasma particles may be written

$$\left(\frac{\partial}{\partial t} + \mathbf{v}\,\frac{\partial}{\partial \mathbf{r}}\right) f_p = \int \Lambda(f_p f_{p_1})\,\sigma(p, p_1)\,dp_1, \qquad (11.4\text{-}4)$$

where f_p is the particle distribution function and v their velocity. The quadratic dependence of the distribution function on the right-hand side is written symbolically $\Lambda(f_p f_{p_1})$, and describes the collisions between particles. An analogous equation may be derived for the interaction of waves in the plasma (Tsytovich, 1970). The kinetic equation for quanta of the transverse waves may be written:

$$\frac{dN(\mathbf{k})^t}{dt} = \int w_t^{ls} N^l(\mathbf{k}) N^s(\mathbf{k}) \frac{d\mathbf{k}_l d\mathbf{k}_s}{(2\pi)^6} - N^t(\mathbf{k}) \int w_t^{ls} (N^l(\mathbf{k}) + N^s(\mathbf{k}) + 1) \frac{d\mathbf{k}_l d\mathbf{k}_s}{(2\pi)^6}, \qquad (11.4\text{-}5)$$

where $N^t(\mathbf{k})$, $N^l(\mathbf{k})$ and $N^s(\mathbf{k})$ are the densities of the number of quanta in the plasma and w_t^{ls} is the conversion probability. If one compares equation (11.4-5) with the radiation transfer equation, the emission coefficient for the conversion is equal to

$$\varepsilon^t(\mathbf{k}) = \int w_t^{ls} N^l(\mathbf{k}) N^s(\mathbf{k}) \frac{d\mathbf{k}_l d\mathbf{k}_s}{(2\pi)^6}$$

and the absorption coefficient is

$$\mu^t(\mathbf{k}) = \frac{1}{v_g^t} \int w_t^{ls} (N^l(\mathbf{k}) + N^s(\mathbf{k}) + 1) \frac{d\mathbf{k}_l d\mathbf{k}_s}{(2\pi)^6},$$

where v_g^t is the group velocity of the electromagnetic waves. From the kinetic equation Vereshkov derived, after som calculations, the following equations for the intensity of the emission I and for the effective temperature T_{eff}:

$$I = \frac{\omega_t k_t^2}{(2\pi)^3} N^l(\mathbf{k}) \left[1 + \frac{4\pi k_s^2}{\tilde{N}^s(\mathbf{k})} (N^l(\mathbf{k}) + 1)\right]^{-1} \qquad (11.4\text{-}6)$$

$$T_{\text{eff}} = \omega_t N^l(\mathbf{k}) \left[1 + \frac{4\pi k_s^2}{\tilde{N}^s(\mathbf{k})} N^l(\mathbf{k}) + 1\right]^{-1}, \qquad (11.4\text{-}7)$$

and the conversion coefficient for electron plasma waves on acoustic waves becomes

$$\eta = \left[1 + \frac{4\pi k_o^2}{\tilde{N}^s(\mathbf{k})} (N^l(\mathbf{k}) + 1)\right]^{-1}, \qquad (11.4\text{-}8)$$

where $\tilde{N}^s(\mathbf{k})$ is the energy density of the acoustic waves determined from

$$N^s(\mathbf{k}) = \tilde{N}^s(\mathbf{k}) \delta(k_x) \delta(k_y), \quad W^s \simeq \frac{1}{(2\pi)^3} \omega_s \tilde{N}^s(\mathbf{k}) \Delta k_s.$$

The probability of the process F may be expressed by

$$F \simeq \frac{2}{3} \omega_p \frac{e^2 M k_s^2 k_l}{m^2 N_e T_e} \qquad (11.4\text{-}9)$$

Application to Type I Bursts

As shown by Trakhtengerts (1966), an electric field is created in the front of the MHD pulse, and electrons are accelerated. Vereshkov derived the electric field in a somewhat different way than made by Trakhtengerts, but the result was in agreement with that obtained by Trakhtengerts as concerns order of magnitude.

Plasma turbulence may now be generated at the front of the MHD wave provided $\ell > d$, where ℓ is the mean free path of electrons and d is the width of the pulse front, $d \simeq r_{Hi}$, the gyro radius of ions. Moreover, the Cerenkov condition must be satisfied, $v_0^2/3v_t^2 > 1$, where v_0 is the velocity of the accelerated electrons. The minimum density of the accelerated cluster of electrons is estimated from the condition that the instability increment is equal to the damping decrement which, according to Kaplan and Tsytovich (1973), occurs for

$$N_0 = \frac{N_e}{N_D}\left(\frac{\delta v_0}{v_0}\right), \qquad (11.4\text{-}10)$$

where N_D is the Debye number ($N_D = N_e(v_t/\omega_p)^3$) and $\delta v_0 \approx v_t$. The maximum value of N_0 is $N_0(\max) = N_e/N_D$ and a larger density of electrons cannot be accelerated in the front of a MHD pulse. When

$$\frac{\delta v_0}{v_0} \approx \frac{v_t}{v_0} > \left(\frac{N_0}{N_e}\right)^{\frac{1}{2}} \approx \left(\frac{1}{N_D}\right)^{\frac{1}{2}},$$

the energy density of the plasma turbulence is approximately (Kaplan and Tsytovich, 1973)

$$W^l = \frac{N_0 m v_0^2}{3} = \frac{N_e T}{N_D}\,\frac{v_0^2}{3v_t^2}. \qquad (11.4\text{-}11)$$

Most of the energy is in the region around the wave number corresponding to the maximum instability increment

$$k_0 = \frac{\omega_p}{v_0} > \frac{\omega_p}{c}\left(\frac{\omega_H/\omega_p}{1+\omega_H/\omega_p}\right)^{\frac{1}{2}}.$$

Therefore $k_0 > k_l \simeq k_t$ introduces difficulties with the conservation laws. Moreover, the turbulence is mainly excited across the magnetic field, whereas waves propagating along the field are of interest for the conversion process. However, the magnetic field has the effect on the plasma turbulence to line up the oscillations along the field direction, and integral transfer (Kaplan and Tsytovich, 1973) brings the wave numbers from $k_0 \approx \omega_p/v_0$ to about $k_t \approx \omega_p/c$. Thus the magnetic field is important for the mechanism to work, but it need not be stronger than to permit quantitative estimates of characteristic times using equations derived for isotropic plasma. For the turning of the wave vectors in the process of differential transfer, scattering of waves on ions is the important factor. In active coronal regions also the integral transfer occurs by scattering on ions, and it can be shown that the following relation between the increments holds:

$$\gamma_{dif} \geq \gamma^{i}_{int} .$$

Three characteristic times are now of importance in the generation of type I bursts:

(a) the instability time

$$\frac{1}{\tau_{in}} = \gamma_{in} = \frac{\pi}{2} \omega_p \frac{N_0}{N_e} \left(\frac{v_0}{\delta v_0}\right)^2 ;$$

(b) the transfer time

$$\frac{1}{\tau_t} = \gamma^{i}_{int} = \frac{(2\pi)^{\frac{1}{2}} T_e T_i k_0 W^l}{8(T_e+T_i)^2 N_e m v_{ti}} ;$$

(c) the conversion time

$$\frac{1}{\tau_c} = \gamma_c = \mu^t_c v^t_g = \frac{2}{3(2\pi)^3} \omega_p \frac{e^2 M k_0 W^s}{m^2 N_e T_e v_s{}^2 \eta} .$$

The above times must be less than the collision time. One then obtains:

$$\left.\begin{aligned}
&\tau_{in} < \tau_{coll} \quad \text{when } v_0^2/3v_t^2 > 1, \\
&\tau_t < \tau_{coll} \quad \text{when } \frac{(2\pi)^{\frac{1}{2}}}{24}\left(\frac{M}{m}\right)^{\frac{1}{2}} \frac{T_e^{3/2} T_i^{\frac{1}{2}}}{(T_e+T_i)^2} \frac{v_0}{v_t} >> 1, \\
&\tau_c < \tau_{coll} \quad \text{when } v^2 > v_s^2\left(\frac{m}{M}\right)^2 \frac{T_e}{e^2 k_l} \frac{\eta}{N_D} 3(2\pi)^3
\end{aligned}\right\} \qquad (11.4\text{-}12)$$

The latter inequality imposes a condition on the characteristic velocity of the displacement of the plasma v.

In order to obtain a sufficient intensity of the emission, the conversion time τ_c must exceed the transfer time τ_t.

$$\frac{2}{3(2\pi)^3}\,\omega_p\,\frac{e^2 M k_l W^s}{m^2 N_e T_e v_s^2 \eta} < \frac{(2\pi)^{\frac{1}{2}}}{8}\,\frac{T_e T_i}{(T_e+T_i)^2}\,\frac{k_0 W^l}{N_e m v_{ti}}. \qquad (11.4\text{-}13)$$

From the inequalities

$$\ell > d,\; v_0^2/3v_t^2 > 1,\; \tau_c < \tau_{coll},\; \tau_c > \tau_t,$$

and the condition for weak damping of the acoustic waves, $\omega_s << \frac{\omega_p}{N_D}\left(\frac{m}{M}\right)^{\frac{1}{2}}$, Vereshkov derived the following relations applicable to type I bursts:

$$B_0 >> 6 \times 10^{-28}\,\frac{N_e T_i^{\frac{1}{2}}}{T_e^2},\quad B_0 v >> 4 \times 10^{12}\, T_e N_e^{\frac{1}{2}},$$

$$v > 1.4 \quad 10 \quad T_e^{\frac{1}{2}}, \qquad \omega_s << 6.5 \times 10^{-24}\,\frac{N_e}{T_e^{3/2}},$$

$$\frac{(B_0 v)^{3/2}}{B_0^2} < \frac{10^{12}}{N_e^{1/4}}\left(\frac{T_e}{T_e+T_i}\right)^2 \left(\frac{T_i}{T_e}\right)^{\frac{1}{2}}.$$

For $N_e \approx 10^8$ cm^{-3} and $T_e \approx 10^6$ K $\approx 10^{-10}$ erg it is found that the inequalities are satisfied by the following values: $B_0 = 30$ G, $v = 10^7$ cm s^{-1}, $\omega_s = 0.02$ s^{-1}, $(v_0/v_t) = 15$.

When $\gamma_{in} > \gamma_{dif} > \gamma^i_{int} > \gamma_c$ the plasma turbulence spectrum W_k^l will be quasi-stationary and T_{eff} and η are determined by the spectrum in the wave-number region $k = k_t$. From this Vereshkov obtained $T_{eff} \sim 10^8 T_e$ and $\eta \approx 1$.

The duration of single bursts is mainly determined by the electron collision frequency.

The wavelength of the accoustic waves is roughly $\lambda = \frac{2\pi v_s}{\omega_s} \approx 3 \times 10^9$ cm. The MHD wave packets overtake the sound waves, and the number of bursts that may be emitted during an intersection amounts to

$$\frac{\lambda}{v_A}\left(\tau_{coll} + \frac{1}{\gamma_{in}} + \frac{1}{\gamma^{i}_{int}} + \frac{1}{\gamma_c}\right)^{-1} \approx 10.$$

Thus the occurrence of small chains of bursts may be explained. The prediction is in good accord with the observational results obtained by Elgarøy and Ugland (1970) and Chernow *et al.* (1972), which were described in section 6.2.

The bandwidth is determined by the distribution of electron densities in the source.

Conclusions

The theory explains the main properties of type I bursts. The suggestion of Trakhtengerts (1966) that the bursts are caused by a current instability excited by hydromagnetic waves is further developed. In particular it is assumed that the conversion from longitudinal to transverse waves occurs when MHD waves overtake acoustic waves propagating in the corona.

One difficulty, as mentioned in connection with the theory of Trakhtengerts, is that the instability excited by the MHD pulse propagating along the magnetic field produces waves of extremely low frequency. No efficient mechanism which can transform these waves into high-frequency electromagnetic waves can be thought of (Mangeney and Veltri, 1976b).

11.5. ZAITSEV'S AND FOMICHEV'S THEORY

It is likely that single type I bursts and type I chains are generated by the same mechanism. Type I chains have properties in common with type II bursts. Coherent plasma radiation from collisionless shock waves in a magnetized plasma has been proposed as a possible mechanism for type II bursts (Wild and Smerd, 1972). The similarity between type II bursts and chains of type I bursts led Zaitsev and Fomichev (1973) to propose collisionless shock waves with a strong magnetic field ($\omega_H \gtrsim \omega_p$) as a possible source for chains.

Radio Emission from Shock Waves

When the collisional frequency in a plasma is low, such as in the corona, a shock front will be an oscillating structure consisting of a series of compressional or rarefaction waves (solitons). The scale of this structure depends on the magnetic Mach number, $M = u_0/v_A$, where u_0 is the velocity of the shock front. It also depends on the angle θ between the plane through the shock front and the direction of the unperturbed magnetic field $\mathbf{B}_0$ and the value of the parameter $\alpha_e = \omega_{H_0}{}^2/\omega_{p_0}{}^2$. The subscript $_0$ indicates values in the unperturbed plasma. For compressional waves propagating transverse to the magnetic field, the scale δ is given by

$$\delta = \frac{c}{\omega_{p_0}} \left(1 + \alpha_e + \alpha_e^2 \frac{m}{M}\right)^{\frac{1}{2}} \left(1 + \alpha_e \frac{m}{M}\right)^{-3/2}. \qquad (11.5\text{-}1)$$

The conditions for exciting plasma waves in a compressional shock are first satisfied in the leading soliton which has the largest amplitude. For $\alpha_e \ll M/m$ it may be written

$$M > M_{cr} = 1 + \frac{3}{4}\left[\frac{8\pi N_{e_0} \kappa T}{B_0^2} (1 + \alpha_e)\right]^{1/3} \qquad (15.5\text{-}2)$$

Provided that condition (11.5-2) is satisfied, the relative drift velocity between electrons and ions in the front of the soliton v_*, which equals

$$v_* = \frac{4 \times 2^{\frac{1}{2}}}{3 \times 3^{\frac{1}{2}}} c \frac{\omega_{H_0}}{(\omega_{p_0}{}^2 + \omega_{H_0}{}^2)^{\frac{1}{2}}} \left(M - 1\right)^{3/2}. \qquad (11.5\text{-}3)$$

becomes larger than the thermal velocity of electrons v_t, and plasma waves may be generated. It is seen from (11.5-2) that M_{cr} decreases with increasing magnetic field strength, and for $\alpha \gg 1$ the value becomes

$$M_{cr} = 1 + \frac{3}{4}\left(2v_t^{\,2}/c^2\right)^{1/3}.$$

The dispersion equation for longitudinal oscillations propagating perpendicular to $\mathbf{B}_0$ may be written

$$1 - \frac{\omega_{p*}^{\,2}}{\omega^2 - \omega_{H_*}^{\,2} - k^2 v_t^{\,2}} - \frac{m}{M}\,\frac{\omega_{p*}^{\,2}}{(\omega - \mathbf{kv}_*)^2 - \Omega_H^{\,2}} = 0, \qquad (11.5\text{-}4)$$

where the coordinate system is referred to the drifting electrons. Values indicated by an asterisk (*) are values in the shock front where the drift velocity reaches a maximum. When $M \leq 1.5$ one has approximately

$$\mathbf{B}_* = \mathbf{B}_0 + \frac{4}{3}\mathbf{B}_0\,(M-1),\quad N_{e_*} = N_{e_0} + \frac{4}{3}N_{e_0}\,(M-1). \qquad (11.5\text{-}5)$$

For $\alpha_e << M/m$ the maximum linear instability growth rate in the direction of the drift becomes

$$\gamma = \frac{3^{\frac{1}{2}}}{2^{\frac{1}{2}}}\left(\frac{m}{M}\right)^{1/3}\omega_{p_*}\left(\frac{\omega_{p*}^{\,2}}{\omega_{p_*}^{\,2} + \omega_{H_*}^{\,2}}\right)^{1/6}\left(1 - \frac{v_t^{\,2}}{v_*^{\,2}}\right)^{1/5}, \qquad (11.5\text{-}5)$$

and in a coordinate system referred to the drifting electrons the frequency of the excited plasma waves will be

$$\omega_0 = \left(\omega_{p_*}^{\,2} + \omega_{H_*}^{\,2} + k_*^{\,2} v_t^{\,2}\right)^{\frac{1}{2}} \approx \left(\omega_{p_*}^{\,2} + \omega_{H_*}^{\,2}\right)^{\frac{1}{2}}$$

where $k_* = \frac{\omega_0}{v_*}$ is the wave number of the plasma waves.

According to Zaitsev (1969) and Stepanov (1970), plasma wave scattering by electromagnetic waves in a collisionless shock front results in electromagnetic radiation at ω_0 and $2\omega_0$ in a coordinate system referred to the drifting electrons.

The electrons drift in opposite directions on adjacent sides of each individual wave (soliton). When the frequency spectrum of the elctromagnetic waves is converted to the laboratory system, the Doppler effect causes a splitting of each band (ω_0 and $2\omega_0$). For the fundamental band one obtains

$$\omega_\pm = \omega_0[1 \pm n_{1,2}(\omega_0)\,\frac{v^*}{c}\,\cos\theta] \qquad (11.5\text{-}8)$$

where θ is the angle between the drift direction and the line of sight and $n_{1,2}(\omega_0)$ is the refractive index for electromagnetic waves in the coordinate system referred to the drifting electrons. The effect of the velocity u_0 of the shock wave is neglected compared with v^*.

The refractive index for o-waves in the quasi-transverse approximation is

$$n_2^2(\omega_0) = 1 - \frac{\omega_p^2}{\omega_0^2} \simeq \frac{\omega_{H*}^2}{\omega_{H*}^2 + \omega_{p*}^2}.$$

The frequency split $\delta\omega$ may then be found from

$$\frac{\delta\omega}{\omega_0} = 2\,\frac{\omega_+ - \omega_-}{\omega_+ + \omega_-}\;\frac{8\sqrt{2}}{3\sqrt{3}}\;\frac{\omega_{H0}\,\omega_{H*}\,(M-1)^{3/2}\cos\theta}{(\omega_{p*}^2 + \omega_{H*}^2)^{\frac{1}{2}}(\omega_{p0}^2 + \omega_{H0}^2)^{\frac{1}{2}}} \qquad (11.5\text{-}9)$$

where $\omega_0 = (\omega_+ + \omega_-)/2$ is the mean frequency of the radio emission in the laboratory coordinate system.

For weak shock waves ($M < 1.3 - 1.5$) and $\alpha_e \geq 1$ the bandwidth of the emission is determined by the growth rate of the plasma wave instability

$$\frac{\Delta\omega}{\omega_0} \simeq \frac{\gamma}{\omega_0} \simeq \frac{3^{1/2}}{2^{4/3}}\left(\frac{m}{M}\right)^{1/3}\left(\frac{\omega_{p*}^2}{\omega_p^2 + \omega_{H*}^2}\right)^{2/3}. \qquad (11.5\text{-}10)$$

The average frequency ω_0 changes because the electron density decreases with increasing height. Therefore a frequency drift occurs:

$$\frac{1}{\omega_0}\left|\frac{d\omega_0}{dt}\right| = \frac{\omega_{p0}^2}{2(\omega_p^2 + \omega_{H0}^2)}\left(\frac{\cos\beta_N}{L_N} + 2\,\frac{\omega_{H0}^2}{\omega_{p0}^2}\,\frac{\cos\beta_H}{L_B}\right)u, \qquad (11.5\text{-}11)$$

where

$$L_N = \frac{|\nabla N_e|}{N_e} \quad \text{and} \quad L_B = \frac{|\nabla B_0|}{B}$$

are the scales of the variation in electron density and magnetic field in the corona and β_N and β_H are the angles between ∇N_e and ∇B_0 and the direction of the source velocity $\mathbf{u}_0$.

Values of the Parameters in the Emission Region

One may use observational results for type I chains to determine the magnetic field and shock-wave parameters. Some simplifying assumptions may be made. Since the shock waves are weak, one may roughly have $\omega_{p*} \simeq \omega_{p0}$ and $\omega_{H*} \simeq \omega_{H0}$. In order to obtain a large frequency split it is assumed that $\cos\theta = 1$ (drift direction parallel to line of sight). Furthermore the radial velocity of the source v_R is estimated under the condition that ∇N_e and ∇B_0 are directed along the radius. Therefore $\beta_N = \beta_B = \beta$ and $v_R = u_0 \cos$.

One may now use (11.5-7) for the centre frequency ω_0, (11.5-9) for the frequency split, (11.5-10) for the bandwidth and (11.5-11) for the frequency drift. From these the following relations are derived between the (unknown) source parameters and the observable data for chains:

$$\omega_{p_0} = \frac{4\pi}{3^{3/2}} f_0\left(\frac{\Delta f}{f_0}\right)^{3/4}\left(\frac{M}{m}\right)^{1/4},$$

$$\omega_{H_0} = 2\pi f_0\left[1 - \frac{4}{3^{3/4}}\left(\frac{\Delta f}{f_0}\right)^{3/2}\left(\frac{M}{m}\right)^{1/2}\right]^{1/2},$$

$$M = 1 + \frac{3}{2^{5/3}}\left(\frac{\frac{f+}{f-} - 1}{\frac{f+}{f-} + 1}\right)^{2/3}\left[1 - \frac{4}{3^{3/4}}\left(\frac{\Delta f}{f_0}\right)^{3/2}\left(\frac{M}{m}\right)^{1/2}\right]^{-2/3},$$

$$v_R = 2\,\frac{(\omega_{p_0}{}^2 + \omega_{H_0}{}^2)^{\frac{1}{2}}}{\omega_{p_0}{}^2}\left|\frac{d\omega_0}{dt}\right|\left(\frac{1}{L_N} + 2\frac{\omega_{H_0}{}^2}{\omega_{p_0}{}^2}\frac{1}{L_B}\right)^{-1}$$

$$u_0 = c\left(\frac{m}{M}\right)^{\frac{1}{2}}\frac{\omega_{H_0}}{\omega_{p_0}} M.$$

Using "typical" data for chains in the 50 - 200 MHz region, Zaitsev and Fomichev obtained the results shown in Table 11.1.

The source height $R/R_\odot$ was determined using electron densities twice those given by the Newkirk model for an active region. The parameter ℓ denotes the scale of irregularities in the region of generation and is equal to the product of the shock velocity u_0 and the characteristic time for intensity fluctuations in a chain (duration of individual bursts in chains).

TABLE 11.1

Source parameters for average type I chains

f_0 (MHz)	200	150	100	50
f_{p0} (MHz)	134	100	67	34
f_{H0} (MHz)	148	111	74	37
B_0 (gauss)	53	40	27	13
$R/R_\odot$	1.38	1.48	1.70	2.23
$L_N/R_\odot$	0.19	0.22	0.29	0.5
$L_B/R_\odot$	0.36	0.48	0.75	0.76
v_R (cm s^{-1})	2.3×10^7	-	4.1×10^7	-
u_0 (cm s^{-1})	-	10^9	-	-
M	-	1.34	-	-
ℓ (cm)	4.2×10^8	4.1×10^8	4.0×10^8	-

Several conclusions were reached from these results.

(a) The emission frequency of type I chains is considerably higher than the plasma frequency in the region where the radiation is generated.

(b) The magnetic field strength in the region where type I chains occur is about one order of magnitude higher than in regions where type II bursts are generated.

(c) Shock waves producing type I chains are weaker ($M \simeq 1.34$) than shocks generating type II bursts ($M \simeq 1.8$). The relatively short lifetime of chains may be connected with this.

(d) The radial component of the shock velocity tends to be lower than the velocity u_0. The shock therefore probably propagates roughly parallel to the solar surface. Take $v_{tang} = u_0 = 10^9$ cm s^{-1} and a chain lifetime of 30 s. Then a horizontal displacement of 6' arc is obtained. This should be compared with accurate observations of possible movements of chains. Such observations should be undertaken.

(e) The radiation is polarized in the o-mode because waves in the x-mode only escape from heights above the level $X = 1 - Y$. This is higher up in the corona than the level at which the chain radiation is generated.

(f) Normal type I bursts and bursts in chains have similar

properties. They are likely to be of the same nature. Chains may be generated by weak shock waves consisting of successive solitons. Type I bursts may be excited by individual (isolated) solitons.

Zaitsev and Fomichev suggested that the most likely factor determining the duration of type I bursts both in chains and outside chains was coronal irregularities in electron density or magnetic field. Perhaps the most likely mechanism for intermittent generation in chains is associated with fluctuations in the direction of the coronal magnetic field relative to the direction of the shock-wave velocity. The most favourable conditions for plasma wave excitation occurs for $u_0 \perp B_0$. A departure of only about 2° from this condition is sufficient to almost stop the generation. According to this the burst duration is determined by the relation $\Delta t = \ell/u_0$ (ℓ is the scale of the irregularities and u_0 is the shock velocity) and not by the characteristic damping time of plasma waves due to collisions.

Radiation is generated only as long as $M > M_{cr}$. This explains that photometer scans along the axis of chains in most cases show a relatively slow decay in intensity ($M > M_{cr}$) before a rapid decline to zero ($M < M_{cr}$).

Conclusions

The theory explains chains and single bursts by the same mechanism. Frequency drift, splitting, bandwidth, polarization and the duration of individual bursts are explained. Since the emission occurs at a frequency well above the plasma frequency, the occurrence of type III bursts "growing out" of chains at lower frequencies may be explained as plasma emission from accelerated electrons escaping along open field lines.

The theory belongs to the class of theories in which plasma waves are generated in shock waves. Wild and Smerd (1972) and Smith and Krall (1974) argued that transverse waves generated in this way have a frequency which is so low that they are unable to escape.

Smith and Krall started with the dispersion relation for plasma waves in the ion frame as derived by Krall and Trivelpiece (1973),

$$1 - \frac{\Omega_p^{\,2}}{\omega_l^{\,2}} \quad \frac{\omega_p^{\,2}}{(\omega_p - k_y U)^2 - k^2_{\,x} v_t^{\,2}} = 0, \tag{11.5-12}$$

where k_x and k_y are the components of the plasma wave vector in the directions along the x- and y-axis (shown in Fig. 11.6) and U is the drift velocity of the electrons relative to the ions; it is assumed that $\omega_H^{\,2} << \omega_p^{\,2}$. Differentiating (11.5-12) and solving for ω_p, Smith and Krall obtained

$$\omega_l \approx \left(\frac{m}{M}\right)^{1/3} \left(\frac{k_y^{\,4} U^4 - 4k_y^{\,2} U^2 k_x^{\,2} v_t^{\,2}}{k_y U}\right)^{1/3} \tag{11.5-13}$$

In deriving (11.5-13) the condition $\omega_l < k_y U$ was used. This is a necessary condition for the generation of the waves. Now, one must also have $U > 3^{\frac{1}{2}} v_t$, and since the growth rate of the waves is largest in the -direction, (11.5-13) may be written approximately as

$$\omega_l \approx \left(\frac{m}{M}\right)^{1/3} k_y U \approx \left(\frac{m}{M}\right)^{1/3} \omega_p, \tag{11.5-14}$$

where $\omega_l' = k_y U \approx \omega_p$ has been applied in the last approximation. This is the condition for excitation of plasma waves by a fast ion in the elctron frame (primed). But this is equivalent with excitation of plasma waves by drifting electrons in the ion frame. Thus (11.5-14) is valid in the ion frame. Since $(m/M)^{1/3} = 0.082$ and the maximum increase of ω_p in the shock is

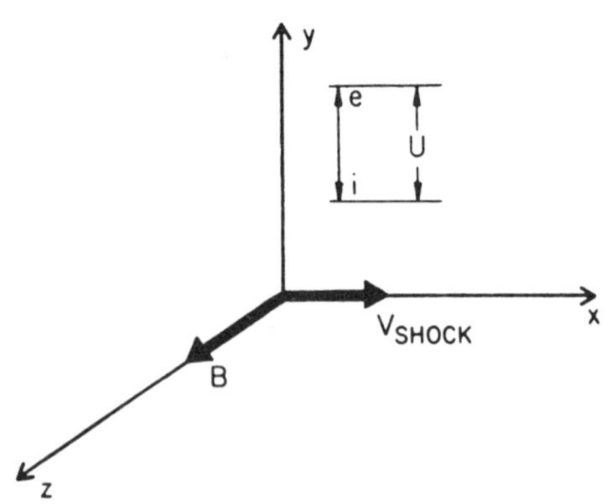

Fig. 11.6. Coordinate system employed by Smith and Krall (1974).

about 3, transverse waves with $\omega_t < 0.03\ \omega_p$ are produced at the fundamental frequency ($\omega_t < 0.6\ \omega_p$ at the second harmonic). In the transformation process from plasma waves to radio waves the frequencies either decrease or remain unchanged. As ω_p refers to the plasma frequency in the ambient medium, the transverse waves generated in the shock cannot escape and will not be observed. For this reason Smith and Krall discard a whole class of theories for type II bursts and denote the application to type I chains undertaken by Zaitsev and Fomichev as unfounded.

The criticism appears to be based on a misunderstanding. The answer to it is as follows: In an equilibrium plasma the scattering of plasma waves occurs on ions, while the scattering on electrons is less important. But the plasma in the shock front is not in equilibrium since the drift velocity of the electrons exceeds their thermal velocity. Then the plasma waves excited by the ion stream are scattered on fast electrons moving with velocity $\mathbf{U}$ in the ions frame of reference. This increases the frequency of the scattered electromagnetic radiation to $\omega_f = \omega_l + \mathbf{kU} \approx \omega_p$. The scattering on ions does not yield radiation of interest. This point was stated by Zheleznyakov (1965) and in subsequent papers on radiation from shock waves in the solar corona.

One may conclude that the theory of Zaitsev and Fomichev shows several attractive features and that no really decisive objection has been raised against it. One difficulty is that it predicts emission at the fundamental and second harmonic frequency, whoreas observations of chains do not show this property to be present (section 6.2).

12. CYCLOTRON RADIATION

The second class of noise storm theories invokes an electromagnetic instability excited by a beam of electrons spiralling around magnetic field lines. A classical objection against theories of this class is that too strong magnetic fields are needed. However, strong magnetic fields may occur in coronal loops and arches.

12.1. FUNG'S AND YIP'S THEORY

Fung and Yip (1966a) suggested that noise storm radiation could be explained as emission from streams of electrons gyrating in the coronal magnetic field above large sunspots.

Coronal Model

Fung and Yip assumed that the electron density in the background corona could be described by the Baumbach - Allen model

$$N_e = 10^8(1.55\rho^{-6} + 2.99\rho^{-16})\ \text{cm}^{-3}, \qquad (12.1\text{-}1)$$

where ρ is the distance from the centre of the sun measured in $R_\odot$. In active regions, the seat of noise storms, electron densities amounting to 5 or 10 times those given by (12.1-1) were assumed.

In the calculations an idealized model of the structure of the sunspot magnetic field was used. In the model a dipole was situated below the photosphere. The orientation and the depth of the dipole below the photosphere were chosen such that the field intensity in the centre of the spot reached a maximum. The detailed geometry is shown in Fig. 12.1. Also a unipolar spot was considered; then the axis of the dipole was directed along the radius vector of the sun and only some lines of force from the pole which was nearest to the photosphere emerged

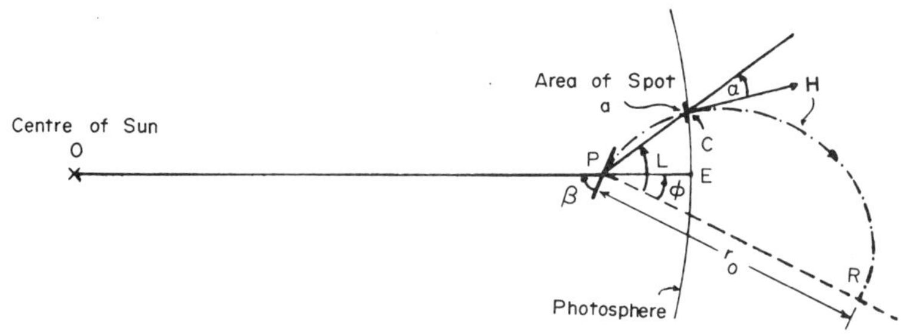

Fig. 12.1. Geometry of the assumed spot field configuarion. P is the midpoint of the dipole. The arrow H represents the field line passing through the centre C of the leading spot. Values $PE = 0.1\ R_{\odot}$, $\beta = 70^{\circ}$, $L = 60^{\circ}$ and $\phi = 20^{\circ}$ are assumed. The field intensity at C is chosen to be 2500 G. (Fung and Yip, 1966a.)

through the spot area.

Emission from Single Electrons

There are good reasons to assume that electrons of energy 10 - 100 keV are present in noise storm regions. Fung and Yip considered such electrons spiralling in the sunspot magnetic field. For electromagnetic waves the refractive index is given by the Appleton - Hartree equation with collisions neglected. The thermal velocity of the electrons in the background corona is small compared with the velocity of light, and the refractive index may be expressed in the following form:

$$n_j^2 = 1 - \frac{A\xi^{-2}(1-A\xi^{-2})}{1-A\xi^{-2}-\frac{1}{2}\xi^{-2}\sin^2\theta \mp [\frac{1}{4}\xi^{-4}\sin^4\theta+(1-A\xi^{-2})^2\xi^{-2}\cos^2\theta]^{\frac{1}{2}}} \qquad (12.1\text{-}2)$$

where $A = \omega_p^{\ 2}/\omega_H^{\ 2}$, θ is the angle between the wave vector $\mathbf{k}$ and the magnetic field $\mathbf{B}$, and $\xi = f/f_H$, where f is the wave frequency.

Heights between 0.2 and 0.5 $R_{\odot}$ above the photosphere are typical for noise storms, and considering a maximum spot field of 2000 G, Fung and Yip found that the value of $A = \omega_p^{\ 2}/\omega_H^{\ 2}$ in the storm region ranged from 0.2 to 5 in their adopted model.

In a reference system tied to the background magnetoactive plasma, the observed frequency of a radiating electron is given by the Doppler equation

$$\xi = \frac{s\gamma}{1-n_j\beta_{\parallel}\cos\theta} \tag{12.1-3}$$

with $\gamma = (1-\beta_{\perp}{}^2-\beta_{\parallel}{}^2)^{\frac{1}{2}}$, $\beta_{\perp} = v_{\perp}/c$, $\beta_{\parallel} = v_{\parallel}/c$ and $v_{\perp}$ and $v_{\parallel}$ denoting the electron velocities perpendicular and parallel to the direction of the magnetic field, and s being the harmonic number. The refractive index n_j must satisfy (12.1-2). In Fig. 12.2 the refractive index n_j is plotted as a function of ξ together with the Doppler equation for two chosen cases.

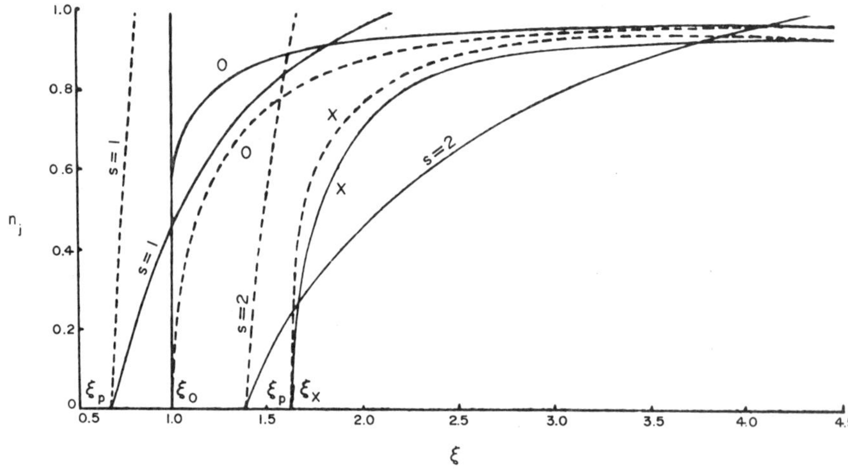

Fig. 12.2. Refractive index n_j for the o-mode and x-mode plotted against wave frequency given by $f/f_H = \xi$ for $A = 1$ and $\theta = 10^\circ$ (full lines) and $\theta = 75^\circ$ (dashed lines). The Doppler equation is plotted for $s = 1$ and $s = 2$ with $\beta_{\perp} = 0.3$ and $\beta_{\parallel} = 0.7$. (Fung and Yip, 1966a.)

As is seen from Fig. 12.2, one or two simultaneous solutions of the dispersion equation (12.1-2) and the Doppler equation (12.1-3) may occur. The distinction between "single" and "double" frequency solutions is important. In the theory of Fung and Yip the double-frequency solution is associated with the burst component, whereas the single-frequency solution leads to an explanation of the background continuum.

When $n_j = 0$, one obtains from (12.1-3), $\xi_D = s\gamma$. Denoting the values of ξ which give $n_j = 0$ in (12.1-2) by $\xi_{o,x}$, the condition for a single-frequency solution may be written $\xi_D = s\gamma \geq \xi_{o,x}$ and the condition for a double-frequency solution is $\xi_D < \xi_{o,x}$.

A gyrating electron emits radiation in both magneto-ionic modes. An expression for the power radiated in both modes by a single electron was derived by Eidman (1958) and completed and slightly corrected by Liemohn (1965). In the case of an electron gyrating with pitch angle $\phi = \mathrm{arctg}\,(\beta_\perp/\beta_\parallel)$ and with the guiding centre moving in the z-direction, the power radiated per unit solid angle in the direction θ in the harmonic with number s amounts to

$$P_s = \frac{e^2}{2\pi c}\,\frac{\omega^2 n_j k^2[-\beta_\perp J_s'(a)+(\alpha_y s\beta_\perp/a+\alpha_z\beta_\parallel)J_s(a)]^2(1+\frac{\omega}{n_j}\frac{\partial n_j}{\partial\omega})}{\left|1-\beta_\parallel n_j\cos\theta\,(1+\frac{\omega}{n_j}\frac{\partial n_j}{\partial\omega}\right|} \qquad (12.1\text{-}4)$$

where J_s and J_s' are Bessel functions and their derivatives with argument $a = n_j\beta_\perp\xi\sin\theta/\gamma$. Moreover, one has

$$\alpha_y = \alpha_\theta\cos\theta + \alpha_K\sin\theta\,, \qquad \alpha_z = \alpha_K\cos\theta - \alpha_\theta\sin\theta\,,$$

$$\alpha_\theta = -\xi\cos\theta/|\xi^2+A/(n_j^2-1)|\,, \qquad \alpha_K = -\xi(n_j^2-1)\sin\theta/(A-\xi^2)$$

$$K = (1+\alpha_\theta^2)^{-\frac{1}{2}}.$$

From a discussion of (12.1-4) using parameters which may be realistic for noise storm sources, Fung and Yip found that:

(a) In the single-frequency solution the power in the x-mode reaches a maximum at angles θ ranging from 15° to 45°, while for the o-mode the angle ranges from 40° to 70°.

(b) For the double-frequency solution the energy reaches a maximum in a direction close to the cut-off angle θ_c (which corresponds to the case when the Doppler equation is a tangent to the dispersion equation). Considerably more power is radiated by an electron when the double-frequency solution exists.

(c) With $\beta_\perp$, $\beta_\parallel$ and s constant, the maximum power in the x-mode exceeds that in the o-mode by one or two orders of magnitude.

(d) In order to have the double-frequency solution, the harmonic number of the x-wave will usually be one larger than that for the wave in the o-mode. Since the power decreases with increasing s, the power radiated in both modes may be of the same order of magnitude for the cases of double-frequency solutions.

The above conclusions regard emission from single electrons. When the radiation from single particles may be combined additively with reabsorption taken into account and intensification of the waves in the system itself is negligible, the emission is by convention denoted incoherent. Ginzburg and Zheleznyakov (1961) showed quite clearly that incoherent cyclotron radiation from electrons is not sufficient to explain the high intensities observed in type I bursts. In order to obtain such intensities amplification of the electromagnetic waves in the stream - plasma system has to take place. The maximum power radiated in the o-mode under the circumstances assumed by Fung and Yip was about 10^{-21} W/steradian, and the power gain necessary to account for a storm burst with maximum flux density of 5×10^{-20} W m^{-2} Hz^{-1}, as observed at the earth was estimated to be of the order of 10^{21}.

Amplification in a Stream - Plasma System

One may now assume that streams of electrons exist, trapped in magnetic fields in active coronal regions. The electrons forming the stream possess a certain momentum distribution, and some care must be exercised at this point. Fung and Yip assumed that all electrons in the stream had the same value of the momentum parallel and perpendicular to the magnetic field. The stream might be regarded as having zero "temperature", and results in $p_\perp = p_{\perp 0}$ and $p_\parallel = p_{\parallel 0}$. The distribution function of the stream is then given by

$$f_0(\mathbf{p})d\mathbf{p} = \left(\frac{1}{2\pi p_{\perp 0}}\right)\delta(p_\perp - p_{\perp 0})\,\delta(p_\parallel - p_{\parallel 0})\,d\mathbf{p}, \tag{12.1-5}$$

where δ is the Dirac delta function. For a stream of this type a large growth rate may be obtained and sufficient amplification to explain type I bursts is achieved. But as pointed out by Melrose (1973a), with the δ distribution assumed for the stream particles, the instability is a "reactive medium", one (Briggs, 1964) which corresponds to a "hydrodynamic" instability in the terminology of Tsytovich (1970). Then the beam is influenced by the waves it excites, and an increase in the velocity spread of the electrons forming the beam occurs. Since the instability has a very large growth rate, the velocity spread takes place rapidly and the distribution given by (12.1-5) is not realistic.

Conclusions

Based on their assumptions and taking the effects of resonance absorption at the first three harmonics into account, Fung and Yip reached several interesting conclusions. Some of them were:

(a) When the source is not too high, the second harmonic wave in the x-mode is either reflected at the level $X = 1 - Y$ or strongly attenuated due to third harmonic resonance absorption. Radiation in the o-mode, both fundamental and second harmonic can be generated when the height exceeds the levels $X = 1$ and $Y = 1$. When the source height is large, radiation in both modes may escape. But the radiation occurs at higher harmonics, and the radiated power is small. Therefore usually only the ordinary mode is observed.

(b) The power radiated in association with the double-frequency solution is much greater than that obtained for the single-frequency solution, and may explain the occurrence of type I bursts. The emission may be of narrow bandwidth and is strongly directive.

(c) When the single-frequency solution is satisfied, the bandwidth is large and the emission polar diagram is wide provided the pitch angle of the stream is not too large. Superposition of several broad-band emissions may form the type I background continuum. As pointed out above, the continuum radiation is also in the ordinary mode.

Fung and Yip reached a surprisingly good agreement between theory and observation. Also the dynamic spectra of type I bursts were explained (Fung and Yip, 1966b). However, as the assumption of a distribution function of the gyrating electrons given by (12.1-5) is untenable, the validity of the results is questionable. This comes out clearly in the treatment of the problem undertaken by Melrose (1973a). Besides this there are also other errors in the theory of Fung and Yip.

12.2. MELROSE'S DISCUSSION OF GYROMAGNETIC EMISSION

Melrose (1973a) considered the problem of coherent gyro emission in the quasi-linear approximation and reached several conclusions which apparently make this radiation mechanism less attractive in the application to solar noise storms. Rather than a δ distribution function, Melrose assumed a bi-Maxwellian distribution of the form

$$f(p,\phi) = \frac{1}{(2\pi)^{3/2} m^3 c^3 \beta_{\perp_0}{}^2 \beta_{\|_0}} \exp\left[-\frac{\beta^2 \sin^2\phi}{2\beta_{\perp_0}{}^2} - \frac{(\beta\cos\phi - \beta_s)^2}{2\beta_{\|_0}{}^2}\right] \qquad (12.2\text{-}1)$$

where βc is the speed of the electron, ϕ is the pitch angle, $\beta_{\perp_0} c$ and $\beta_{\|_0} c$ are thermal velocities, and $\beta_s c$ is the streaming velocity. The results of the analysis undertaken by Melrose are not much dependent on the specific form chosen for the distribution function.

The quasi-linear equations may be given in terms of the emissivity ε^σ (mean power lost by an electron per unit solid angle about the direction of the wave normal per frequency unit), the absorption coefficient γ^σ and a quantity W^σ describing the energy density per unit solid angle about the wave normal direction per frequency unit. When all variations are assumed to be temporal, one may write for waves of mode σ

$$\frac{\partial W^\sigma(\omega,\theta)}{\partial t} = \varepsilon^\sigma(\omega,\theta) - \gamma^\sigma(\omega,\theta) W^\sigma(\omega,\theta). \qquad (12.2\text{-}2)$$

The different terms in (12.2-2) are quite lengthy functions and will not be written down explicitly. Using the bi-Maxwellian streaming function (12.2-1) and considering growth rates, limitations on the maser mechanism and competing instabilities, Melrose arrived at the following results for coherent gyro emission by non-relativistic electrons:

(a) Coherent emission of waves able to escape occurs only for $\beta_{\perp 0}{}^2 \gg \beta_{\| 0}$ and, for even more extreme anisotropies, for $n_j \ll 1$ or $|\cos\theta| \ll 1$.

(b) Coherent emission in the x-mode is favoured over coherent emission in the o-mode at the same harmonic s. For $s \geq 2$ the growth rate decreases with increasing s as $\beta_{\perp 0}{}^{2s}$.

(c) Coherent emission at $s = 1$ in the o-mode has a growth comparable with that at $s = 3$ in the o-mode, i.e. less than that at $s = 2$ in the o-mode and also less than that at $s = 2$ in the x-mode.

(d) Coherent emission at $s = 1$ in the x-mode is possible only for a rapidly streaming distribution ($|\beta_s| \geq \beta_{\parallel 0}$) in a very low density plasma with $\omega_p << \omega_H$.

According to this the mechanism should produce radiation polarized in the x-mode and occur at the frequency $\omega \approx 2\omega_H$ except when the condition $\omega \approx \omega_H >> \omega_p$ is satisfied. Since the available evidence quite strongly indicates that the noise storm radiation occurs in the o-mode, coherent gyro emission from electron streams in the corona is less likely to be the mechanism producing noise storm emission.

12.3. MANGENEY'S AND VELTRI'S THEORY

Further work on the mechanism investigated by Fung and Yip (1966a), i.e. an electromagnetic instability excited by a beam of electrons spiralling around magnetic field lines was undertaken by Mangeney and Veltri (1976a, b). They showed that nonlinear coupling of linearly unstable whistler and high frequency x-mode waves with low-frequency MHD waves provides an effective stabilization mechanism for these two modes for low-beam densities. Quasi-linear relaxation of the remaining o-mode wave instability appeared rewarding when it comes to explain polarization, bandwidth and directivity of type I bursts. In this way one escapes from the difficulties pointed out by Melrose (1973a).

Turbulent Fluctuations in Background Plasma

Mangeney and Veltri considered the stability of a beam of mildly relativistic electrons spiralling along magnetic field lines in a plasma which was assumed to be cold and turbulent. The density of the beam electrons was low, and the dispersion of momenta was small. The pitch angle was different from zero.

Turbulent fluctuations in noise storm regions are quite likely. An enhanced flux of hydromagnetic waves may supply the energy which is lost by the active region through radiation and heat conduction. But the spectrum of the hydromagnetic waves is not known. Mangeney and Veltri (1976a) assumed that $(0.01 - 0.1\ \text{Hz}) \lesssim \omega \lesssim (\omega_H \Omega_H)^{\frac{1}{2}}$. The lower limit is determined by the generating photospheric motions. The high-frequency limit occurs because at higher frequencies non-linear steepening may be balanced by dispersive effects (Sagdeev and Galeev, 1969).

With a direct view to application of the results to noise storms the following assumptions were made:

(a) A statistically uniform and stationary background plasma permeated by a magnetic field so that $1 < A < 20$ with $A = \omega_p{}^2/\omega_H{}^2$.

(b) The spectrum of the turbulent fluctuations is broad with a high frequency cut-off at $\omega^* \simeq (\omega_H \Omega_H)^{\frac{1}{2}}$.

(c) A uniform electron beam of infinite extent. The density of the beam electrons $n_b \equiv qN_e$ is very low (N_e is the electron density of the background plasma). Furthermore the dispersions $\Gamma p_{\perp 0}$ and $\Gamma p_{\parallel 0}$ of the perpendicular and parallel electron momenta of the beam electrons around their most probable values $p_{\perp 0}$ and $p_{\parallel 0}$ are very small. Therefore one has $q \ll 1$, $0 < \Gamma \ll 1$.

(d) Electrons in the beam have an energy of the order of 50 - 250 keV.

The form of the distribution function used for calculations was given by

$$f(\mathbf{p}) = a \exp\left[\left(\frac{p_\perp - p_{\perp 0}}{\Gamma p_{\perp 0}}\right)^2 + \left(\frac{p_\parallel - p_{\parallel 0}}{\Gamma p_{\parallel 0}}\right)^2\right] \qquad (12.3\text{-}1)$$

As long as $\Gamma \ll 1$ was satisfied, the results were independent of the detailed mathematical form of the distribution function.

Effect of Turbulence on Beam - Plasma Instability

In a quiescent plasma a beam with a distribution as given by (12.3-1) is highly unstable. But plasma turbulence has a strong stabilizing influence.

Two main effects may be expected. Firstly, long wavelength fluctuations, $\omega \ll (\omega_H \Omega_H)^{\frac{1}{2}}$, are seen by the unstable waves as random modulations of the background plasma parameters. The equation for the number $N^{\sigma}(\mathbf{k})$ of plasmons of wave mode σ with wave vector $\mathbf{k}$ for a slightly inhomogeneous plasma may be written (Tsytovich, 1970)

$$\frac{\partial N_{\sigma}}{\partial t} + \mathbf{v}_g^{\sigma} \frac{\partial N_{\sigma}}{\partial \mathbf{r}} = - \gamma^{\sigma}(\mathbf{k}) N^{\sigma}(\mathbf{k}) + \eta^{\sigma}(\mathbf{k}), \tag{12.3-2}$$

where $\eta^{\sigma}(\mathbf{k})$ is the emissivity, $\gamma^{\sigma}(\mathbf{k})$ twice the linear damping rate and $\mathbf{v}_g^{\sigma}$ the group velocity of the wave. Equation (12.3-2) must now be considered as an equation with random coefficients. Both $\gamma^{\sigma}(\mathbf{k})$ and $\eta^{\sigma}(\mathbf{k})$ can be split into random and non-random components, $\gamma^{\sigma}(\mathbf{k}) = \overline{\gamma}^{\sigma}(\mathbf{k}) + \delta\gamma^{\sigma}(\mathbf{k})$, $\overline{\delta\gamma}^{\sigma} = 0$. (The bar signifies an ensemble average.)

When the turbulent fluctuations have a sufficiently small r.m.s. amplitude and correlation length, the average plasmon number $\overline{N}^{\sigma}(\mathbf{k})$ follows the equation

$$\frac{\partial}{\partial t}\overline{N}^{\sigma}(\mathbf{k}) = -\gamma^{\sigma}_{\text{eff}}\,\overline{N}^{\sigma}(\mathbf{k}) + \eta^{\sigma}_{\text{eff}}\,Q^{\sigma}(\mathbf{k}) + \left(\frac{\partial \overline{N}^{\sigma}}{\partial t}\right)_{\text{diff}}, \tag{12.3-3}$$

where $\gamma^{\sigma}_{\text{eff}} = \overline{\gamma}^{\sigma}(\mathbf{k})$ and $\eta^{\sigma}_{\text{eff}} = \eta^{\sigma}(\mathbf{k})$ plus terms which are of second order in $\overline{\delta B}/B_0$ and may be neglected. $Q^{\sigma}(\mathbf{k},t)$ is the probability that a plasmon emitted with arbitrary wave vector has a given wave vector $\mathbf{k}$ at time t. The term $\left(\frac{\partial \overline{N}^{\sigma}}{\partial t}\right)_{\text{diff}}$ represents the diffusion resulting from scattering, and may be neglected except when the gradients in $\mathbf{k}$-space are large.

Secondly, short wavelength (high frequency, $\omega \leq (\omega_H \Omega_H)^{\frac{1}{2}}$) fluctuations may interact non-linearly with waves which are driven unstable by the beam. These interactions are described by using the standard balance equation (Tsytovich, 1970)

$$\left(\frac{\partial \overline{N}^{\sigma}(\mathbf{k})}{\partial t}\right)_{NL} = \sum_{\sigma_1} \int \frac{d\mathbf{k}_1}{(2\pi)^3} \frac{d\mathbf{k}_f}{(2\pi)^3} W_f^{\sigma_1,\sigma}(\mathbf{k},\mathbf{k}_1,\mathbf{k}_f) \times$$

$$\left\{ N^f(\mathbf{k}_f)[N^{\sigma_1}(\mathbf{k}_1) - N^{\sigma}(\mathbf{k})] - N^{\sigma}(\mathbf{k}) N^{\sigma_1}(\mathbf{k}_1) \right\} \tag{12.3-4}$$

where $N^f(\mathbf{k}_f)$ is the number of waves of wave vedtor $\mathbf{k}_f$ which belong to the high-frequency part of the turbulent spectrum, σ_1 denotes the wave mode which may be non-linearly coupled with the wave modes σ and f, and $W_f^{\sigma_1,\sigma}(\mathbf{k},\mathbf{k}_1,\mathbf{k}_f)$ is the "transition probability" for this coupling.

In the situations under discussion, $N^f >> N^\sigma >> N^{\sigma_1}$

The wave σ is linearly unstable whereas the wave σ_1 is linearly stable or has a much smaller growth rate. Mangeney and Veltri then defined a non-linear damping rate $\gamma^\sigma_{NL}(\mathbf{k})$ such that one may write

$$\frac{\partial \overline{N}^\sigma(\mathbf{k})}{\partial t} = -\gamma^\sigma_{NL}(\mathbf{k})\overline{N}^\sigma(\mathbf{k}) \qquad (12.3\text{-}5)$$

and

$$\gamma^\sigma_{NL}(\mathbf{k}) = \sum_{\sigma_1}\int \frac{d\mathbf{k}_1}{(2\pi)^3}\frac{d\mathbf{k}_f}{(2\pi)^3}N^f(\mathbf{k}_f)W_f^{\sigma_1,\sigma}(\mathbf{k},\mathbf{k}_1,\mathbf{k}_f). \qquad (12.3\text{-}6)$$

In order to determine whether a wave mode σ will grow or be damped, the net growth rate $\gamma^\sigma_T(\mathbf{k})$ must be calculated

$$\gamma^\sigma_T(\mathbf{k}) = -\overline{\gamma}^\sigma(\mathbf{k}) + \gamma^\sigma_{NL}(\mathbf{k}).$$

When $\gamma^\sigma_T(\mathbf{k}) > 0$ damping occurs, while the condition for amplification is $\gamma^\sigma_T(\mathbf{k}) < 0$.

Linear Growth Rate

The very low frequency fluctuations will affect the resonance between the electromagnetic waves and the beam electrons. This must be taken into account in the calculation of the average linear growth rate $\overline{\gamma}^\sigma(\mathbf{k})$.

Consider an electron of momentum $\mathbf{p}$ and energy equal to $E = (m^2c^4 + c^2p^2)^{\frac{1}{2}}$ and a wave σ of frequency $\omega^\sigma = \xi\omega_H$ propagating at an angle $\theta^\sigma = \text{arc}\cos(u)$ with respect to the magnetic field. The resonance condition may then be expressed by

$$\xi\left(1 - \frac{\mu_\sigma u v_{/\!/}}{c}\right) - \frac{\nu mc^2}{E} = 0, \qquad (12.3\text{-}7a)$$

$$\mu_\sigma = \mu_\sigma(A,\xi,u). \tag{12.3-7b}$$

The refractive index $\mu_\sigma = ck/\omega_\sigma$, ν is the harmonic number and $v_\parallel$ is the electron velocity parallel to the magnetic field.

Solving (12.3-7a,b) with respect to u one obtains the solution

$$u = U_\nu^{\ \sigma}(\mathbf{p},\xi,A). \tag{12.3-8}$$

Using this solution, two characteristic angular widths of the resonant domain in **k**-space may be defined:

(a) a "thermal width" $(\Delta u)_1$ due to the existence of a small dispersion in the beam momenta;

(b) a "turbulent width" $(\Delta u)_2$ due to the turbulent fluctuations of the magnetic field and the cold plasma density.

For $1 \leq A \leq 20$, $0.1 \leq \xi \leq 10$, $p_{\parallel 0} \geq 0.3$ mc and $p_{\perp 0} \leq 0.9$ mc, the turbulent width $(\Delta u)_2$ was found to be larger than the thermal width $(\Delta u)_1$ for all modes predicted by cold plasma theory provided

$$\frac{\Delta\omega_H}{\omega_H} \approx \frac{\overline{\delta B}}{B_0} > \Gamma. \tag{12.3-9}$$

Mangeney and Veltri now assumed that (12.3-9) was valid. Then the effect of the turbulence is to increase the angular width of the resonances to $(\Delta u)_2$ which is larger than the thermal width $(\Delta u)_1$ corresponding to the dispersion in electron momenta. It can then be shown that the average growth rate $\overline{\gamma}^\sigma(\mathbf{k})$ can be approxiamted by the mean value of $\gamma^\sigma(\xi,u)$ over the cosine u of the angle of propagation θ^σ:

$$\overline{\gamma}^\sigma(\xi,u) \approx \frac{1}{\Delta u}\int_{-1}^{1} du\,\gamma^\sigma(\xi,u) \tag{12.3-10}$$

for

$$|u - U_\nu^{\ \sigma}(A,\xi,\mathbf{p}_0)| \leq \Delta u,$$

where Δu is a typical value of the turbulent width $(\Delta u)_2$ of the resonance.

Mangeney and Veltri derived an expression for the average linear damping rate for each harmonic, $\overline{\gamma}_\nu{}^\sigma(\xi)$. Then

$$\overline{\gamma}^\sigma(\xi) = \sum_{\nu=-\infty}^{+\infty} \overline{\gamma}_\nu{}^\sigma(\xi). \tag{12.3-11}$$

Using $1 \leq A \leq 20$, $\Gamma = 10^{-2}$, $p_{\perp 0} \geq 0.3$ mc, $p_{\parallel 0} \leq 0.9$ mc numerical calculations were performed. The most important results for the high-frequency modes ($\xi > A^{\frac{1}{2}}, \mu_\sigma \leq 1$) were:

(a) For each harmonic ν there is a range of values of A where this harmonic has the largest growth rate. In this range of A the growth rate has one maximum at an angle $\theta_\nu{}^\sigma$ and a frequency $\xi_\nu{}^\sigma$. The unstable domain has an angular width $\Delta\theta$ at half-height of about $20^\circ - 25^\circ$ and a bandwidth of the order of $0.04\ \xi_\nu{}^\sigma - 0.08\ \xi_\nu{}^\sigma$.

(b) When the harmonic number increases by one, the maximum growth rate decreases by roughly one order of magnitude.

(c) In a range of A where the o-mode is most unstable on the νth harmonic, only the $(\nu+1)$th harmonic of the x-mode is unstable with a growth rate of the same order of magnitude as that of the o-mode, at least for pitch angles which are not too close to $\pi/2$.

(d) The growth rate decreases very rapidly with increasing electron pitch angle. This indicates that the main cause of instability is the existence of an excess of perpendicular momentum. The growth rate increases with increasing electron energy.

The whistler mode is the most unstable cold plasma mode. Whatever the value of A, there is a first-order resonance ($\nu=\pm 1$) between the beam electrons and the whistler mode. In the range of parameters used, the resonant whistlers have frequencies $0.1 < \xi < 0.9$. Therefore both Cerenkov and resonance absorption are ineffective.

The longitudinal modes are stable. When $(\Delta\omega_p/\omega_p) > \Gamma$ Mangeney and Veltri showed that the beam is stable against the usual electrostatic beam - plasma instability.

This occurs because a fluctuation $\delta\omega_p$ of the plasma frequency induces a fluctuation $\delta v_\phi = \delta\omega_p/k$ of the phase velocity of an unstable wave with wave vector k. When the condition $\Delta\omega_p/\omega_p > \Gamma$

is satisfied, this particular wave will be successively in resonance with nearly all particles of the beam. Since the higher phase velocities are of greatest importance for the growth rate of the instability, the net result is that the average damping rate is positive. Thus the beam is stable against the usual electrostatic beam plasma instability.

Non-linear Growth Rate

There are several possibilities for non-linear interactions between the unstable waves and the short wavelength MHD fluctuations. A discussion of the various couplings shows that:

(a) The o-mode may couple with itself ($o + \mathrm{MHD} \Leftrightarrow o$). The process involves magneto-sonic waves with small wave vectors. The frequency of the scattered wave remains in the vicinity of that of the primary wave and does not leave the resonant domain.

(b) For the x-mode the coupling process $x + \mathrm{MHD} \rightarrow o$ is always possible for the x-mode waves driven unstable by the beam. The resulting o-mode wave is not in resonance with the beam.

(c) For the whistler mode $w + \mathrm{MHD} \rightarrow w$ is always possible. The scattered whistler remains within the resonant domain, but the frequency jump [ω(primary) - ω(scattered)] is greater or of the same order of magnitude as the bandwidth of the unstable domain and the growth rate is reduced.

It now follows that for a particular interval of A, $A_\nu^{\,1} < A < A_\nu^{\,2}$, the o-mode waves driven unstable by a νth order resonance with the beam electrons are not significantly affected by non-linear coupling with short wavelength flucutations. But all other unstable waves are scattered by this process into waves which are either stable or have a much lower growth rate. To answer the question whether these waves are stabilized or not it is necessary to determine how rapidly they are scattered, i.e. γ_{NL}^{σ} must be determined. Mangeney and Veltri then found that the whistler mode and the x-mode were both stabilized when the beam density was less than a critical value $q < q_c$ ($q = n_b/N_e$; beam density relative to background plasma).

Quasi-linear Relaxation of the Beam Plasma Instability

It is now only necessary to take the high-frequency ordinary modes into account, since the other modes are stabilized or have a lower growth rate. The instability is esentially a form of maser effect; it is due to an inversion in energy level populations. The excitation of waves leads to a fast flow of energy from the exciting particles into waves. This causes a fast "equalization" of the inversion in the level populations until the cause of the instability disappears. This process is the quasi-linear relaxation. It is therefore necessary to study the equation describing the simultaneous evolution of the plasmon number $N^{\sigma}(\mathbf{k})$ and of the electron distribution function $f(\mathbf{p})$. In their analysis Mangeney and Veltri showed that during the quasi-linear relaxation the energy and pitch angle of the electrons changed simultaneously. The quenching of the instability was related to the decrease of the pitch angle of the beam electrons. The emitted frequency of a particular harmonic $\bar{\nu}$ increases by about 10% while the relative bandwidth decreases to zero. The maximum energy density of the waves $W_{\max}$ becomes

$$W_{\max} \simeq 0.1\, n_b \frac{\Delta u}{u} mc^2 . \tag{12.3-12}$$

The duration of the quasi-linear process (the time necessary to amplify at the linear rate the spontaneous emission of the electrons up to the saturation level) is

$$t_{QL}^{\bar{\nu}} \simeq 20 - 30\, |\gamma_{\bar{\nu},\max}|^{-1} \tag{12.3-13}$$

When the linear growth of the wave intensity causes the frequency angular spectrum of the unstable waves to become too narrow, the gradients in $\mathbf{k}$ are quite large and diffusion spreads the spectrum to a width determined by the intensity of the fluctuations

$$\delta\xi \sim \delta u \sim \frac{D}{|\gamma_{\bar{\nu},\max}|}, \tag{12.3-14}$$

where $D \simeq \frac{\omega_H}{k l^*}\left(\frac{\delta B}{B_0}\right)^2$, where l^* is a characteristic wavelength of the turbulent fluctuation.

Summary of Theoretical Results

The most important results obtained by Mangeney and Veltri (1976a) were that the hydromagnetic fluctuations of long wavelength stabilize the electrostatic waves. This occurs mainly because of a large broadening of the resonances between waves and beam electrons. If long wavelength fluctuations only are considered, the linear theory predicts that most of the free energy in the beam will be radiated in the whistler mode. But the short wavelength fluctuations couple non-linearly with the whistler waves and provide a stabilization mechanism for sufficiently low beam densities. The physical process is that an unstable whistler wave excited by the beam couples resonantly with a low-frequency hydromagnetic wave and another whistler wave which is stable or has a low rate of growth. The result is a fast transfer of energy from unstable to stable waves, which prevents energy building up in the resonant whistler wave.

This mechanism also serves to stabilize the high-frequency x-mode, while the o-mode waves excited by the electrons in the beam are nearly unaffected.

Melrose (1973a) considered the same type of instabilities, but assumed a non-turbulent background plasma. It was then found that large anisotropies were required to destabilize high-frequency modes. Competing instabilities (whistlers, electrostatic waves) might have growth rates of the same order of magnitude or larger than those of the high-frequency electromagnetic modes. The results of Melrose may now be generalized. It seems clear that an electron beam radiates coherent electromagnetic waves of high frequency only when the electrons in the beam have large anisotropies or they do not move parallel to the magnetic field. Moreover, a stabilization mechanism must be acting which serves to suppress competing instabilities. In a turbulent plasma such a mechanism exists. It affects all wave modes which are excited by the beam except the o-mode.

The o-mode waves generated under such circumstances have narrow bandwidth and the emission is very directive.

Therefore the inclusion of wide-band hydromagnetic turbulence in the background plasma has effects which make coherent gyro emission from electrons in the corona a possible mechanism for the production of type I bursts.

Application to Type I bursts

The theoretical results were used to construct a model for type I bursts (Mangeney and Veltri, 1976b).

Several assumptions were then made:

(a) The magnetic field has a strength such ahat $1 \leq A \leq 25$. The reason for this choice is that the intensity of the radiation decreases rapidly with increasing harmonic numbers ν. The condition for radiation to escape is $\nu\omega_H > \omega_p$. Take $\nu = 1$ to 5, then the above range of A occurs. Magnetic fields equal to 10 - 30 G are required in the source regions.

(b) The beam electrons have an anisotropic velocity distribution.

(c) A broad band of hydromagnetic fluctuations exists in the background plasma. Moreover, the relative amplitude of the electron gyro frequency fluctuations, due to the long wavelength hydromagnetic fluctuations, must satisfy the condition $(\Delta\omega_H/\omega_H) > \Gamma$. This may occur for $\Gamma \simeq 10^{-2}$. Moreover, the beam density must be small, $q \leq 10^{-5}$.

A Model for the Beam

The electrons in the beam must have an average pitch angle different from zero, and the dispersion in momenta (Γ) must be very small ($\Gamma \simeq 10^{-2}$). Mangeney and Veltri (1976a,b) did not solve the problem of the generation of a beam of this type. But they showed that if the acceleration mechanism mainly increases the perpenducular momentum of the electrons ($p_\perp$), and the electrons propagate a distance along a decreasing magnetic field, then a beam of the required properties may be formed.

Assume a magnetic field following the relation

$$B_0^{\,0} = B_0^{\,0}(\ell_0/\ell)^2, \qquad (12.3\text{-}15)$$

where ℓ is the distance along the field line and B_0^0 the field at the reference level (ℓ_0) where the particles are accelerated. Then the scale height becomes $H_B = -B_0\left(\frac{\partial B_0}{\partial \ell}\right)^{-1}$, which equals $\ell/2$.

Using the adiabatic approximation the free propagation of the electrons is described by the Vlasov equation

$$\frac{\partial f}{\partial t} + v_{\parallel}\frac{\partial f}{\partial \ell} = Q(\ell,J,E;t), \tag{12.3-16}$$

where $f(\ell,J,E;t)$ is the distribution function of the particles at ℓ with energy E and magnetic momentum $J = p_{\perp}{}^2/m^2B_0$. Q is a source function describing the injection of energetic electrons by the acceleration process, which is assumed to be instantaneous. The acceleration was assumed to increase mainly the perpendicular momentum. Hence $p_{\parallel} \approx mv_t$ where v_t is the thermal velocity of the coronal electrons. The perpendicular momentum may be of the order $p_{\perp} \simeq mc$. Therefore the pitch angle becomes $\alpha = \pi/2$ with a dispersion $\delta\alpha \simeq$ arctg $c/v_t \simeq 10^{-2}$. The solution of the Vlasov equation may be written.

$$f(\ell,J,E;t) = \int_{-\infty}^{t} dt'Q[\ell(t'),J,E;t'], \tag{12.3-17}$$

where $\ell(t')$ is the solution of the equation of motion,

$$\frac{d\ell}{dt} = v_{\parallel} = [v^2 - JB_0(1 - v^2/c^2)]^{\frac{1}{2}} \tag{12.3-18}$$

satisfying

$$\ell(t') = \ell \text{ for } t' = t,$$

leading to

$$\ell(t') = \frac{\ell}{p}\left\{[p_{\parallel} - \frac{p^2}{E\ell}(t-t')]^2 + p_{\perp}{}^2\right\}^{\frac{1}{2}}, \tag{12.3-19}$$

where $p = (p_{\perp}{}^2 + p_{\parallel}{}^2)^{\frac{1}{2}}$.

In the case of simultaneous acceleration of all beam electrons,

$$Q = F(\ell,J,E)\delta(t). \tag{12.3-20}$$

Mangeney and Veltri assumed a particularly convenient form of F for the sake of illustration:

$$F = F_0 \exp[-(\ell-\ell_0)^2/L^2](p/p^*)^2\exp-(p/p^*)^2\exp-(\cos^2\alpha/\Delta^2), \tag{12.3-21}$$

where L is the linear dimension of the acceleration region, p^* is the characteristic total momentum of the accelerated particles and Δ is a measure of the angular width of the distribution of the pitch angles of the electrons.

Using (12.3-17), (12.3-19) and (12.3-20) the distribution function at time t at coordinate ℓ was obtained:

$$f(\ell, J, E; t) = F(\ell(t=0), J, E) \tag{12.3-22}$$

Numerical values of this function were calculated assuming

$$L = 0.02\,\ell_0, \quad \Delta = 0.05, \quad E^* = (m^2c^4 + c^2p^{*2})^{\frac{1}{2}} = 1.3\ mc^2.$$

It was found that when the electrons have travelled a distance equal to half of the scale height of the magnetic field, the momentum dispersion of the beam at the maximum of intensity is smaller than 5%.

The beam density versus time at two levels $\ell - \ell_0 = 0.25\ \ell_0$ and $0.5\ \ell_0$ was determined. The width of the profiles was of the order of the propagation time, $t_p \simeq H_B/v_\parallel$. The form of the curves depended on the form of F.

The electrons propagate freely provided no serious instabilities arise, which requires a beam to plasma density ratio q satisfying $q < 10^{-5}$. Then the low-frequency turbulence stabilizes the whistler instability.

The growth rate of the high-frequency o-mode decreases rapidly as A increases. Thus it may occur that the instability has insufficient time to grow up and free propagation of the electrons is secured in regions where A is high.

Variation of A with Height

It has been assumed that the electrons propagate freely between the acceleration region and the place where the instability occurs. In order to obtain this A must decrease with increasing height in this height interval. The reason is that in the region where the acceleration occurs, A must be sufficiently high to suppress instability. But during propagation (upwards) the perpendicular momentum of the beam particles decreases, and the beam becomes less unstable. Therefore A must be lower at the height of emission than in the region of acceleration in order to allow the instability to occur.

Is there any justification for the assumption of decreasing A with increasing height in the low corona?

$$A \propto N_e/B_0^{\ 2},$$

$$\frac{1}{A}\frac{\partial A}{\partial z} = \frac{1}{N_e}\frac{\partial N_e}{\partial z} - \frac{2}{B_0}\frac{\partial B_0}{\partial z} = -H_{N_e}^{-1} + 2H_B^{\ -1},$$

where H_{N_e} and H_B are the scale heights for the electron density and the magnetic field.

Assuming hydrostatic equilibrium one has

$$H_{N_e} = c_s^{\ 2}/g,$$

where c_s is the velocity of sound and g the acceleration of gravity. In the corona one may have $T \simeq 10^6$ K and $g \simeq 2.7 \times 10^4$ cm s^{-2}. Then $H_{N_e} \simeq 0.07$ $R_\odot$ and for A to decrease with height one must have $H_B > 2H_{N_e} \simeq 0.14$ $R_\odot$. This last condition is not unreasonable. Previously Fung and Yip (1966a) and Sy (1973b) assumed that A increases with height through the whole active region. Their conclusion was supported by reference to the Baumbach - Allen and Newkirk models of coronal densities. But the application of these models to localized regions of the low corona is uncertain. The behaviour of A in the noise storm region is not known, and there are arguments in favour of both models.

Interaction of the Beam with the Plasma

After acceleration at $z = z_0$ a small group of electrons propagate outwards. At the height z one has $\mathbf{p}(z) = [p_\perp(z), p_\parallel(z)]$. Assume that the particles are in resonance with an o-mode wave of frequency ω and angle of propagation $\theta = \arccos(u)$ with respect to the magnetic field at height z. If ν is the harmonic number of the resonance this implies:

$$|u - U_\nu(\mathbf{p},\omega,\omega_p,\omega_H)| < \Delta u,$$

where U_ν is the solution of the resonance equation and Δu is the effective width of the resonance due to the presence of the turbulent hydromagnetic fluctuations.

As the particles propagate, $\mathbf{p}$, ω_p and ω_H change and they leave the region of resonance. If the effective length of the resonance region is Δz_ν, it can be shown that $\Delta z_\nu << H_B \simeq H_{N_e}$. Therefore the ambient plasma and magnetic field may be regarded as homogeneous in the interaction region.

The particles are in the resonance region for a time of about $\Delta z_\nu / v_\parallel$ and the waves for a time $\Delta z_\nu / v_{g_\parallel}$ (where $\mathbf{v}_g$ is the group of velocity of the waves). The resonance between waves and particles occurs during a time

$$t_R^\nu \simeq \min(\Delta z_\nu / v_{g_\parallel}, \Delta z_\nu / v_\parallel)$$

The characteristic time scale of the beam due to propagation is of the order of

$$t_p \simeq H_B / v_\parallel >> t_R^\nu$$

In consequence the beam may be considered as being of infinite extent during the time waves and particles are in resonance.

For the quasi-linear relaxation to proceed one must have

$$t_R^\nu \gtrsim t_{QL}^\nu ,$$

where t_{QL}^ν is the time necessary to amplify the spontaneous emission of the electrons up to the saturation level.

If this relation does not hold, only a very small amount of energy will be extracted from the beam in the form of electromagnetic waves.

A Model for Type I Bursts Emission

Assume that a beam of electrons of small density (the whistler mode is stabilized) is accelerated perpendicularly to the magnetic field in a region where A is sufficiently high for the beam to travel undisturbed. But A decreases with height, and at an altitude z_{em} one may have

$$t_{QL}^{\nu} \simeq t_{R}^{\nu} << t_{p}.$$

The beam then undergoes a rapid quasi-linear relaxation and a burst of electromagnetic waves is emitted. When the beam leaves the emission region the particles have lost most of their perpendicular momentum.

The burst of radio waves has some characteristic properties:

Polarization

The x-mode is stabilized together with the whistler mode. The radio waves are therefore polarized in the o-mode.

Emitted Frequencies

The emission occurs for a harmonic ν which depends on the value of A. But the frequency is smaller than the x-mode cut-off frequency. The angle of emission varies from $\simeq 0^{\circ}$ $(\nu = 1)$ to $\simeq 40^{\circ}$ $(\nu = 6)$.

Bandwidth and Directivity

The bandwidth $\delta\xi$ of the amplified radiation is mainly determined by the turbulent fluctuations in the source. An order of magnitude estimate shows that

$$\frac{\overline{\delta\xi}}{\xi} \simeq C\frac{\overline{\delta B}}{B_0},$$

where C is a factor varying from 0.1 to 1. Provided $(\overline{\delta B}/B_0) \geq 10^{-2}$, this agrees with observed bandwidths.

It was shown by Mangeney and Veltri (1976a) that the emission occurred within an opening angle which was related to the bandwidth. A relative bandwidth of 3 - 4% corresponded to an opening angle $\overline{\delta\theta} = 10^{\circ} - 12^{\circ}$ at $\theta_{\nu} \simeq 0^{\circ}$ and $\overline{\delta\theta} \simeq 4^{\circ} - 5^{\circ}$ at $\theta_{\nu} \simeq 30^{\circ} - 40^{\circ}$. The values obtained here are smaller than the upper limit of 25° derived by Steinberg *et al.* (1974).

Time Profile and Duration

At the height where the instability develops ($z = z_{em}$), the beam varies during a time $t_p \simeq H_B/v_{\parallel}$ which is much longer than the time t^{ν}_{QL}. Therefore the quasi-linear relaxation can be considered as instantaneous. The radiated energy was shown by Mangeney and Veltri (1976a) to be proportional to the beam density. Therefore the time profile of the type I bursts will correspond to that of the beam density at the level $z = z_{em}$.

The duration of a burst may then be estimated:

$$t \simeq t_p \simeq H_B/v_{\parallel} \simeq 0.6\ s$$

for $H_B \simeq 10^5$ km, $v_{\parallel} = 0.5$ c.

Frequency Drift

In most cases the beam electrons have pitch angles $\alpha \leq 45^{\circ}$. Then the burst will show no frequency drift or a drift towards lower frequencies as the source drifts towards higher altitudes. When $\alpha > 45^{\circ}$, a drift towards higher frequencies is observed. This occurs because at a level z the maximum in the electron distribution function drifts towards the origin while the pitch angle remains constant. Considering the emitted frequency as a function of the particle energy for constant pitch angle, it was found (Mangeney and Veltri, 1976a) that the frequency increased when the electron energy decreased (for $\alpha \geq 45^{\circ}$).

Intensity and Source Dimensions

The maximum energy of the waves was given by (12.3-12), and the intensity observed at the earth may amount to

$$I \simeq (10^{-12} - 10^{-11}) K n_b L^2 \text{ W m}^{-2},$$

where K is an attenuation factor representing reabsorption in the source, n_b is the beam density (m^{-3}) and L the transverse linear dimension of the source (in $R_{\odot}$). Take $n_b = 10^8$ m^{-3} and $L \simeq 10^{-3}$ $R_{\odot}$. Then

$$I = (10^{-10} - 10^{-9}) K \text{ W m}^{-2}$$

and $K \simeq 10^{-5} - 10^{-3}$ in order to explain the observed intensities.

The source size $L \simeq 10^{-3}\ R_{\odot}$ is considerably smaller than indicated by observations ($L_{obs} \simeq 0.1 - 0.2\ R_{\odot}$). If this is an effect of scattering, the scattering must take place far from the source, or it is impossible to explain the high directivity of type I bursts (Steinberg *et al.*, 1974).

Model calculations undertaken by Mangeney and Veltri (1976b) showed that for source heights in the range 0.3 to 0.64 $R_{\odot}$ and magnetic fields decreasing from 24 to 8 G, bursts of duration 0.1 - 3 s could be emitted over a band from 300 - 30 MHz. The emitted frequency f lies in the range

$$f_p \leq f \leq 1.5\ f_p$$

which may agree with observations.

In order to explain chains, Mangeney and Veltri suggested repetitive production of beams in the acceleration region. If the acceleration region emits successive beams with exponentially increasing (or decreasing) energy at a given level, the (slow) frequency drift of chains may be explained. Another possibility is that the acceleration region itself drifts upwards or downwards with the Alfvén speed emitting beams of the same energy.

Conclusions

Summarizing their results, Mangeney and Veltri concluded:

(a) The emission occurs in regions of strong magnetic fields with a vertical scale height greater than twice the density scale height. This requirement may explain why only a small number of spots are associated with noise storms.

(b) A particular acceleration process is required which increases the transverse angular momentum of a fraction of the coronal electrons by a factor of 10 while their parallel momentum remains roughly constant. The betatron mechanism acts in this way, but the acceleration problem needs to be solved.

(c) The background continuum emission is not explained by the theory of Mangeney and Veltri, but they pointed out one possibility to solve the problem. One may suppose that the acceleration mechanism works regularly for a long period and that very short perturbations occur. The bulk of the accelerated electrons radiate the continuum by some unknown process, while the electrons accelerated during the perturbations form beams with small dispersion in momenta, which generate the bursts.

One may remark that the theoretical results obtained by Mangeney and Veltri depend on the existence of a broad-band turbulence spectrum in the corona. Unfortunately almost nothing is known about coronal turbulence from observations, in particular as regards the high-frequency part. Moreover, the acceleration problem remains to be solved.

13. SOME THEORIES WITH POSSIBLE APPLICATION TO STORM EMISSION

13.1. PLASMA TURBULENCE AND RADIO EMISSION

Since about 1960 a rapid development in experimental and theoretical plasma physics occurred. Concepts such as plasma turbulence and non-linear interaction of waves were introduced. Previous results, which had been of a qualitative nature, and problems which had been left unanswered, were treated in a quantitative way. The results were of prime importance for astrophysics, as is seen from the expositions of Tsytovich (1970), Kaplan and Tsytovich (1973), and from the large number of papers in which theoretical results in plasma physics have been applied on astrophysical problems.

The state of a plasma in which waves of one or more modes are excited in a wide range of frequencies and wave numbers is called plasma turbulence. The excitation is furnished by some source for the turbulence energy. It may be suggested that type I bursts, which occur frequently, are connected with a major channel of turbulence supplying energy to the corona (Gordon, 1970).

Spectrum of Plasma Turbulence

It is useful to devote some attention to the spectrum of plasma turbulence, as was done by Pikelner and Tsytovich (1969) and by Kaplan *et al.* (1972).

Longitudinal waves have a frequency close to the plasma frequency ω_p. Their wavelength is determined by the wave number k which is related to the phase velocity through $v_\phi = \omega_p/k$. Particles with velocity v will interact with waves who have phase velocities $v_\phi \leq v$. The interaction is the strongest when $v_\phi \simeq v$. It is possible to determine the effect which plasma turbulence will have on particles, but it is then necessary to know the spectrum of the turbulence which gives the relative energy of the waves as a function of k.

The function $W^l(k)$ is an expression for the energy of plasma turbulence in 1 cm^3 of the plasma within the interval dk. For the total energy of the turbulence one then has $W^l = \int_0^\infty W^l(k)dk$.

In some respects problems encountered in connection with the turbulence spectrum in a fluid have a similarity to those encountered when plasma turbulence is treated. One difference should be pointed out: in a fluid the energy is supplied on the largest scale and is transmitted to the smallest scales where viscosity takes care of it. In a plasma the flow is in the opposite direction, i.e. from large to small wave numbers.

Plasma waves may be generated in different ways. Of particular interest in solar physics are waves generated by particle beams and low-frequency waves such as Alfvén waves.

It is easy to find the maximum wave number in plasma turbulence since plasma waves with phase velocities which are equal to the thermal velocity of electrons are strongly damped. One may regard the value $k = k_{max} = \omega_p/v_t$ as an upper limit. In the region below this value, generation of plasma turbulence takes place. From the generation region the energy flows towards longer wavlengths (smaller wave numbers) because of different non-linear processes. This is also a flow towards larger phase velocities. The increase in energy in the region of generation is compensated for by a non-linear outflow of energy. In regions of wave numbers outside the excitation region a kind of quasi-stationary spectrum is formed. The inflow of energy is compensated for by non-linear outflow. The energy flux decreases with decreasing k and the energy $W^l(k)$ increases. For some value k_0 the accretion of energy because of non-linear interactions becomes comparable with the energy damping because of collisions with fast particles. The form of the spectrum is illustrated in Fig. 13.1.

When strong damping occurs due to coulomb collisions and acceleration of suprathermal particles, the spectrum changes in form, the part for lower wave numbers being supressed.

Two consequences of plasma turbulence are very important for solar radio astronomy: particles may be accelerated and generation of transverse electromagnetic waves may occur. The acceleration

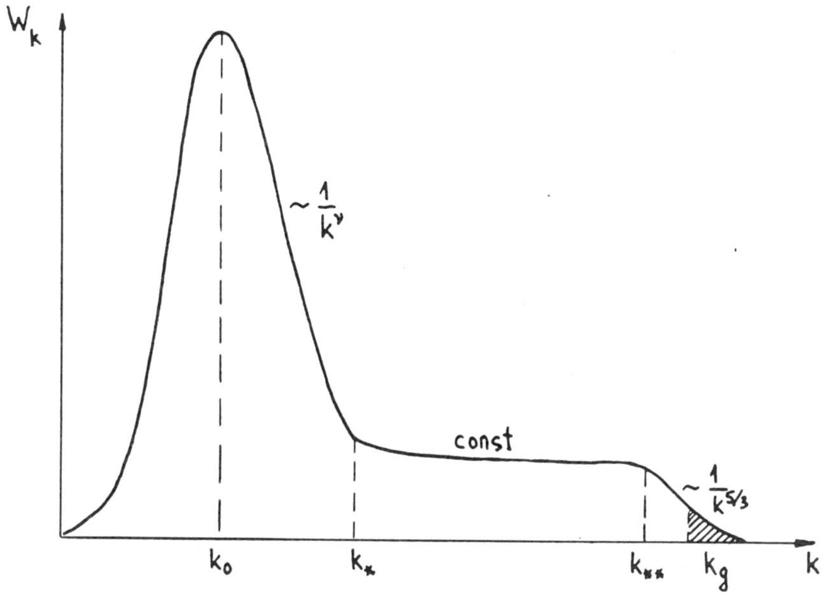

Fig. 13.1. Spectrum of plasma turbulence. The region of generation of waves is k_g, and k_0 is the region of accumulation of energy of the waves. $k_* = (\omega_p/v_t)(m/9M)^{\frac{1}{2}}$, $k_{**} = (\omega_p/v_t)(m/M)^{1/5}$. (Kaplan *et al.*, 1972.)

takes place because energy flows from waves with high to waves with low wave numbers (high phase velocities) and exchange of energy with the particles of the plasma occurs.

Non-linear transfer of energy along the plasma wave spectrum is accompanied by simultaneous conversion of longitudinal waves into transverse waves with frequency close to ω_p.

Acceleration of Particles

Gordon (1970) suggested that the relation between type I and type III bursts might be explained by non-linear processes in the corona. The observation of chains of bursts provides evidence for drift velocities which are about one order of magnitude less than v_t. It appears reasonable to associate the drift velocity of chains with the velocity of propagation of hydromagnetic waves. In the front of the wave electrons may be accelerated, resulting in the production of type I bursts. But there is also a relation between chains and type III bursts, as was illustrated in Fig. 6-12. Furthermore, Malville (1962) found that the starting frequencies of type III bursts usually were lower than the upper frequencies covered by noise storm emission. It should also be remembered that dekametric storms reveal the existence of electron energies in the range of tens of keV. Gordon proposed that the energetic electrons producing type III bursts were accelerated by non-linear processes in the type I source. He assumed that the initial phase velocity of the excited plasma waves was $v_\phi \approx 2v_t$. For the temperature in the active region he assumed a range from 10^6 to 2×10^7 K. The initial phase velocity of the excited plasma waves lies in a rapidly rising part of the plasma turbulence spectrum where one has

$$W^l(k) \propto k^{-5/2} \tag{13.1-1}$$

and this rising part extends to a value of k equal to $(\omega_p/v_t)/5.6$ where the slope of the spectrum changes. Denoting ω_p/v_t by k_D one finds $k_D = 1.1\ \text{cm}^{-1}$ and $0.25\ \text{cm}^{-1}$ for temperatures of 10^6 and 2×10^7 K respectively, when the frequency is assumed equal to 100 MHz. In the first case the turnover point in the spectrum corresponds to a phase velocity of about 0.1 c and in the second case to roughly 0.5 c. Acceleration of electrons may thus occur up to velocities which are required to explain type III bursts and dekametric type III storms. The characteristic time taken to transform the spectrum of plasma waves involving a change in wave number of $\Delta k \simeq k$ has been derived by Tsytovich (1970)

$$\frac{1}{\tau} = \omega_p \left(\frac{v_t}{v_\phi}\right)^3 \frac{W_e^l}{N_e m v_t^2} \tag{13.1-2}$$

where $W_e^l = \int W^l(k)\,dk$ is the energy density of the plasma waves. For ratios of $W_e^l/N_e m v_t^2 \approx 10^{-4}$ or somewhat larger, one obtains transformation times τ considerably less than 1 s.

Conclusion

In accordance with the theory of Gordon, electron beams of velocity in the range 0.1 - 0.5 c are created when type I bursts are excited. Thus one may explain the "on-fringe" type III bursts described in section 10.2. It is more difficult to explain the observations of Stewart and Labrum (1972) with this theory since the sources of the noise storm and the type III bursts were 0.5 $R_\odot$ apart (Fig. 8.7). There are obviously different mechanisms which may release type III activity. The type III bursts observed by Stewart and Labrum were possibly of the "off-fringe" type and may be associated with another acceleration mechanism in which particles generated by flares had direct access to open field lines.

13.2. MODE COUPLING IN A WARM PLASMA

An attempt to explain enhanced, directive, circularly polarized radiation in the *o*-mode from the sun by means of mode coupling in a warm plasma was made by Mollwo (1970).

Propagation in a Warm Plasma

Consider waves propagating in a warm (non-relativistic) plasma. A linearized kinetic theory of plane waves in an infinite, warm, magnetized plasma has been derived by Sitenko and Stepanov (1957). The problem has also been treated by Fried and Conte (1961), Akhiezer *et al.* (1967) and several others (cf. Bekefi, 1966).

Using a coordinate system with the z-axis in the direction of the magnetic field B_0 and the x-axis taken such that the wave propagates in the x-z-plane, the effective conductivity tensor σ of the plasma with collisions neglected may be written

$$\sigma = -\frac{i\omega_p^{\,2}\zeta_0}{4\pi\omega}\exp(-\eta)\mathbf{S}. \qquad (13.2\text{-}1)$$

In (13.2-1) $\mathbf{S}$ is a tensor and depends on ζ_s and η where

$$\zeta_s = \frac{\omega - s\omega_H}{\omega n\beta\cos\theta} \quad \text{and} \quad \eta = \tfrac{1}{2}\left(\frac{\omega}{\omega_H}n\beta\sin\theta\right)^2$$

$$\beta = \frac{v_t}{c} = \left(\frac{2T}{mc^2}\right)^{\frac{1}{2}} \quad (T \text{ in ergs}).$$

The complex refractive index is $n = \mu + i\chi$, and the dielectric tensor ε is given by $\varepsilon = 1 + \frac{4\pi i}{\omega}\sigma$. The calculations of Mollwo were restricted to the case $X = \omega_p{}^2/\omega^2 \approx 1$, $\sin^2\theta << 1$ and $Y = \omega_H/\omega \approx 1$.

For the tensor resistivity σ^{-1} one gets

$$\sigma^{-1} = \frac{1}{-\frac{i\omega_p{}^2\zeta_0}{4\pi\omega}\det[\exp(-\eta)\,\mathbf{S}]}\,\mathbf{A}. \tag{13.2-2}$$

Changing coordinate system by turning around the y-axis an angle θ so that the z-axis is parallel to the wave propagation direction and the magnetic field is in the x-z-plane, one has $\mathbf{B} = \mathbf{T}^{-1}\mathbf{A}\mathbf{T}$ where

$$\mathbf{T} = \begin{bmatrix} \cos\theta & 0 & -\sin\theta \\ 0 & 1 & 0 \\ \sin\theta & 0 & -\cos\theta \end{bmatrix}$$

Calculating the determinant $\det[\exp(-\eta)\mathbf{S}]$ and the matrix explicitly, Mollwo obtained

$$\bar{\sigma}^{-1} = \frac{1}{-\frac{i\omega}{4\pi}Xb_0}\begin{bmatrix} b_{11} & ib_{12}\cos\theta & b_{13}\sin\theta\cos\theta \\ -ib_{12}\cos\theta & b_{22} & ib_{23}\sin\theta \\ b_{13}\sin\theta\cos\theta & -ib_{23}\sin\theta & b_{33} \end{bmatrix} \tag{13.2-3}$$

where $b_0 = 2\zeta_0 Z\left(\zeta_0\frac{\omega+\omega_H}{\omega}\right)$, b_{ik} are functions of $Z(\zeta_n)$, Y and ζ_0, and $Z(\zeta_n)$ is the plasma dispersion function tabulated by Fried and Conte (1961).

The expression for the conductivity may be formally written (Heald and Wharton, 1965)

$$\bar{\sigma} = -\frac{i\omega}{4\pi}\begin{bmatrix} M^2 & 0 & 0 \\ 0 & M^2 & 0 \\ 0 & 0 & -1 \end{bmatrix} \tag{13.2-4}$$

where $M^2 = n^2 - 1$.

The general relation $\sigma\mathbf{E} = \mathbf{J}$ and $\sigma^{-1}\mathbf{J} = \mathbf{E}$ are consistent if $(\sigma^{-1}\sigma - 1)\mathbf{E} = 0$. Accordingly $n^2 - 1 = M^2$ can be derived from

$$\bar{\sigma}^{-1}\bar{\sigma} - 1 = \frac{M^2}{b_0 X}\begin{bmatrix} b_{11} - b_0\frac{X}{M^2} & ib_{12}\cos\theta & -b_{13}\frac{1}{M^2}\sin\theta\cos\theta \\ -ib_{12}\cos\theta & b_{11} - b_0\frac{X}{M^2} & -ib_{23}\frac{1}{M^2}\sin\theta \\ b_{13}\sin\theta\cos\theta & -ib_{23}\sin\theta & -(b_{33} + b_0 X)\frac{1}{M^2} \end{bmatrix}$$

(13.2-5)

The solutions for different wave modes are found when the determinant of the matrix is set equal to zero. This leads to the following characteristic equation

$$\left[(b_{11} - b_0\frac{X}{M^2})^2 - b_{12}\cos^2\theta\right](\frac{b_{33}}{b_0} + X)$$

$$-\left[(b_{11} - b_0\frac{X}{M^2})(\frac{b_{13}^2}{b_0}\cos^2\theta + \frac{b_{23}}{b_0}) + 2b_{12}\frac{b_{13}b_{23}}{b_0}\cos^2\theta\right]\sin^2\theta = 0. \quad (13.2\text{-}6)$$

Evaluation and discussion of (13.2-6) is rather complicated. In the cold plasma approximation the dispersion equation may be written

$$A k^4 + B k^2 + C = 0 \quad (13.2\text{-}7)$$

But the warm plasma dispersion equation is no longer biquadratic because A, B and C are also functions of k as well as of ω and θ. In the present context it appears justified to proceed directly to some of the results obtained by Mollwo.

For $\theta \neq 0$ two characteristic wave modes, the ordinary and the extraordinary, occur. In Fig. 13.2 $Re\, n^2$ is shown and in Fig. 13.3 $Im\, n^2$ for $X \approx 1$, small angles θ and $Y \approx 1$. The imaginary part of n^2 represents cyclotron damping, and it is seen that the damping decreases with decreasing θ.

It is important to note from the two figures that the x-wave and the o-wave have almost the same value of the real part of the refractive index ($Re\, n^2 \simeq 0.5$) for small values of θ near $1 - X = 3(\beta/2)^2$ and that the imaginary part of n^2 is negligible. Moreover, both waves have roughly the same polarization coefficient in this region. This means that one here has a coupling point, and exchange of energy between both wave modes occurs. In the cold plasma approximation this does not occur.

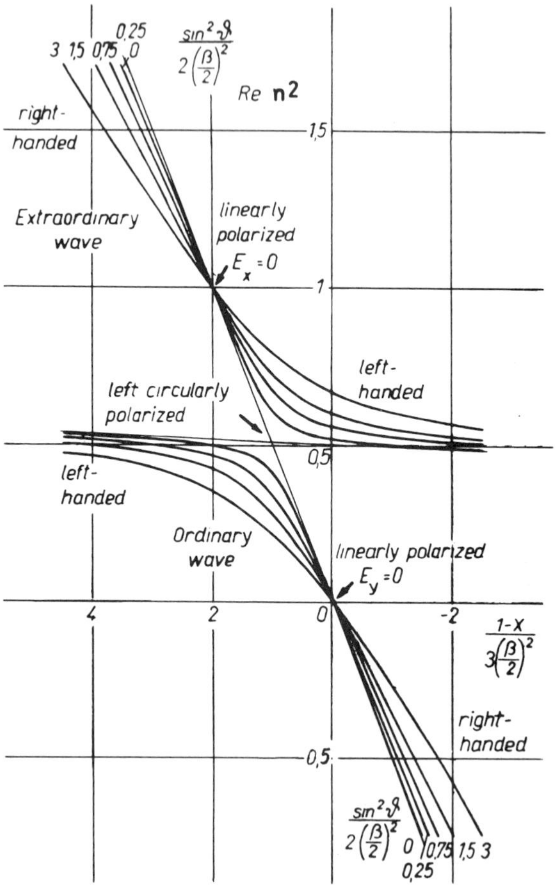

Fig. 13.2. $Re\, n^2$ for a warm plasma. $Y = 1$, $X \propto$ electron density, θ is the angle between the magnetic field and the direction of wave propagation, $\beta^2 = (v_t/c)^2$ is proportional to the electron temperature. (Mollwo, 1970.)

A larger portion of the dispersion curves is given in Fig. 13.4 for $\omega/\omega_H = 0.9$ and 1.1 and a small value of θ. As appears from Fig. 13.4 the extraordinary wave with $\omega/\omega_H = 1.1$ and the ordinary wave with $\omega/\omega_H = 0.9$ join the electron plasma waves for the respective ω/ω_H and θ towards large values of n^2 in agreement with the results of Denisse and Delcroix (1961) and Allis *et al.* (1963).

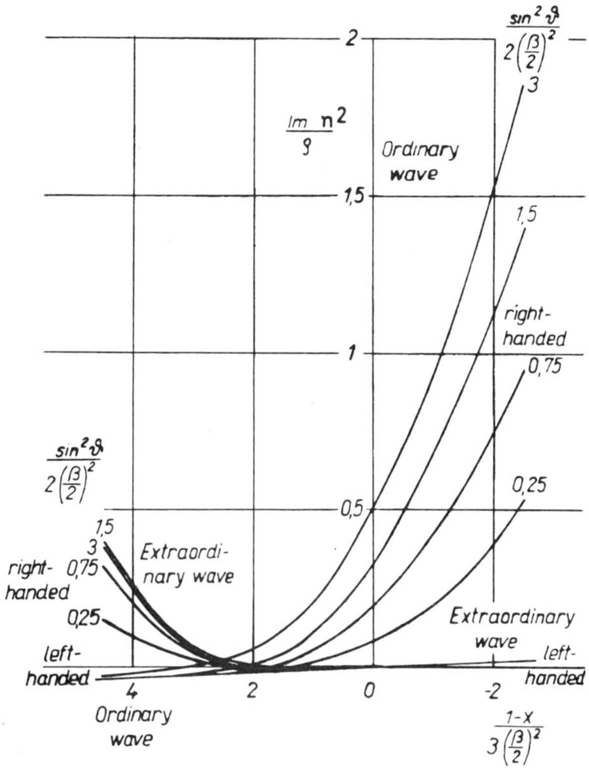

Fig. 13.3. *Im* n^2 for a warm plasma. The plasma parameters are as in Fig. 13.2; $\rho = \beta n/2\pi^{\frac{1}{2}}$. The curve for the left-handed circularly polarized wave is the asymptote of the other curves. The slope of this curve ($\theta = 0$) depends on β. (Mollwo, 1970.)

According to Bernstein (1958) asymptotes of the dispersion curves exist. These are nearly parallel to the dispersion curve of the plasma wave with $\theta = 0$ and cross the abscissa where the magneto-ionic theory gives poles of the refractive index. In Fig. 13.4 Mollwo therefore assumed such parallels as asymptotes and tangents respectively to the points of inflection. Where the dispersion curves approach the points for $\omega/\omega_H = 0.9$, strong imaginary parts of n^2 exist.

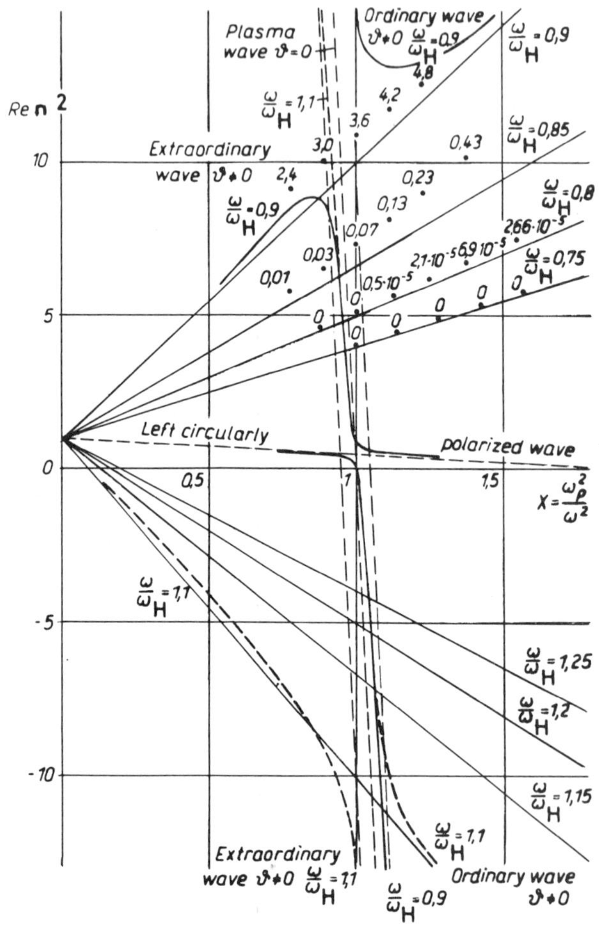

Fig. 13.4. Dispersion curves in a warm plasma for very small angles θ and $1/Y = 1.1$ and 0.9. Thin lines represent dispersion curves for $\theta = 0$ after magnetoionic theory for several values of ω/ω_H, and results from kinetic theory are given by distinct points for $\beta = 0.0269$ and $T = 2.14$ million degrees. Imaginary parts of n^2 are added to each point. (Mollwo, 1970.)

Application to Storm Radiation

As described in section 8.3, the stationary metre-wave continuum occurring during the later stage of type IV bursts seems to be of the same nature as the noise storm background continuum emission. Both phenomena may therefore be generated in the same way. From the properties of the warm plasma dispersion curves, Mollwo proposed an interpretation of the two phenomena. It was assumed that the primary emission mechanism was gyro-radiation from electrons with speeds of the order of 0.5 c and with negligible velocity component in the direction of the field lines. In the source region $Y \simeq 1$ and X may be slightly less than 1. According to Fig. 13.4 propagation is then possible for very small angles θ ($Im\, n^2$ is small and $Re\, n^2 < 10$). The waves are polarized in the right-handed sense and have a large longitudinal component. Since the corona is inhomogeneous, the waves may encounter a region of increased electron density. Therefore X may become equal to 1 or a little more than 1 before the general decrease of X with increasing distance from the solar centre dominates. During the propagation through the inhomogeneity the waves have to pass twice through a coupling point and energy is transferred to the ordinary mode. The ordinary waves from the first coupling point are reflected at $X = 1$, travel towards the solar centre and may be reflected once more. The energy delivered to the ordinary mode at the second coupling point propagates to the observer and suffers no important damping. The energy remaining in the original wave mode is absorbed and the waves are unable to escape. Thus the ordinary mode only is observed, and because the coupling mechanism works for very small values of θ, a strong directivity is expected. With the assumed electron speed of about 0.5 c an effective temperature approaching 10^9 K may be obtained provided the energy is about equally divided between the wave modes in a coupling point. One may argue against the mechanism suggested by Mollwo that this value is somewhat on the lower side in order to account for the phenomena in question.

In order to explain type I bursts Mollwo assumed that the primary radiation was waves excited by the Cerenkov effect by electrons with velocity about 0.5 c travelling along the magnetic field lines in the same region as discussed above ($Y \simeq 1$, X somewhat less than 1). The waves make very small angles with respect to the direction of the magnetic field lines, they have right-handed polarization and a large longitudinal component. The coupling process is as described before. Mollwo suggests that the short duration and narrow bandwidth of the bursts are effects of local conditions which must be such that standing waves arise. The coupling process requires an electron density profile which leads to reflection of one electromagnetic mode and of the plasma mode. In addition a reflection point must exist for the waves propagating in the direction of the solar centre. Then a building-up process may take place and coherent waves be emitted.

Conclusions

From his investigations Mollwo concluded that the radio emission has a frequency close to the electron plasma frequency. The ordinary wave has a strong directivity and originates from the extraordinary wave by an efficient process of mode coupling. Since the emitted frequency is no harmonic of the gyro frequency, a relative low energy of the emitting electrons is sufficient. The type IVmB bursts and noise storms are generated by electrons with small velocity in the direction of the magnetic field. This explains the long durations of these phenomena. The explanation of type I bursts is based on the Cerenkov effect and a building-up process of standing waves.

In order to account for the prevailing polarization in the *o*-mode, asymmetry of the magnetic field structure in the active region is necessary. This question needs further consideration.

The conditions for the mechanism to work are relatively strict. There is as yet no proof of the importance of the different propagation processes in the solar corona. Moreover, the condition $Y \simeq 1$ requires strong magnetic fields at considerable heights above the solar surface.

13.3. THE ROLE OF WHISTLERS FOR THE PRODUCTION OF RADIO NOISE

Whistlers were first detected as an ionospheric phenomenon. Being waves of very low, audible frequencies, the swishing sounds they produced in the receivers gave rise to their name. But whistlers are not confined to the ionosphere. They are transverse waves with frequencies below the electron gyro frequency, and may readily be generated in the solar corona. As shown by Chiu (1970), Chin (1972) and Kuipers (1975) an anisotropic distribution of electrons trapped in coronal magnetic loops easily generates a non-thermal level of whistler turbulence. Since non-thermal plasma waves (Langmuir waves) should be present at the same time, one could hope that a whistler wave w coupled with a plasma wave l to produce a transverse wave t above the plasma frequency: $l + w \rightarrow t$.

It was suggested by Chiu that whistler wave trains executing bouncing motion in bipolar coronal magnetic fields could produce double, sympathetic sources of type III bursts and microwave pulsations. The process of coalescence of whistlers and coherent plasma waves to produce transverse waves with frequency near the plasma frequency was further studied by Chiu in connection with type III bursts. Kuipers (1975) applied the process for the explanation of intermediate drift bursts, which may occur as fine structure in type IV continuum bursts. In view of the periodic phenomena described in section 7.3, which show variations on a time scale in the range of seconds, fusion of whistlers and plasma waves appear to be of interest in connection with noise storms. The pulsation periods predicted by Chiu fall in the correct range.

Fusion of Longitudinal Waves and Whistlers into o-mode Waves

When electron plasma waves and whistlers coalesce to form transverse waves, the following resonance conditions apply:

$$\mathbf{k}_t = \mathbf{k}_l \pm \mathbf{k}_w, \qquad \omega_t = \omega_l \pm \omega_w \tag{13.3-1}$$

where the subscripts signify the different wave modes.

When whistlers propagate at an angle θ with the magnetic field, the dispersion equation in the cold plasma approximation is (Kennel and Scarf, 1969)

$$v_\phi = c\,\frac{\omega_H}{\omega_l}\left[\frac{\omega_w}{\omega_H}\left(\cos\theta - \frac{\omega_w}{\omega_H}\right)\right]^{\frac{1}{2}} \tag{13.3-2}$$

For the wave number k_w one then obtains ($\omega_l = \omega_p$)

$$k_w = \frac{\omega_p}{c}\left(\frac{\omega_w}{\omega_H\cos\theta - \omega_w}\right)^{\frac{1}{2}} \tag{13.3-3}$$

The typical frequency range of whistlers falls below the electron gyro frequency, and one has $\omega_w << \omega_H|\cos\theta|$. From (13.3-3) one then obtains

$$k_w << \frac{\omega_p}{c} \tag{13.3-4}$$

For the escaping transverse waves the dispersion equation in the case $\omega_H << \omega_p$ is well known,

$$n = \frac{ck_t}{\omega} = \left(1 - \frac{\omega_p}{\omega_t}\right)^{\frac{1}{2}}$$

giving

$$k_t = \frac{(\omega_t^2 - \omega_p^2)^{\frac{1}{2}}}{c} << \frac{\omega_p}{c}. \tag{13.3-5}$$

The longitudinal waves generated by particles with velocity v have wave numbers

$$k_l = \frac{\omega_p}{v} > \frac{\omega_p}{c}. \tag{13.3-6}$$

It is seen, as was pointed out by Melrose (1975a), that there is no way in which (13.3-4), (13.3-5) and (13.3-6) can be made compatible with the resonance conditions (13.3-1).

There appears to be two ways to escape from the difficulty: to allow higher whistler frequencies, or to invoke plasma waves with phase velocities greater than the velocity of light.

Kuipers made use of the first-mentioned possibility. The damping of whistlers propagating along the magnetic field was determined using the dispersion equation given by Scharer and Trivelpiece (1967) for a finite temperature plasma

$$n^2 = \frac{c^2k^2}{\omega^2} = 1 + \frac{\omega_p^{\ 2}}{2^{\frac{1}{2}}kv_t} Z_t, \qquad (13.3\text{-}7)$$

where $v_t = (\kappa T/m)^{\frac{1}{2}}$, and $Z_t \equiv Z(\psi_t)$ is the plasma dispersion function (Fried and Conte, 1961) with argument $\psi_t = (\omega - \omega_H)/2^{\frac{1}{2}}kv_t$. Using parameters which are typical for the corona, $T = 10^6$ K, f_p = 300 MHz and ω_p/ω_H = 30 - 10, Kuipers found that cyclotron damping was very important for frequencies above 0.5 ω_H. At frequencies below 0.5 ω_H whistlers can propagate with slight damping. It was also shown that the group velocities were directed within a small cone around the magnetic field. Provided the Alfvén velocity decreases along the path, whistlers focus towards the direction of the magnetic field. When the group velocity equals the phase velocity, discrete wave packets, solitons, may propagate along the field. This occurs for a velocity of 21.5 v_A and whistler frequencies between 0.25 and 0.5 ω_H.

In the search for possible solutions of the coupling problem, Kuipers followed a method described by Høijer and Wilhelmsson (1970). Using parameters $T = 10^6$ K and $\omega_p/\omega_H = 30$, it was found that the whistler frequency should not exceed 0.37 ω_H in order to avoid too strong damping, and the maximum wave number was 0.056 cm^{-1}. Coupling could then occur with plasma waves having small wave numbers, i.e. $k_l \simeq 0.06$ cm^{-1} and $\theta_l \leq 10^o$ (θ_l is the angle between the magnetic field and the plasma wave vector). As a result, transverse electromagnetic waves in the ordinary mode would be produced. It should be mentioned that in order to obtain $k_l \simeq 0.06$ cm^{-1}, the plasma waves must be excited by very fast particles.

Fusion of Z-Mode Waves with Whistlers

The possibility to satisfy conditions (13.3-1) by considering plasma waves with phase velocities exceeding the speed of light was explored by Melrose. Fusion and decay processes (involving two l-waves and one w-wave) may pump plasma waves from the region with $v_\phi < c$ into the region $v_\phi > c$. But simultaneously there occurs a change in the properties of the plasma waves which cannot any longer be regarded as longitudinal. They are nearly transverse waves for sufficiently small wave numbers and may be treated as Z-mode waves, whish are known from ionospheric sounding experiments (Ratcliffe, 1959). Melrose therefore investigated the possible fusion of Z-mode waves with whistlers and found that the resonance conditions could only be satisfied for

$$\omega_w \simeq \omega_p - \omega_Z \gg \omega_t - \omega_p, \qquad k_w \simeq k_Z \gg k_t$$

and with $\omega_w \simeq \omega_H \left(\frac{\omega_H}{2\omega_p}\right)^{\frac{1}{2}}$.

Since a specific whistler frequency was required, it was suggested that the process was probably not of physical significance. It appears that a more interesting effect of whistlers is the pumping of longitudinal plasma waves into Z-mode waves. If the Z-mode waves can be effectively converted into radiation, the mechanism may be important for the interpretation of noise storm radiation.

Scattering of Electrons by Whistlers

The role of whistlers for the production of non-thermal radio emission may also be manifested in another way. When energetic electrons are injected in a magnetic structure, electrons with $v > 43\, v_A$ are strongly scattered by whistlers, whereas those with lower velocities move freely (Melrose, 1974c). In this way a spatial separation of fast and slow particles is obtained, and a sharply defined "gap distribution" of electrons is formed.

13.4. PLASMA EMISSION DUE TO FAST, ISOTROPIC ELECTRONS

The properties of plasma emission from an isotropic distribution of fast electrons were explored by Melrose (1975c) and a theory of type I bursts was given in outline. Emission from a power law distribution could not explain the high brightness temperatures observed for type I bursts. Therefore gap distributions were investigated.

Gap Distributions

Melrose considered gap distributions with the following properties: (1) the distribution is isotropic; (2) the density of particles with $v \simeq v_0$ is sufficiently high for these particles to dominate over the thermal particles in the emission and absorption of plasma waves with $v_\phi \simeq v_0$; (3) in the velocity range $v_t << v << v_0$ the number density of electrons is sufficiently low that absorption of plasma waves with phase velocities v_ϕ in the interval $v_t << v_\phi << v_0$ is dominated by the thermal particles, while the emission of plasma waves in this range is dominated by the particles with $v \simeq v_0$.

An electron distribution with the properties specified above gives the highest possible brightness temperature $T^l(k)$ for plasma waves when the number density and energy density of the isotropic suprathermal particles are given.

Gap distributions may be formed as mentioned in the last part of section 13.3. Such distributions, although less sharply defined, can also be created through the effect of collisions, since the slower electrons trapped in a coronal region are more quickly slowed down than the faster ones.

Emission

The predicted effective temperature of plasma waves excited by a gap distribution becomes (Melrose, 1975c)

$$T^l = \frac{9.0 \times 10^{20}}{f_p} \frac{N_1}{N_e} \left(\frac{v_0}{c}\right)^2 \left(\frac{v_\phi}{v_0}\right)^3 \qquad (13.4\text{-}1)$$

where N_1 is the density of suprathermal electrons and f_p is the plasma frequency in megahertz. The brightness temperature of radio waves T_1^t at the fundamental frequency, ignoring induced scattering and assuming a coronal temperature of 1.5×10^6 K and a scale length of 3×10^5 km, is given by

$$T_1^t = 1.5 \times 10^{12} \frac{N_1}{N_e} f_p \left(\frac{v_0}{c}\right)^{-1} \tag{13.4-2}$$

For the second harmonic one has

$$T_2^t = 1.8 \times 10^{22} \left(\frac{N_1}{N_e}\right)^2 \frac{v_0}{c}. \tag{13.4-3}$$

Induced scattering is important for the fundamental when $v_\phi = v_0$ and

$$\frac{N_1}{N_e} \geq \frac{2 \times 10^{-3}}{f_p} \left(\frac{v_0}{c}\right)^2$$

The highest brightness temperature is obtained at the fundamental and is due to induced scattering,

$$\left(T_1^t\right)_{\max} = \frac{1}{24\sqrt{3}\pi} \frac{N_1}{N_e} N_e \lambda_D{}^3 T_e \left(\frac{v_0}{c}\right)^2 \left(\frac{c}{v_t}\right)^5 \ln\frac{v_0}{v_t}, \tag{13.4-5}$$

where λ_D is the Debye length and T_e the electron temperature. Incerting numerical values, (13.4-5) may be written

$$\left(T_1^t\right)_{\max} = \frac{6 \times 10^{23}}{f_p} \frac{N_1}{N_e} \left(\frac{v_0}{c}\right)^2$$

Assuming f_p = 100 MHz and $v_0 = 10^5$ km s^{-1}, the maximum value of T_1^t occurs for $N_1/N_e = 3 \times 10^{-5}$ and amounts to $T_1^t = 2 \times 10^{16}$ K. For the second harmonic Melrose obtained $T_2^t = 3 \times 10^{13}$ K as a maximum value.

Application to Type I Bursts

In the application to type I bursts Melrose pointed out some favourable properties of the mechanism: (1) the radiation occurs at the fundamental, and a relatively weak magnetic field produces 100% polarization in the ordinary mode because the emission frequency becomes less than the cut-off frequency of the

x-mode; (2) high brightness temperatures can be explained; (3) the brightness temperature of the fundamental is orders of magnitude larger than that of the harmonic and harmonic structure in noise storms is ordinarily not expected; (4) since the mechanism involves amplification, short-duration bursts may be expected. The shortest time scale in bursts should not be shorter than the characteristic time for amplification. The e-folding amplification time as estimated by Melrose (1975c) amounts to

$$t_{\text{growth}} \simeq \left(2 \times 10^4 \frac{N_1}{N_e} f_p \frac{c}{v_0}\right)^{-1} \qquad (13.4\text{-}6)$$

and may be of the order of milliseconds for a reasonable choice of parameters.

The lifetime of the gap distribution is

$$t_{\text{coll}} = \frac{1.9 \times 10^9}{f_p^{\,2}} \left(\frac{v_0}{c}\right)^3 \qquad (13.4\text{-}7)$$

Values in the range of hours may be found for t_{coll}.

The above time scales overlap those which are associated with type I bursts.

It is difficult to explain the background continuum in this theory. One possibility pointed out by Melrose is that the continuum is made up of many small unresolved sources. But the dimensions of such sources cannot be smaller than the path length over which the fundamental is amplified, about 10^3 km, and it is not obvious that contribution from such subsources would merge into a continuum.

13.5. CONSIDERATIONS CONCERNING THE BACKGROUND CONTINUUM

In the previous sections special attention has been paid to the explanation of the burst component of noise storms, although possible interpretations of the background continuum have been pointed out in several instances. But some special remarks concerning the generation of the background continuum are appropriate.

As previously stated, there appears to be no distinction between type IV continuum storms (type IVmB) and noise storm continuum. They are both highly polarized in the ordinary sense; the source diameters are small, the intensity may reach high values, the radiation is directive, it covers a relatively wide band at metric wavelengths and there is definite evidence of an origin near the critical (plasma) frequency in the corona.

The characteristics mentioned above rules out an explanation by simple synchrotron radiation as proposed by Takakura (1962). In order to explain the polarization in the ordinary mode, Takakura suggested that the radiation in fact might be produced in the extraordinary mode in a source region near the following spot in a group. If the polarization then was thought to be associated with the leading spot, which has the opposite magnetic polarity, one would find that the ordinary mode was emitted. However, it is now definitely established that this possibility is quite unlikely, as, for instance, is borne out from observations of bipolar sources.

Simple Estimates of Radiation from Coherent Plasma Waves

Kai (1965) suggested that during the first part of a type IV event the magnetic field became disordered because of the impact of energetic particles. Afterwards the field recovered and assumed a regular structure and radiation was emitted in the ordinary mode. The time taken for the field to become regular is of the order

$$t \simeq \frac{L}{v_A}, \qquad (13.5\text{-}1)$$

where L is the scale of the linear dimension of the source and v_A the Alfvén velocity. Taking $N \simeq 5 \times 10^8$ cm^{-3} and $L \simeq 10^{10}$ cm one obtains $t \simeq 10^3/B$, and values $1 < B \leq 50$ G give values of t in accordance with the observations. The radiation mechanism is not quite clear, but coherent radiation from plasma waves may be considered as proposed by Ginzburg and Zheleznyakov (1958). This is in principle the same mechanism as was suggested by Takakura (1963) to account for the type I bursts.

The amplitude of coherent plasma waves excited by an electron beam was estimated by Bohm and Gross (1949b) under the assumption that the rate of growth of the plasma waves was balanced by the damping rate due to collisions with the ambient particles and that the electron beam was in a steady state,

$$\frac{eE_0}{k} = \frac{m}{2}\left(\frac{\omega}{k}\right)\left[\frac{16}{3}\,\frac{\omega_p^{\,2}}{k^2}\,\frac{\tau_s}{\tau_f} f'\left(\frac{\omega}{k}\right)\right]^2, \qquad (13.5\text{-}2)$$

where E_0 is the amplitude of the electric field, k is the wavenumber, τ_s is the collision time for slow particles, τ_f the collision time for fast particles, and $f'(\frac{\omega}{k})$ is the slope of the velocity distribution $f(v)$ at $v = v_\phi$ = phase velocity of the plasma wave. According to Kai $f'(\omega/k) = (1/v_0{}^2)(N_f/N_0)$, where v_0 is the velocity of the beam, N_f is the number of fast electrons and N_0 is the number of thermal electrons. Substituting this into (13.5-2) and using $v_0 \simeq \omega/k$ one obtains

$$\frac{eE_0}{k} \simeq \frac{m}{2}v_0{}^2\left[\frac{16}{3}\,\frac{\tau_s}{\tau_f}\left(\frac{v_0}{v_1}\right)\,\frac{N_f}{N_0}\right]^2, \qquad (13.5\text{-}3)$$

where v_1 is the velocity dispersion of the beam particles.

If the coherent plasma waves are transformed to transverse waves by scattering on thermal density fluctuations (spontaneous emission), the emitted energy P is given by (Ginzburg and Zheleznyakov, 1958)

$$P = \frac{ne^4 N_{sc} V_{sc}}{6m^2 c^3}\, E_0{}^2, \qquad (13.5\text{-}4)$$

where n is the refractive index of the electromagnetic waves, N_{sc} the density in the scattering region and V_{sc} the volume of the scattering region. The effective temperature becomes

$$T_{\text{eff}} = \frac{c^2 P}{2\kappa f^2 S\Delta f} = 8.4\times 10^{12}\,\omega_p{}^2\, n\frac{N_{sc} L v_0{}^{10}}{f^2\Delta f v_1{}^8}\left(\frac{\tau_s}{\tau_f}\right)\left(\frac{N_f}{N_0}\right), \qquad (13.5\text{-}5)$$

where S is the area of the scattering region, L the height interval of the region and Δf the bandwidth of the radiation. For $v_0 > 2v_t$ one has (Spitzer, 1956) $\tau_s/\tau_f = 2.6(v_t/v_0)^3$. Assuming f = 200 MHz, Δf = 10 MHz, $N_0 = 5\times 10^8$ cm^{-3}, $L \simeq 10^{10}$ cm and

$v_1 \simeq v_t$, the expression (10.9-5) may be written

$$\frac{N_f}{N_0} \simeq \frac{(v_0/v_t)^{\frac{1}{2}}}{10^{3/4} n^{1/4} v^{1/2}} T_{\text{eff}}^{1/4} \qquad (13.5\text{-}6)$$

Take $v_0/v_t = 25$, $n^{1/4} \simeq 1$, $v_t = 5\times10^9$ cm s^{-1} and $T_{\text{eff}} = 10^{10}$ K, then $N_f/N_0 \simeq 1/300$.

In connection with the above estimate it may be remarked that the effective temperature may be higher by several orders of magnitude than assumed here and that the source volume may be considerably smaller. It is quite likely that induced effects are important. The above considerations should only be regarded as a first approach to a solution of the problem of continuum emission at metre wavelengths.

Radiation Generated in Flux Tubes

Observational evidence of noise storm radiation suggests that the generation occurs in magnetic flux tubes (magnetic arches) connecting regions of opposite polarity. The existence of such structures is evident from *XUV* observations undertaken outside the earth's atmosphere (Sheeley *et al.*, 1975). One may then possibly arrive at a realistic model by considering high-frequency instabilities in magnetic tubes. Assume that electrons of energy 10 - 100 keV are injected repeatedly into a closed magnetic configuration. The velocity distribution at the time of injection may be isotropic, but will rapidly evolve into a "loss cone" distribution. This occurs because the fast particles have their mirror points at low heights and their energy is lost through collisions. Within the loss cone the distribution consisting of a thermal, isotropic, "cold" plasma and a superposed loss cone distribution of fast particles (Fig. 13.5) can be unstable. This was studied by Kuipers (1975) in order to explain continuum emission and fine structure in type IV bursts.

Consider a magnetic arch which at each point is filled with a "cold" ($T = 10^6$ K) isotropic background component and a superposed loss cone distribution of fast electrons (>30 keV). The half-aperture θ of the loss ocne is given at every point by

$$\frac{\sin^2\theta}{B} = \frac{1}{B_c}, \qquad (13.5\text{-}7)$$

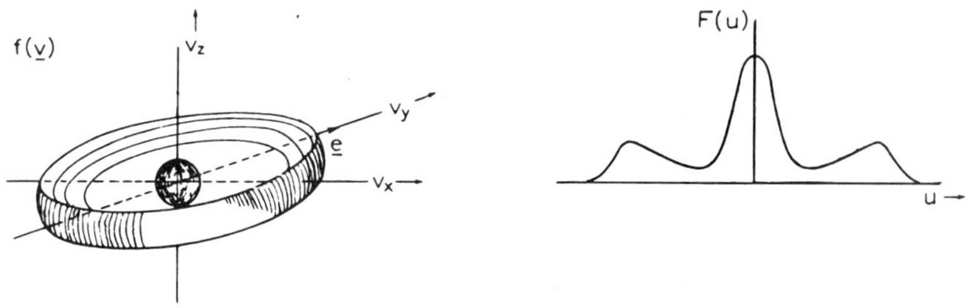

Fig. 13.5. Sketch of the particle distribution in velocity space together with the integrated function $F(u)$ for a ring distribution superposed on a cold background. The direction of the unit vector **e** along which the velocity component u is measured, is indicated. (Kuipers, 1975.)

where B_c is the value of the magnetic field at the level where collisions are dominating. In the source region one may have $\omega_H/\omega_p << 1$, and the fractional density of fast particles $q = N_{\text{hot}}/(N_{\text{hot}} + N_{\text{cold}}) << 1$.

Under such circumstances one may apply the instability criterion for the non-magnetic case

$$\left.\frac{\partial E}{\partial u}\right|_{u = \omega/k} > 0, \qquad (13.5\text{-}8)$$

where $F(u) = \int f(\mathbf{v})\,\delta(u-\mathbf{kv}/k)\,d\mathbf{v}$ and $f(\mathbf{v})$ is the velocity distribution for the electrons. As an example Kuipers found that coherent plasma waves could be generated for values of θ larger than 7° and up to 80° for beam velocities between $3v_t$ and extending to 5×10^9 cm s^{-1}. The waves are generated at frequencies near the upper hybrid frequency, $\omega_{UH} = (\omega_p{}^2 + \omega_H{}^2)^{\frac{1}{2}} \simeq \omega_p$, preferably in a direction perpenducular to the magnetic field.

Conversion to Transverse Waves

The longitudinal waves must be transformed to electromagnetic waves For spontaneous processes the conversion efficiency is low, in general less than 10^{-4}. But by the process of induced scattering the temperature of the transverse waves may become comparable to the temperature of the longitudinal waves. The most important process appears to be induced scattering of plasma waves off thermal ions. The characteristic transfer time in this case amounts to (Tsytovich, 1970, Kuipers, 1975)

$$\tau_i \simeq \frac{N_e m v_t^2}{W^l} \left(\frac{M}{m}\right)^2 \left(\frac{v_t}{v_\phi^l}\right)^3 \frac{1}{\omega_p} \qquad (13.5\text{-}9)$$

provided the condition $v_\phi^l << 3\left(\frac{M}{m}\right)^{1/2} v_t \simeq 130 v_t$ is satisfied. W^l is the total energy density of the plasma waves and v_ϕ^l their phase velocity.

Induced scattering into transverse waves occurs when

$$\tau_i v_g^t < L, \quad \tau_i << 1/\nu_c, \qquad (13.5\text{-}10)$$

where v_g^t is the group velocity of transverse waves and ν_c the collision frequency for electrons and ions in the background plasma. L is the linear dimension of the scattering region. It does not appear difficult to find physical parameters for the source region which satisfies the conditions for induced scattering to be accomplished.

Polarization

The escaping waves are polarized in the ordinary mode provided the longitudinal waves are confined to a narrow band above the plasma frequency, since one may then have

$$\omega_t < \omega_p + \frac{1}{2}\omega_H \qquad (13.5\text{-}11)$$

which implies that the x-mode waves encounter a stop region.

Melrose and Sy (1972) considered plasma emission processes in magneto-active plasma and found that in the weak field limit the condition for (13.5-11) to be satisfied was that

$$v_\phi \geq \max\left[v_t(3\omega_p/\omega_H)^{1/2},\ v_t\left(\frac{m}{M}\right)^{1/2}(2\omega_p/\omega_H)\right]. \qquad (13.5\text{-}12)$$

For $(\omega_p/\omega_H) \simeq 10 - 30$ this implies $v_\phi > 20 - 40\ 000$ km s^{-1}.

Further studies of instabilities connected with loss cone distributions in magnetic traps appear rewarding for a better theoretical understanding of the noise storm radiation.

Accumulation of Energetic Particles

The long duration of noise storms requires the presence of energetic electrons for hours and days. If electrons are instantaneously injected into the source region, they will loose energy through collisions with thermal plasma particles. The energy loss time t_E of electrons in the energy range 20 - 100 keV is given by (Brown, 1972)

$$t_E \simeq \frac{1.5 \times 10^8 \; E^{3/2}}{N_e} \; s \qquad (13.5\text{-}13)$$

where E is the kinetic energy of the fast electrons in keV and N_e is the number of thermal electrons per cubic centimetre. Take E = 100 keV, $N_e = 10^8$ cm^{-3}, one then finds $t_E \simeq 2.5$ min. In accordance with this result electrons must be produced continuously in the source or be injected repeatedly.

One possibility for accumulation of energetic particles in solar magnetic traps was pointed out by Kaplan and Tsytovich (1973), who suggested that strong ion-sound turbulence would fill the traps with particles. The mechanism is as follows.

Ion-sound waves can be changed into whistlers through nonlinear scattering by ions and electrons. This process occurs quite easily. Next the whistlers are absorbed by particles increasing the particle energy in the transverse direction to the magnetic field. Therefore a particle which at first was moving mainly along the magnetic field can acquire a sufficient velocity component perpendicular to the field to become trapped (reflected near the ends of the magnetic trap). The process is obstructed by ion-sound waves which serve to make the particle velocity distribution isotropic. When the condition

$$\frac{W^w}{W^s} >> \frac{\omega_p^{\,2}}{\omega_H^{\,2}} \frac{v_t}{v} \qquad (13.5\text{-}14)$$

is satisfied, the accumulation of energetic particles in traps is possible.

If the size of the trap is R and the change in transverse

energy during the flight through the trap is of the same order of magnitude as the total energy, the condition for confinement may be written (Tsytovich and Kaplan, 1973)

$$\frac{\omega_p v^3}{\omega_H{}^2 R v_t{}^2} \leq \frac{W^\omega}{N_e T_e} \leq \frac{W^s}{N_e T_e} \leq \frac{v^4}{\omega_p R v_t{}^3}. \qquad (13.5\text{-}15)$$

If $R \simeq 10^{10}$ cm, $\omega_p/\omega_H = 10$, $\omega_p = 5 \times 10^8$ s^{-1}, the relations (13.5-15) are satisfied for $v \simeq 100\ v_t$ and $W^\omega \simeq 10^{-2} N_e T_e$. The condition $v \simeq 100\ v_t$ is rather stringent, but the estimate demonstrates that there exists a possibility for trapping of fast electrons in magnetic arches by means of ion sound waves.

Conclusions

In some of the theories for type I bursts possible explanations of the storm continuum have been proposed. This is true for cyclotron as well as plasma wave theories. In this section som possibilities for the explanation of continuum emission in terms of plasma radiation have been considered. The long duration of storms necessitates either continuous supply of energetic particles or storing of energy which is slowly released. Further attention should be devoted to the storm continuum, both observationally and theoretically.

CONCLUDING REMARKS

The journey through a considerable mass of observations, interpretations and theories regarding noise storms is brought to an end. For the first time the noise storm has been in focus on all pages, whereas little or no attention has been paid to other types of solar emissions.

Noise storms present an entangled complex of problems to the investigator. Some solar radio emissions are easier to understand. However, progress in observation and theory of the simpler phenomena are not without interest for the understanding of noise storms because links and cross-relations exist between the different phenomena. It is frequently useful to explore the surroundings before directly attacking the central target. To mention a specific example: progress in observation and theory of stationary type IV continuum bursts is valuable for the understanding of the background continuum of noise storms since the physical processes responsible for the radiation are likely to be the same in the two cases.

Looking back in time it is seen that the first qualitative phase in the history of noise storms, a period which brought much excitement, has been passed. One is at present in the transition region between a semi-quantitative and a quantitative stage. Observations are leading somewhat in comparison with theory.

Even though the storm phenomenon is complex, one can certainly record progress in the observations and understanding of storms. Information has been gained about the relation between storms and optically observable phenomena, the variation of source height with observing frequency is established and general features of bursts and background continuum have been determined. There is convincing evidence that the emission occurs in the ordinary mode and that the observed direction of rotation corresponds to that at the source: no reversal takes place. Ordering of the activity in the form of chains has been detected and frequency splitting, time splitting and periodic phenomena are observed. Type I bursts are strongly directive, as has been established through combined observations from space and the ground. The effects of coronal irregularities in electron density on observed properties of bursts under various suppositions have been calculated. Observational evidence for the predicted effects is being searched for. Relations between noise storms and other forms of solar radio emission have been established. Considerable progress can be noted concerning the interrelation between metric and dekametric storms.

Theories for noise storms were at first diffuse and tentative. Step by step the situation improved, mainly because better models of the storm region were derived from observations and because theoretical plasma physics experienced an almost explosive development.

At present there is a need for combination of data from different types of instruments. Records from wide-band radio spectrographs should be used together with observations undertaken with high resolution spectrographs, and it is a matter of interest to compare the spectral information in detail with determinations of positions and spatial source structure. Spatial and polarization measurements at multiple, closely spaced frequencies with high time resolution may provide valuable information. Unpolarized limb storms deserve special attention. The time-consuming work to study the structure and evolution of different storms also appears necessary.

The background continuum has been offered little interest for a long time. The spectrum and changes with time in the spectrum of this component of the storm emission is almost unknown and should be investigated. Quasi-periodic variations in background continuum have been analysed in some cases, but more systematic work should be carried out. Also possible ionospheric and tropospheric influences on the observations should be further explored.

Models of storm-emitting regions on the sun have certainly become more refined. Radio astronomical investigations cannot be isolated from observations undertaken in other parts of the electromagnetic spectrum and intensified cooperation between radio and optical solar astronomers is called for. Studies of optical, ultra-violet and X-ray observations coupled with simultaneous radio observations, may provide better information on the coronal structures which are the seat of noise storms.

Simultaneous observations at the ground and from space probes have the potentiality to give new information on the directivity of storm emission, and it is to be hoped that further progress may occur with new experiments.

Observations with a higher degree of precision than is commonly obtained are needed. Quantitative data on the dynamic spectra of storm bursts, time profile, frequency profile and frequency drift are needed. This may be obtained from high resolution radio spectrographs of good sensitivity supplied with modern data acquisition systems. A few spectrographs of this type operating in some selected frequency bands and at suitable locations are wanted and calls for international cooperation.

As is the case in so many other branches of science, instruments of increasing sophistication and costs are prerequisite for new developments. Many solar radio astronomers dream of new radio heliographs with high resolution in time and space and operating at many frequencies between dekametric and centimetric wavelengths becoming available. It is a sad state of affairs that most solar interferometers operate in one dimension and with inadequate resolution.

Scattering calculations have been performed, but clarification of the degree to which scattering influences observations of type I emissions is still needed.

Interest is also connected with further researches into the relations which exist between metric and dekametric storms.

For a long time noise storms had little appeal to plasma physicists and other theoreticians. Together with the great advances which have taken place in plasma theory, an increased interest in noise storms can be recorded. One is now at least able to ask relatively precise questions and to give semi-quantitative answers. The theoretical tools necessary to explain the noise storm phenomenon is probably at hand, but the application to storms is not straightforward because the physical conditions in the storm region are complex and not sufficiently well known. Developments of new instruments and methods of observation, together with closer cooperation with theoreticians and more effective application and adaption of theoretical results which already exist, may lead to an understanding of the noise storm phenomenon. The only way to understanding is hard work and constant improvements both of observations and theory.

The sun is the closest star and noise storms the most frequently occurring radio phenomenon. This provides some of the background for the fight for a continued progress in the investigations of storms. Furthermore, the processes in the regions where storms occur on the sun lead to phenomena which extend even to our earth, and thus is important to all of us.

The complexity of noise storms may be frustrating, but it is also rewarding because there is always the chances of finding something new. That should appeal to everybody.

REFERENCES

Abrami, A. (1971) *Proc. of the 1st CESRA Meeting, Utrecht*, p. 31.

Abrami, A. (1974) *Proc. of the 4th CESRA Meeting, Berne*, (ed. E. Schanda), p. 123.

Akhiezer, A.I., Akhiezer, I.A., Polovin, R.V., Sitenko, A.G., Stepanov, K.N. (1967) *Collective Oscillations in Plasma* (MIT Press, Cambridge, Mass.).

Allen, C.W. (1947) *Mon. Not. R. Astr. Soc.* 107, 426.

Allis, W.P., Buchsbaum, S.J., Bers, A. (1963) *Waves in Anisotropic Plasmas* (MIT Press, Cambridge, Mass.).

Aubier, M., Boischot, A. (1972) *Astron. Astrophys.* 19, 343.

Axisa, F., Avignon, Y., Martres, M.J., Pick, M., Simon, P. (1971) *Solar Phys.* 19, 110.

Bailey, V.A. (1950) *Phys. Rev.* 78, 428.

Barrow, C.H., Saunders, H. (1972) *USNC-URSI Meeting, Washington DC.*

Baselyan, L.L., Goncharov, N.Yu., Zaitsev, V.V., Zinichev, V.A., Rapoport, V.O., Tsybko, Ya.G. (1974a) *Solar Phys.* 39, 213.

Baselyan, L.L., Goncharov, N.Yu., Zaitsev, V.V., Zinichev, V.A., Rapoport, V.O., Tsybko, Ya.G. (1974b) *Solar Phys.* 39, 223.

Bekefi, G. (1966) *Radiation Processes in Plasma* (Wiley, New York).

Bernstein, I.B. (1958) *Phys. Rev.* 109, 10.

Berz, F. (1956) *Proc. Phys. Soc., London* B 69, 939.

Blackwell, D.E. (1956) *Mon. Not. R. Astr. Soc.* 116, 56.

Blum, E.T. (1953) *Ann. Univ. Paris* 23, 136.

Bohm, D., Gross, E.P. (1949a) *Phys. Rev.* 75, 1851.

Bohm, D., Gross, E.P. (1949b) *Phys. Rev.* 75, 1864.

Bohm, D., Gross, E.P. (1950) *Phys. Rev.* 79, 992.

Boischot, A. (1958) *Ann. d'Astrophys.* 21, 273.

Boischot, A., de la Noë, J., Möller-Pedersen, B. (1970) *Astron. Astrophys.* 4, 159.

Boischot, A., de la Noë, J., du Chaffaut, M., Rosolen, C. (1971) *C.R. Acad. Sci. Paris* B 272, 166.

Booker, H.G. (1936) *Proc. R. Soc. London* A 155, 235.

Bougeret, J.L. (1972) *Doctoral thesis* (Univ. Paris)

Bougeret, J.L. (1973a) *Proc. of the 3rd CESRA Meeting, Bordeaux*, (ed. J. Delannoy and F. Poumeyrol), p. 182.

Bougeret, J.L. (1973b) *Astron. Astrophys.* 24, 53.

Briggs, R.J. (1964) *Electron-stream Interaction with Plasmas* (MIT, Cambridge, Mass.).

Brown, J.C. (1972) *Solar Phys.* 25, 158.

Budden, K.G. (1952) *Proc. R. Soc. London* A 215, 215.

Buneman, O. (1960) *Nature* 165, 474.

Caroubalos, C., Steinberg, J.L. (1974) *Astron. Astrophys.* 32, 245.

Caroubalos, C., Aubier, M., Leblanc, Y., Steinberg, J.L. (1972) *Astron. Astrophys.* 16, 374.

Caroubalos, C., Pick, M., Trottet, G., Heyvaerts, J. (1974) *IAU Symp. no. 57, Coronal Disturbances* (ed. G. Newkirk), p. 243.

Chambe, C., Lantos, P. (1971) *Solar Phys.* 17, 197.

Chandrasekhar, S. (1952) *Mon. Not. R. Astr. Soc.* 112, 475.

Chernov, G.P., Chertok, I.M., Fomichev, V.V., Markeev, A.K. (1972) *Solar Phys.* 24, 215.

Chin, Y.C. (1972) *Planet. Space Sci.* 20, 711.

Chiu, Y.T. (1970) *Solar Phys.* 13, 420.

Clavelier, B. (1967) *Ann. d'Astrophys.* 30, 895.

Clavelier, B. (1968) *IAU Symp. no. 35, Structure and Development of Solar Active Regions* (ed. K.O. Kiepenheuer), p. 556.

Clavelier, B., Jarry, M.F., Pick, M. (1968) *Ann. d'Astrophys.* 31, 523.

Cohen, M.H. (1958) *Proc. IRE* 46, 172.

Cohen, M.H. (1959) *Ap. J.* 130, 221.

Daigne, G. (1968) *Nature* 220, 567.

Daigne, G. (1975) *Astron. Astrophys.* 38, 141.

Daigne, G., Lantos-Jarry, M.F., Pick, M. (1971) *IAU Symp. no. 43, Solar Magnetic Fields* (ed. R. Howard), p. 609.

de Groot, T. (1959) *IAU-URSI Paris Symposium on Radio Astronomy* (ed. R.N. Bracewell), p. 245.

de Groot, T. (1960) *Bull. Astr. Inst. Netherl.* 15, 229.

de Groot, T. (1962) *Inf. Bull. of Solar Radio Obs. no. 4.*

de Groot, T. (1966) *Rech. Astr. Obs. Utrecht* 18, 1.

de Groot, T., van Nieuwkoop, J. (1968) *Solar Phys.* 4, 332.

de Jager, C., van't Veer, F. (1958) *Rech. Astr. Obs. Utrecht* 14, 1.

de la Noë, J. (1972) *Proc. of the CESRA Summer School on Plasma Physics and Solar Radioastronomy* (ed. A. Mangeney, Paris), p. 260.

de la Noë, J., Boischot, A. (1972) *Astron. Astrophys.* 20, 55.

Denisse, J.F. (1960) *Inf. Bull. Solar Radio Obs. no. 4* and *URSI 13th General Assembly, London 1960.*

Denisse, J.F., Delcroix, J.L. (1961) *Theorie des Ondes dans les Plasmas* (Dunod, Paris).

Denisse, J.F., Rocard, Y. (1951) *J. de Phys. et le Radium* 12, 58.

Diemel, W.E. (1969) *Inf. Bull Solar Radio Obs. no. 25.*

Dodson, H.W. (1958) *Proc. IRE* 46, 149.

Dodson, H.W., Hedeman, E.R. (1960) *McMath-Hulbert working list of IGY flares, IGY Solar Activity Rep. Ser. no. 12.*

Dodson, H.W., Hedeman, E.R., Owren, L. (1953) *Ap. J.* 118, 169.

Dröge, F. (1967) *Z. Astrophys.* 66, 200.

Dulk, G.A., Altschuler, M., Smerd, S.F. (1971) *Astrophys. Letters* 8, 235.

Dulk, G.A., Nelson, G.J. (1973) *Proc. ASA* 2, 211.

Eckhoff, H.K. (1966) *Report no. 18, Inst. of Theoret. Astrophys. Oslo.*

Eidman, V.J. (1958) *Soviet Phys. JETP* 7, 91.

Elgarøy, Ø. (1961) *Astrophys. Norv.* 7, 123.

Elgarøy, Ø. (1962) *Inf. Bull. Solar Radio Obs. no. 9.*

Elgarøy, Ø. (1965) *Solar System Radio Astronomy* (ed. J. Aarons, Plenum Press, New York), p. 201.

Elgarøy, Ø. (1967) *Astrophys. Letters* 1, 13.

Elgarøy, Ø. (1968) *Astrophys. Letters* 3, 39.

Elgarøy, Ø., Hauge, Ø. (1958) *Astrophys. Norv.* 6, 85.

Elgarøy, Ø., Sveen, O.P. (1973) *Solar Phys.* 32, 231.

Elgarøy, Ø., Ugland, O. (1966) unpublished.

Elgarøy, Ø., Ugland, O. (1970) *Astron. Astrophys.* 5, 372.

Ellis, G.R.A. (1969) *Aust. J. Phys.* 22, 177.

Ellis, G.R.A., McCulloch, P.M. (1966) *Nature* 211, 1070.

Ellis, G.R.A., McCulloch, P.M. (1967) *Aust. J. Phys.* 20, 583.

Eliseyev, G.E., Lazareva, L.F., Chistjakov, V.F. (1974) *Coleznye Dannye no. 5*, p. 88.

Erickson, W.C. (1961) *J. Geophys. Res.* 66, 1773.

Fainberg, J., Stone, R.G. (1970a) *Solar Phys.* 15, 222.

Fainberg, J., Stone, R.G. (1970b) *Solar Phys.* 15, 433.

Fokker, A.D. (1957) *Rep. no. 12, Sect. IRA, Netherl. PTT.*

Fokker, A.D. (1960) *Doctoral thesis* (Univ. Leiden).

Fokker, A.D. (1963) *Bull. Astr. Inst. Netherl.* 17, 214.

Fokker, A.D. (1965a) *Solar System Radio Astronomy* (ed. J. Aarons, Plenum Press, New York), p. 171.

Fokker, A.D. (1965b) *Bull. Astr. Inst. Netherl.* 18, 111.

Fokker, A.D. (1971) *Solar Phys.* 19, 472.

Fried, D.B., Conte, S.D. (1961) *The Plasma Dispersion Function* (Academic Press, New York).

Fung, P.C.W. (1966) *Aust. J. Phys.* 19, 489.

Fung, P.C.W., Yip, W.K. (1966a) *Aust. J. Phys.* 19, 759.

Fung, P.C.W., Yip, W.K. (1966b) *Planet. Space Sci.* 14, 1139.

Försterling, K. (1942) *Z. Hochfrequenztechn. u. Elektroakust.* 59, 10.

Gergely, T.E., Kundu, M.R. (1975) *Solar Phys.* 41, 163.

Gibbons, J.J., Neetney, R.J. (1952) *J. Geophys. Res.* 57, 323.

Ginzburg, V.L., Zheleznyakov, V.V. (1958) *Soviet Astr.* 2, 653.

Ginzburg, V.L., Zheleznyakov, V.V. (1959a) *Soviet Astr.* 3, 235.

Ginzburg, V.L., Zheleznyakov, V.V. (1959b) *IAU-URSI Paris Symposium on Radio Astronomy* (ed. R.N. Bracewell), p. 574.

Ginzburg, V.L., Zheleznyakov, V.V. (1961) *Soviet Astr.* 5, 1.

Gnezdilov, A.A. (1970) *Soviet Astr.* 14, 59.

Gnezdilov, A.A. (1972) *Soviet Astr.* 16, 546.

Gnezdilov, A.A. (1974) *Soviet Astr.* 18, 334.

Goldstein, S. (1959) *Ap. J.* 130, 393.

Gordon, I.M. (1970) *Astrophys. Letters* 5, 251.

Grognard, R.J.M., McLean, D.J. (1973) *Solar Phys.* 29, 149.

Gutmann, M., Steinberg, J.L. (1959) *IAU-URSI Paris Symposium on Radio Astronomy* (ed. R.N. Bracewell), p. 123.

Haeff, A.V. (1949) *Phys. Rev.* 75, 1546.

Hanasz, J. (1966) *Aust. J. Phys.* 19, 635.

Hansen, R.J., Garcia, C.J., Hansen, S.F., Loomis, H.G. (1969) *Solar Phys.* 7, 417.

Hatanaka, T. (1956) *Publ. Astr. Soc. Japan* 8, 73.

Hauge, Ø. (1956) *Astrophys. Norv.* 5, 227.

Hauge, Ø. (1958) *Astrophys. Norv.* 6, 43.

Heald, M.A., Wharton, C.B. (1965) *Plasma Diagnostics with Microwaves* (Wiley, New York).

Hewish, A. (1965) *Solar System Radio Astronomy* (ed. J. Aarons, Plenum Press, New York), p. 255.

Hewish, A., Wyndham, J.D. (1963) *Mon. Not. R. Astr. Soc.* 126, 469.

Hey, J.S. (1946) *Nature* 157, 47.

Hey, J.S. (1973) *The Evolution of Radio Astronomy* (Elek. Science, London).

Hey, J.S., Parsons, S.J., Phillips, J.W. (1948) *Mon. Not. R. Astr. Soc.* 108, 354.

Heyvaerts, J. (1975) *Astron. Astrophys.* 38, 45.

Hollweg, J.V. (1967) *Doctoral thesis* (MIT, Center for Space Research Document CSRT-67-6).

Howard, R. (1959) *Ap. J.* 133, 983.

Högbom, J.A. (1959) *Doctoral thesis* (Cambridge Univ.).

Høijer, S., Wilhelmsson, H. (1970) *Plasma Phys.* 12, 585.

Jaeger, J.C., Westfold, K.C. (1949) *Aust. J. Sci. Res.* A2, 322.

Jaeger, J.C., Westfold, K.C. (1950) *Aust. J. Sci. Res.* A3, 376.

James, J.C. (1966) *Ap. J.* 146, 356.

Jorand, M. (1955) *Ann. d'Astrophys.* 18, 180.

Kai, K. (1962) *Publ. Astr. Soc. Japan* 14, 1.

Kai, K. (1965a) *Publ. Astr. Soc. Japan* 17, 294.

Kai, K. (1965b) *Publ. Astr. Soc. Japan* 17, 309.

Kai, K. (1970) *Solar Phys.* 11, 456.

Kai, K., Nakajima, H. (1974) *Publ. Astr. Soc. Japan* 26, 379.

Kai, K., Sekiguchi, H. (1973) *Proc. ASA* 2, 217.

Kai, K., Sheridan, K.V. (1974) *Solar Phys.* 35, 181.

Kai, K., McLean, D.J., Payton, W.J., Sheridan, K.V., Wild, J.P. (1968) *Proc. ASA* 1, 79.

Kakinuma, T., Swarup, G. (1962) *Ap. J.* 136, 975.

Kaplan, S.A., Pikelner, S.B., Tsytovich, V.N. (1972) *Comments on Astrophys. and Space Phys.* 4, 71.

Kaplan, S.A., Tsytovich, V.N. (1969) *Soviet Phys. Usp.* 12, 42.

Kaplan, S.A., Tsytovich, V.N. (1973) *Plasma Astrophysics* (Pergamon Press, Oxford).

Kennel, C.F., Scarf, F.L. (1969) *Plasma waves in Space and in the Laboratory* (ed. J.O. Thomas and B.J. Landmark, University Press, Edinburgh) vol. 1, p. 451.

Kerdraon, A. (1973) *Astron. Astrophys.* 27, 361.

Kiepenheuer, K.O. (1946) *Nature* 158, 340.

Krall, N.A., Trivelpiece, A.W. (1973) *Principles of Plasma Physics* (McGraw-Hill, New York).

Krüger, A. (1972) *Physics of Solar Continuum Radio Bursts* (Aakademie-Verlag Berlin).

Kruse, U.E., Marshall, L., Platt, J.R. (1956) *Ap. J.* 124, 601.

Kuipers, J. (1975) *Doctoral thesis* (Utrecht).

Kundu, M.R. (1965) *Solar Radio Astronomy* (Interscience Publishers, New York).

Körper, K. (1957) *Z. Naturforsch.* 12a, 815.

Lantos-Jarry, M.F. (1970) *Solar Phys.* 15, 40.

Leblanc, Y. (1972) *Proc. of the CESRA Summer School on Plasma Physics and Solar Radio-Astronomy* (ed. A. Mangeney, Paris) p. 312.

Leblanc, Y. (1973) *Astrophys. Letters* 14, 41.

Le Squeren, A.M. (1963) *Ann. d'Astrophys.* 26, 97.

Letfus, V. (1965) *Bull. Astr. Inst. Czech.* 16, 231.

Liemohn, H.B. (1965) *Radio Sci.* 69D, 741.

Lin, R.P. (1970) *Solar Phys.* 12, 266.

Lin, R.P., Anderson, K.A. (1967) *Solar Phys.* 1, 446.

Lin, R.P., Evans, L.G., Fainberg, J., (1973) *Astrophys. Letters* 14, 191.

Little, A.G., Payne-Scott, R. (1951) *Aust. J. Sci. Res.* A4, 508.

Loonen, J.P. (1971) *Proc. of the 1st CESRA Meeting, Utrecht*, p. 39.

Machin, K.E., O'Brien, P.A. (1954) *Phil. Mag. Ser. 7*, 45, 973.

McCready, L.L., Pawsey, J.L., Payne-Scott, R. (1947) *Proc. R. Soc. London*, A 190, 357.

McLean, D.J. (1969) *Proc. ASA* 1, 188.

McLean, D.J., Sheridan, K.V. (1972) *Solar Phys.* 26, 176.

Maltby, P. (1958) *Astrophys. Norv.* 6, 75.

Maltby, P. (1960) *Astrophys. Norv.* 6, 147.

Maltby, P., Steen, O. (1960) *Astrophys. Norv.* 7, 1.

Malville, J.M. (1961) *Doctoral thesis* (Univ. Colorado).

Malville, J.M. (1962) *Ap. J.* 136, 266.

Malville, J.M., Aller, A.D., Jensen, C.J. (1967) *Ap. J.* 147, 711.

Mangeney, A., Veltri, P. (1976a) *Astron. Astrophys.* 47, 165.

Mangeney, A., Veltri, P. (1976b) *Astron. Astrophys.* 47, 181.

Markeev, A.K., Chernov, G.P. (1971) *Soviet Astr.* 14, 835.

Martres, M., Pick, M., Parks, G.K. (1970) *Solar Phys.* 15, 48.

Martyn, D.F. (1946) *Nature* 158, 308.

Martyn, D.F. (1947) *Nature* 159, 26.

Melrose, D.B. (1973a) *Aust. J. Phys.* 26, 229.

Melrose, D.B. (1973b) *Proc. ASA* 2, 208.

Melrose, D.B. (1974a) *Aust. J. Phys.* 27, 31.

Melrose, D.B. (1974b) *Aust. J. Phys.* 27, 43.

Melrose, D.B. (1974c) *Solar Phys.* 37, 353.

Melrose, D.B. (1975a) *Aust. J. Phys.* 28, 101.

Melrose, D.B. (1975b) *Solar Phys.* 43, 79.

Melrose, D.B. (1975c) *Solar Phys.* 43, 211.

Melrose, D.B., Sy, W.N. (1971) *Proc. ASA* 2, 56

Melrose, D.B., Sy, W.N. (1972) *Aust. J. Phys.* 25, 387.

Mercier, C. (1973) *Solar Phys.* 33, 177.

Mercier, C., Rosenberg, H. (1974) *Solar Phys.* 39, 193.

Mollwo, L. (1970) *Solar Phys.* 12, 125.

Morimoto, M. (1963) *Publ. Astr. Soc. Japan* 15, 46.

Morimoto, M., Kai, K. (1961) *Publ. Astr. Soc. Japan* 13, 294.

Mosier, S.R., Fainberg, J. (1972) *USNC-URSI Meeting, Washington DC.*

Mosier, S.R., Fainberg, J. (1975) *Solar Phys.* 40, 501.

Möller-Pedersen, B. (1974) *Astron. Astrophys.* 17, 163.

Ness, N.F., Wilcox, J.M. (1967) *Solar Phys.* 2, 351.

Newkirk, G. (1961) *Ap. J.* 133, 983.

Newkirk, G. (1971) *IAU Symp. no. 43, Solar Magnetic Fields* (ed. R. Howard), p. 547.

Owren, L. (1954) *Radio Astr. Rep. no. 15, Cornell Univ.*

Palmer, I.O., Lin, R.P. (1972) *Proc. ASA* 2, 101.

Payne-Scott, R., Little, A.G. (1951) *Aust. J. Sci. Res.* A4, 508.

Peersen, T. (1970) *Thesis* (Oslo Univ.).

Petschek, H.E. (1966) *Solar Wind* (ed. R.J. Mackin Jr. and M. Neugebauer, Pergamon Press), p. 257.

Pick, M. (1961) *Ann. d'Astrophys.* 24, 183.

Piddington, J.H., Minnett, H.C. (1951) *Aust. J. Sci. Res.* A4, 131.

Pikelner, S.B., Tsytovich, V.N. (1969) *Soviet Astr.* 13, 5.

Pneuman, G.W. (1972) *Solar Phys.* 23, 223.

Ratcliffe, J.A. (1959) *The Magneto-ionic Theory and Its Applications to the Ionosphere* (Cambridge Univ. Press).

Reber, G. (1946) *Nature* 158, 945.

Riddle, A.C. (1972) *Proc. ASA* 2, 98.

Riddle, A.C. (1974) *Solar Phys.* 35, 153.

Roberts, J.A. (1958) *Aust. J. Phys.* 11, 215.

Roberts, J.A. (1959) *Aust. J. Phys.* 12, 327.

Rosenberg, H. (1973) *Doctoral thesis* (Univ. Utrecht).

Ryle, M. (1952) *Proc. R. Soc.* A 211, 351.

Ryle, M., Vonberg, D.D. (1946) *Nature* 158, 339.

Ryle, M., Vonberg, D.D. (1948) *Proc. R. Soc.* A 193, 98.

Rydbeck, O.E.H. (1950) *J. Appl. Phys.* 21, 1205.

Sagdeev, R.Z., Galeev, A.A. (1969) *Non-linear Plasma Theory* (W.A. Benjamin, Inc., New York).

Sakurai, K. (1967) *Publ. Astr. Soc. Japan* 19, 316.

Sakurai, K. (1971a) *Solar Phys.* 16, 125.

Sakurai, K. (1971b) *Solar Phys.* 16, 198.

Sakurai, K., Stone, R.G. (1971) *Solar Phys.* 19, 247.

Sastry, CH.V. (1969) *Solar Phys.* 10, 429.

Sastry, CH.V. (1971) *Astrophys. Letters* 8, 115.

Sastry, CH.V. (1972) *Astrophys. Letters* 11, 47.

Scharer, J.E., Trivelpiece, A.W. (1967) *Phys. Fluids* 10, 591.

Scheuer, P.G. (1960) *Mon. Not. R. Astr. Soc.* 120, 180.

Sheeley Jr., N.R., Bohlin, J.D., Brueckner, G.E., Purcell, J.D., Scherrer, V.E., Tousey, R. (1975) *Solar Phys.* 40, 103.

Sheridan, K.V., Labrum, N.R., Payten, W.J. (1972) *Nature* 238, 115.

Sheridan, K.V., McLean, D.J., Smerd, S.F. (1973) *Astrophys Letters* 15, 139.

Shklovsky, J.S. (1946) *Astr. J. USSR* 23, 333.

Sitenko, A.G., Stepanov, K.N. (1957) *Soviet Physics JETP* 4, 512.

Slottje, C. (1972) *Proc. of the CESRA Summer School on Plasma Physics and Solar Radio Astronomy* (ed. Mangeney, Paris) p. 245.

Smerd, S.F. (1964) *Ann. of the IGY* 34, 331.

Smerd, S.F. (1965) *The Solar Spectrum* (ed. C. de Jager, Reidel Publ., Dordrecht).

Smerd, S.F. (1968) *Proc. ASA* 1, 124.

Smerd, S.F., Sheridan, K.V., Stewart, R.T. (1973) *Astrophys. Letters* 16, 23.

Smith, D.F., Krall, N.A. (1974) *Ap. J.* 194, L 163.

Smith, D.F., Pneuman, G.W. (1972) *Solar Phys.* 25, 461.

Smith, D.F., Sturrock, P.A. (1971) *Astrophys. Space Sci.* 12, 411.

Smith, E.v.P., McIntosh, P.S. (1962) *J. Geophys. Res.* 67, 1013.

Spitzer, L. (1956) *Physics of Fully Ionized Gases* (Interscience Publishers, Inc., New York).

Steinberg, J.L., Caroubalos, C. (1970) *Astron. Astrophys.* 9, 329.

Steinberg, J.L., Caroubalos, C., Bougeret, J.L. (1974) *Astron. Astrophys.* 37, 109.

Steinberg, J.L., Aubier-Girard, M., Leblanc, Y., Boischot, A. (1971) *Astron. Astrophys.* 10, 362.

Stepanov, A.V. (1970) *Radiophys. and Quantum Electronics* 13, 1034.

Stewart, R.T. (1975) *Solar Phys.* 40, 417.

Stewart, R.T., Labrum, N.R. (1972) *Solar Phys.* 27, 192.

Stewart, R.T., Sheridan, K.V. (1971) *Proc. ASA* 2, 60.

Suzuki, S. (1961) *Ann. Tokyo Astr. Obs.* 7, 75.

Sveen, O.P. (1971) *Proc. of the 1st CESRA Meeting, Utrecht*, p. 44.

Sy, W.N. (1973a) *Proc. ASA* 2, 215.

Sy, W.N. (1973b) *Doctoral thesis* (Austr. National Univ.).

Takakura, T. (1956) *Publ. Astr. Soc. Japan* 8, 182.

Takakura, T. (1959) *Publ. Astr. Soc. Japan* 11, 55.

Takakura, T. (1962) *J. Phys. Soc. Japan* 17, Suppl. A-II, 243.

Takakura, T. (1963) *Publ. Astr. Soc. Japan* 15, 462.

Takakura, T., Yousef, S. (1975) *Solar Phys.* 40, 421.

Tarnstrom, G.L., Philip, K.W. (1972) *Astron. Astrophys.* 16, 21.

Tidman, D.A., Dupree, T.H. (1965) *Phys. Fluids* 8, 1860.

Tidman, D.A., Birmingham, T.J., Stainier, H.M. (1966) *Ap. J.* 146, 207.

Titulaer, C. (1967) *Bull. Astr. Inst. Netherl.* 19, 82.

Trakhtengerts, V.Yu. (1966) *Soviet Astr.* 10, 281.

Tsuchiya, A. (1963) *Publ. Astr. Soc. Japan* 15, 368.

Tsytovich, V.N. (1970) *Nonlinear Effects in Plasma* (Plenum Press, New York).

Tsytovich, V.N. (1972) *An Introduction to the Theory of Plasma Turbulence* (Pergamon Press, Oxford).

Twiss, R.Q. (1958) *Aust. J. Phys.* 11, 564.

Twiss, R.Q., Roberts, J.A. (1958) *Aust. J. Phys.* 11, 424.

Ugland. O. (1966) unpublished.

van de Hulst, H.C. (1950) *Bull. Astr. Inst. Netherl.* 11, 135.

Vereshkov, G.M. (1974) *Soviet Astr.* 18, 152.

Vinokur, M. (1968) *Ann. d'Astrophys.* 31, 457.

Vitkevich, V.V. (1955) *Dokl. Akad. Nauk. USSR* 101, 229.

Vitkevich, V.V. (1957) *IAU Symp. no. 4, Radio Astronomy* (ed. H.C. van de Hulst) p. 363.

Vitkevich, V.V., Gorelova, M.V. (1961) *Soviet Astr.* 4, 595.

Warwick, J.W., Dulk, G.A. (1969) *Ap. J. Letters* 158, L 123.

Weber, W.J., Rosenberg, H. (1974) *Solar Phys.* 37, 409.

Westfold, K.C. (1949) *Aust. J. Sci. Res.* A2, 169.

Wilcox, J.M., Ness, N.F. (1965) *J. Geophys. Res.* 70, 5793.

Wild, J.P. (1950) *Aust. J. Sci. Res.* A3, 399.

Wild, J.P. (1951) *Aust. J. Sci. Res.* A4, 36.

Wild, J.P. (1957) *IAU Symp. no. 4, Radio Astronomy* (ed. H.C. van de Hulst), p. 521.

Wild, J.P. (1967a) *Proc. ASA* 1, 38.

Wild, J.P. (1967b) *Proc. IREE Austral.* 28, no. 9.

Wild, J.P. (1969a) *Proc. ASA* 1, 181.

Wild, J.P. (1969b) *Solar Phys.* 9, 260.

Wild, J.P., McCready, L.L. (1950) *Aust. J. Sci. Res.* A3, 387.

Wild, J.P., Sheridan, K.V. (1958) *Proc. IRE* 46, 160.

Wild, J.P., Sheridan, K.V., Neylan, A.A. (1959) *Aust. J. Phys.* 12, 369.

Wild, J.P., Smerd, S.F. (1972) *Ann. Rev. Astron. Astrophys.* 10, 159.

Wild, J.P., Smerd, S.F., Weiss, A.A. (1963) *Ann. Rev. Astr. Astrophys.* 1, 291.

Wild, J.P., Tlamicha, A. (1965) *Bull. Astr. Soc. Czech.* 16, 73.

Wild, J.P., Zirin, H. (1956) *Aust. J. Phys.* 9, 315.

Yoh, P., James, J.C. (1967) *Ap. J.* 149, 441.

Zaitsev, V.V. (1969) *Doctoral thesis* (Univ. Gorkii).

Zaitsev, V.V., Fomichev, V.V. (1973) *Soviet Astr.* 16, 666.

Zheleznyakov, V.V. (1965) *Soviet Astr.* 9, 191.

Zheleznyakov, V.V. (1970) *Radio Emission of the Sun and Planets* (Pergamon Press, Oxford).

Zheleznyakov, V.V., Zlotnik, E.Ya. (1963) *Soviet Astr.* 7, 485.

Zlobec, P. (1972) *Proc. of the 2nd Meeting of CESRA, Trieste.*

Zlobec, P. (1973) *Proc. of the 3rd Meeting of CESRA, Bordeaux* (ed. I. Delannoy and F. Pouneyrol) p. 151.

AUTHOR INDEX

SUBJECT INDEX